Acid and Sour Gas Treating Processes

Latest data and methods for designing and operating today's gas treating facilities

Stephen A. Newman, *Editor*

Acid and Sour Gas
Treating Processes

Library of Congress Cataloging in Publication Data
Main entry under title:

Acid and sour gas treating processes.

Includes index.
1. Gases—Purification. 2. Gas, Natural—Sweetening.
I. Newman, Stephen A., 1938–
TP754.A25 1985 665.7′3 84-25339
ISBN 0-87201-839-3

I would like to acknowledge the assistance and cooperation of all individuals who contributed to the development of this publication. A special thanks to Mrs. Mary Ellen Newman and the staff of Gulf Publishing Company for their assistance with the manuscript.

Stephen A. Newman

Cover photograph and title page photograph of carbon dioxide removal unit in an ammonia plant are courtesy of Foster Wheeler Energy Corporation.

Contents

Preface

Most of the papers in this text are based upon material presented at a recent national conference on acid and sour gas treating processes organized and chaired by the editor. Many of the papers presented at the conference were subsequently published in several well-known journals and permission has been obtained to incorporate the manuscripts within this text. All other conference papers, as well as additional papers that have been selected to extend the coverage of the subject, have also been included in this compendium.

The engineer required to select an acid gas removal process for a given application often confronts a bewildering array of available processes. Each year additional alternatives are offered, increasing the maze of possibilities, but also presenting greater opportunities for attaining clean-up goals with greater confidence and more economy. Most of the papers in this text deal with the latest information on widely used commercial gas treating processes and will interest engineers who have to select, design, and/or operate these units. Some of the papers describe recent data and models used in the design of the treating units. The papers have been written by individuals who have been at the forefront in this important process area. The papers provide the latest available information, state the problems, present the anticipated goals, and relate the costs necessary to accomplish these goals.

Contributions to the text provided by many of the leading developers and licensors describe the latest technological advancements in the gas processing field. Gas treating units are incorporated in many process plants and are very expensive; in fact, they are quite often the most capital-intensive unit in the entire plant. For this reason there is substantial interest in the proper selection and economic design of acid and sour gas treating facilities. Most of the processes described in the text are well-established mainstays of the gas processing industry. Others are still in the developmental or early commercial stages, but have the potential of offering unique features that complement, and in some cases, supplant known routes. The nature of the solvent is fundamental to the successful operation of the configurations common to many clean-up processes. Several papers in the text address this vital consideration of the design process. Most of these papers, relating to vapor-liquid-equilibria and solubility data, have been written by individuals who have been pioneers in providing high-quality data used in the design of many recent non-proprietary gas treating facilities.

Stephen A. Newman

Acid and Sour Gas Treating Processes

1

Gas Treating with a Physical Solvent

John W. Sweny
Market Manager, Selexol® Process
Norton Chemical Process Products

Treating of gases today is more complex than in the past, as shown by examples taken from applications in enhanced oil recovery and in offshore gas treating.

Successful handling of such projects requires not only a well-developed process, but a considerable backlog of commercial experience. Basic principles of treating with physical solvents is explained, followed by accounts of the growth of one such process from simple applications to the complex applications of today.

In recent years, gas treating demands have become more exacting because of higher fuel costs, environmental protection, and the appearance of CO_2 as a large volume product, needed now in the new techniques of enhanced oil recovery. Three recent projects illustrate these new demands. These will all employ a physical solvent, the dimethyl ether of polyethylene glycol (DMPEG), whose proprietary technology is marketed by the Norton Company, U.S.A.

Outlined in Table 1 are the bare process requirements for the three projects. The raw gas for Project No. 1 was discovered in 1962. Despite the huge size of the deposit (capable of 2-3 billion SCFD raw feed), this gas has not been produced because of low heating value, associated treatment difficulties, depth of formation, and remoteness. Now, however, since physical solvent treating has been developed and CO_2 has become an important by-product for enhanced oil recovery (EOR), this gas is about to be utilized. Corporations are considering construction of plants to produce CH_4 and CO_2 at high purity and

Table 1
Recent Significant Applications of Treating with DMPEG

Example 1

PROJECT NO. 1: Economically recovers CO_2 sulfur and CH_4 as separate products with full protection of the environment.

Feed		Products
□CH_4, C_2, N_2	32.2 vol%	CH_4:H_2S, 4ppmv; CO_2, 100ppmv
□CO_2	65.3 vol%	CO_2:97 vol% +; H_2S, 10ppmv
□H_2S	4.5 vol%	Sulfur:99.8% recovery; 99.5% purity; less than 10 MT/yr. total S to atmosphere
□COS	170 ppmv	

Example 2

PROJECT NO. 2: Removes H_2S from CO_2-rich natural gas for enhanced oil recovery.

Feed		Products
□CH_4, C_2, N_2	9.0 vol%	CH_4, CO_2:H_2S, 25ppmv
□CO_2	90.6 vol%	
□H_2S	0.4 vol%	Waste:CO_2, H_2S to flare

Example 3

PROJECT NO. 3: Removes CO_2 and H_2S in an offshore location without use of heat in regeneration.

Feed		Products
□CH_4, C_2, N_2	63.7 vol%	CH_4:H_2S, 2ppmv; CO_2, 1.2 vol%
□CO_2	36.3 vol%	
□H_2S	90 ppmv	Waste:CO_2, H_2S to flare

recovery, and to recover sulfur to a very high degree as a matter of preserving this environment, which has had no exposure to industrial contamination.

Project No. 2 is land-based in a remote area. The feed gas for this project is about 90% CO_2, and is, therefore, suitable for EOR, but the 4000 ppmv of H_2S it contains must be removed to permit transmission to the oil field for injection. The high selectivity of physical solvent treating makes this possible.

Project No. 3 is located offshore. This plant has been constructed and is presently being started up. The feed gas contains 36% CO_2 and a minor amount of H_2S. These need to be removed offshore with economy, preferably without heating for regeneration. This was made possible through physical solvent treating.

These three projects deal with treating tasks which did not exist only a short time ago. Before treating with physical solvent was developed, these projects would have been economically unattractive or technically unfeasible. Although physical solvents left the pilot plant stage about 20-25 years ago, applications such as these, critical to their owners, require demonstration in full-scale plants before acceptance. A process needs to grow commercially through simple applications, uncovering all the operating and economic problems, before the critical applications can be undertaken. These three applications will be discussed in this paper in greater detail, showing how earlier applications made it possible to confidently cope with the challenges of these recent applications.

Selective Removal of Sulfur: Project No. 1

If the H_2S and CO_2 contained in the raw feed were removed together in a non-selective manner, as they would be with a chemical solvent such as DEA or hot potassium carbonate, the acid gas would contain only about 6 vol. % H_2S, which would be too dilute for processing in a Claus plant, which usually requires 20 vol. % for a split-flow plant and 50% for a straight-through plant. The answer to this problem is selective removal of H_2S with a physical solvent, which can, on this feed, reach 30% to 70%, depending on the process flow scheme chosen.

Absorption can be made selective if there is a sufficiently large difference in solubility between key components in a multi-component system. In the situation here, we wish to remove H_2S while taking a minimum of CO_2. The differences in solubility we have to work with is shown on Fig. 1, which gives relative solubilities for a number of gaseous components in SELEXOL® Solvent. (These relative solubilities cannot be used for design, since the solubilities will change if the temperatures, pressures and compositions differ from the standardized ones used for Fig. 1). Fig. 1 shows that the relative solubility between CO_2 and H_2S is 134/15.0, or almost 9/1, which offers a good basis for separating the two key components.

The separation is effected by restricting the solvent flow to a level high enough to remove all the H_2S, but to a level too low to take more than

Component	$R = \frac{K'\,CH_4}{K'\,Component}$	Component	$R = \frac{K'\,CH_4}{K'\,Component}$
H_2	0.20	NH_3	73.
CO	0.43	→ nC_5	83.
→ C_1	1.	→ H_2S	134.
C_2	6.4	→ C_6	165.
C_2H_4	7.3	CH_3SH	340.
→ CO_2	15.0	→ C_7	360.
C_3	15.3	CS_2	360.
iC_4	28.	SO_2	1400.
nC_4	35.	C_6H_6	3800.
→ COS	35.	C_4H_4S	8100.
iC_5	67.	→ H_2O	11000.
C_2H_2	68.	HCN	18000.

Figure 1. Relative solubilities of various gases in SELEXOL® solvent.

a portion of the CO_2. If one uses an abundance of solvent, all the H_2S and CO_2, indeed all of the gas including even the lightest components, will be taken up by the solvent. If one reduces solvent flow in stages, the less soluble materials will be left unabsorbed, the amount depending on how much flow is reduced. An adjustment of the flow can be made to take all the solubles from, say, H_2S down to HCN, while leaving nC_5 to H_2 in the gas phase.

This separation will not be perfectly sharp. Virtually all the H_2S to HCN can be removed, but some portion of each of the lights will be removed too; that is, the solvent will contain some C_5, CO_2, and C_1. This co-absorption can be minimized by reducing the solvent flow as far as possible, which is done by using a tall absorber, one with enough theoretical stages to remove all the key component with a minimum of solvent.

All this can be seen clearly on the Kremser chart shown in Fig. 2. The abcissa is the absorption factor, $A = L/KV$, in which L and V are

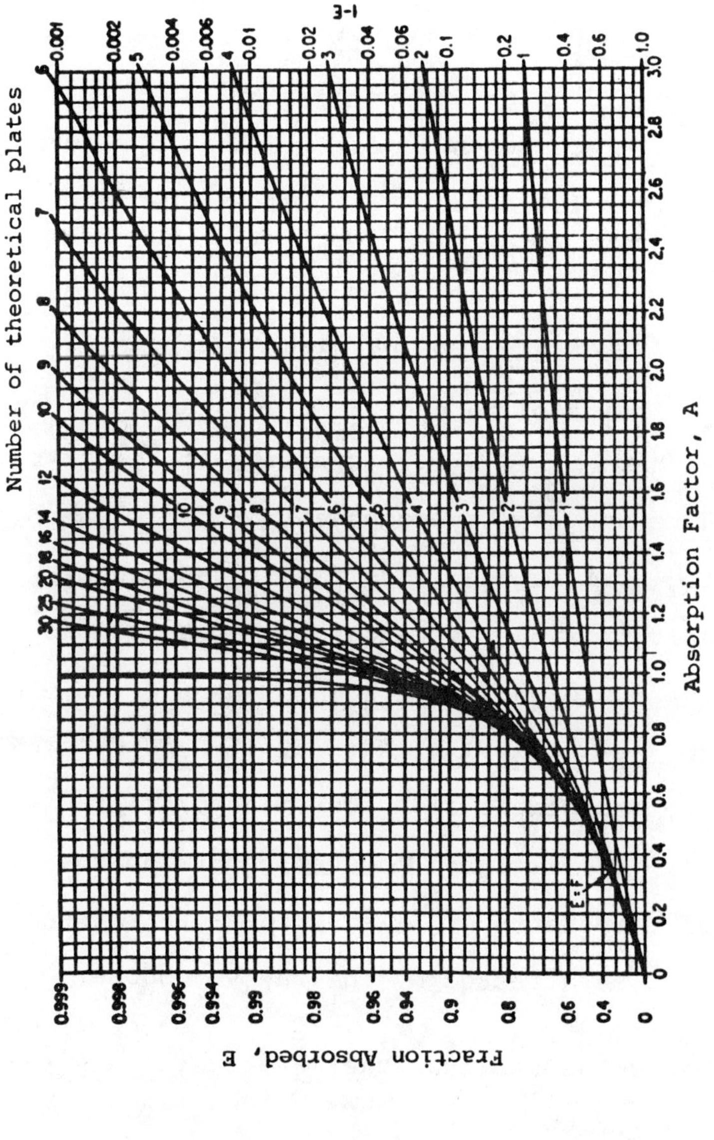

Figure 2. Theoretical stages by Kremser method.

liquid and vapor rates in moles per unit time, and K is the equilibrium vaporization constant, K = y/x, which expresses the solubility of a component in the solvent. The ordinate E is the degree of removal of a component, provided that the lean solvent is solute-free. The number of theoretical stages is n.

Let us assume that we wish to remove 99.8% of the H_2S from a mixture of H_2S, CO_2, and CH_4, and that we have fixed V, the quantity of the mixture. We can remove the key component, H_2S, in two ways: by using more solvent with few stages, or by using less solvent with many stages. We can remove 99.8% of H_2S with A at 2.8 and n at 6 or we can do the same with A at 1.4 and n at 15. Either way would be acceptable from the standpoint of H_2S removal. Our mixture, however, contains the lighter components CO_2 and CH_4, and their absorption will be affected by the way we choose.

Let us assume that the K-value for H_2S is 1 and that for CO_2 is 5.

If the A-factor for H_2S is 2.8, the A factor for CO_2, since L and V are constant, will be 2.8/5 = 0.56, which according to the chart would lead to 55% or so removal of CO_2. On the other hand, if the A factor for H_2S is 1.4, the removal appears to be only 28%, a substantial reduction in the amount of CO_2 absorbed with the H_2S.

Thus, if we wish to absorb selectively, we require some difference in solubility between the keys to work with, and then, using an efficient absorber, we restrict the flow of solvent as much as we can while still removing the key component to the required degree.

Efficient absorption is the basic tool for exploiting natural selectivity, but it will not prevent the co-absorption of lighter-than-key components; some CO_2 for example, will always be co-absorbed with the H_2S. The deabsorption process, which necessarily follows the absorption process, does, however, offer means of making further separation between the heavies and lights.

The most common and most economical way to remove lights is to recycle the gases given off in flashing. Solutes have been absorbed under pressure; as the pressure is relieved, solutes come out of solution. As pressure is reduced, first to atmospheric, and then perhaps to vacuum, all the solutes come off and the solvent is regenerated. If one stops at an intermediate pressure, the light materials will have come off and the heavies will be left in solution. The lights will contain some heavies, which can be removed in the main absorber when the flash gas is recompressed and added to the fresh feed.

This flowscheme is shown in Fig. 3, which shows the favored scheme if recycling were to be used. This scheme is not capable of sharp separations, since only one, two, or three stages of flashing are practical. It is capable, with the given raw feed and conditions, of producing an acid gas feed to the Claus plant of 30-35% H_2S concentration.

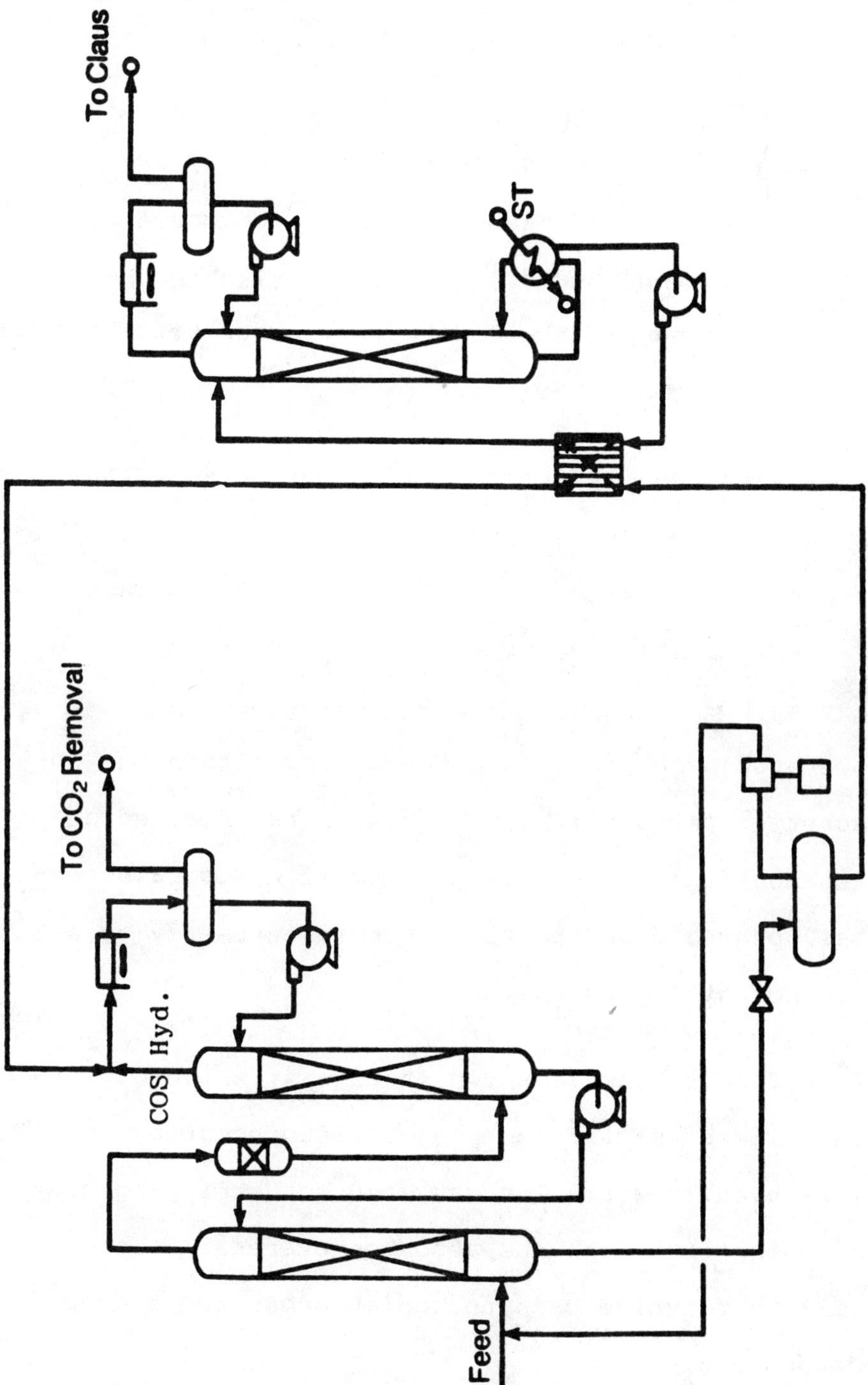

Figure 3. Project No. 1—H_2S removal with flash-recycle.

Much better results than this are possible. One could conceive in theory a series of recompressions and reabsorptions which would completely separate H_2S and CO_2 in mixture, but the resulting plant would be very uneconomic both in equipment and utilities.

A reasonably economical way of introducing staging into the desorption process is partial stripping or enrichment. Instead of flash tanks, a tower can be used to carry out the desorption of excess CO_2 from rich liquid. The tower is operated at a pressure somewhat lower than the absorption pressure; rich liquid, first flashed at the top of the tower or perhaps in a separate vessel, passes down the tower countercurrently to a flow of stripping gas.

This scheme is shown in Fig. 4, which depicts the flowscheme for the project incorporating an enrichment tower. The gas used for stripping must be compatible with the absorber overhead, since it will be recycled back to the absorber to recover desorbed H_2S.

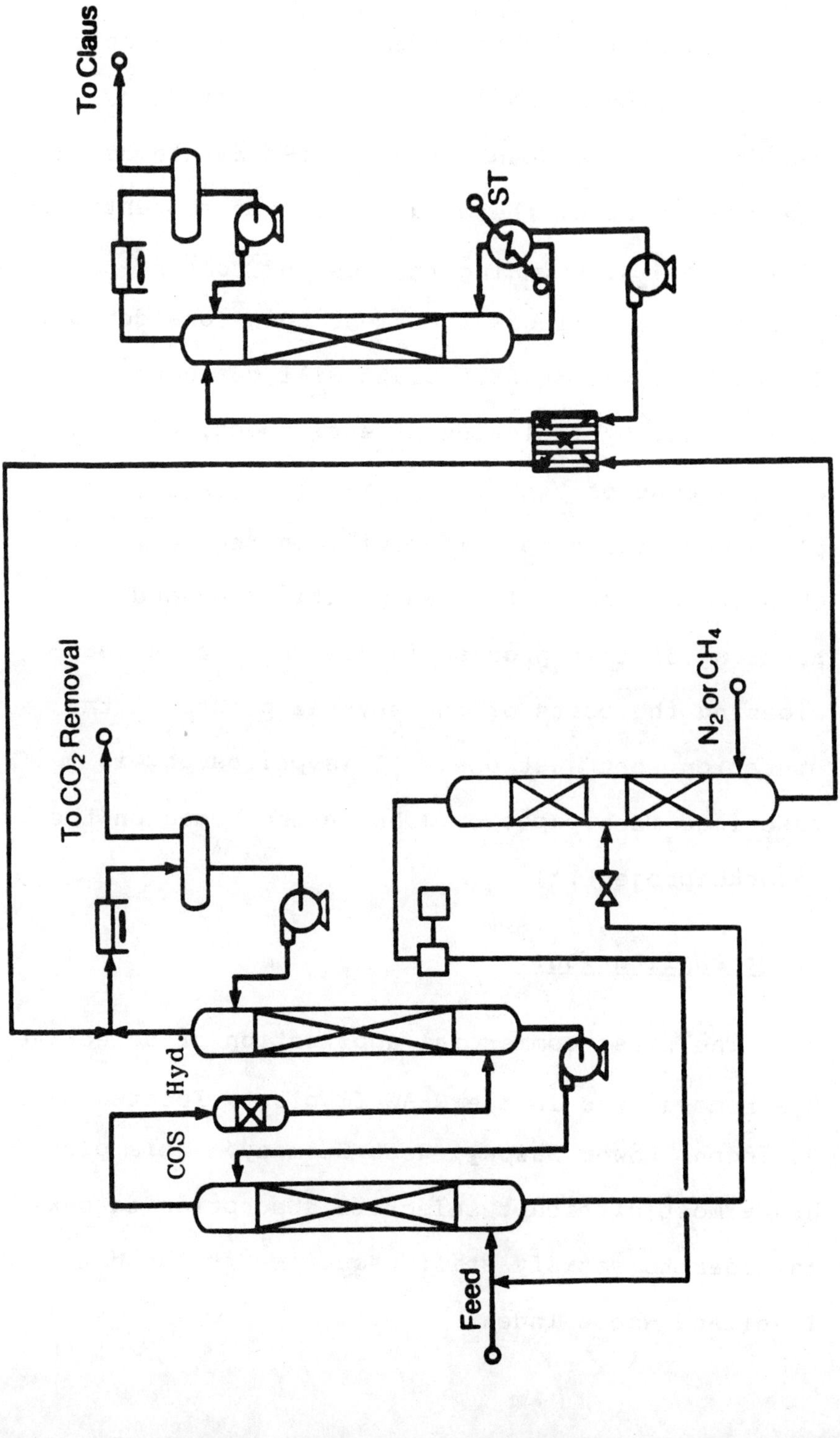

Figure 4. Project No. 1—H_2S removal with partial stripping.

This device, which introduces staging into the desorption process, is very effective in producing well-concentrated Claus-plant feeds: in the case in hand, the acid gas can be concentrated to 70-90% H_2S, bringing the cost of sulfur recovery and tail-gas treatment down considerably. The cost of the SELEXOL plant will be increased, it is true, by inclusion of enrichment, but the overall cost of H_2S removal and its virtually complete conversion to sulfur will be reduced. Choosing a process for gas purification and the handling of that process in design involves considering the costs of the several plants in the operation, not just one. (This philosophy is described in a paper by Johnson and Homme on the LaBarge project.[1])

Prior Applications:

The first commercial application of selective H_2S removal was in the NEAG II plant, located near Sulingen, Lower Saxony, West Germany. This plant had a most difficult selective absorption to make in order to satisfy stiff requirements for H_2S level and Wobbe Index.

The feed gas composition is shown on the simplified flow diagram of NEAG given in Fig. 5. The main components besides CH_4 were H_2S at 9.0 vol. % and CO_2 at 9.5 vol. %. The H_2S had to be reduced from 90,000 ppmv to 2 ppmv while retaining about 80% of the CO_2 in order to satisfy Wobbe Index and sulfur requirements for town gas in the North German distribution system; if these requirements could not be met, the gas would have had to be sold for a modest price as power-plant fuel instead of a high price as town gas.

As reported in a paper by Michaeli, Stein, and Unland[2], a system involving four separate processes was installed at first. This system did not succeed: it could meet sulfur specification but not Wobbe Index, or it would meet Wobbe Index but not the sulfur specification. We were asked if the SELEXOL process could bring the operation to success. At the time only laboratory and pilot plant data were available to encourage us and our client to make the try - there was no commercial plant yet operating in 1970. Only minimum retrofitting was done because of the urgency of the situation, but the revamped plant did succeed. The plant has gone through further revampings since 1971, and is operating today in accordance with the flowsheet of Fig. 5. Throughout this development, the client and ourselves worked closely for our mutual benefit; the client could not do without our process, but certainly we could not have succeeded without the client's contributions.

This experience proved to others, and most importantly to ourselves, that DMPEG had a major position in the field of selective absorption. Our process was soon afterward applied to selective applications in the Dueste I, Dueste II, and Ruetenbrock plants of Wintershall AG, also located in Lower Saxony[3]. It was in these plants that the technique of steam stripping was first developed and applied.

At the time NEAG was revamped, regeneration was carried out only by flashing and gas stripping. As Fig. 5 shows, stripping at NEAG was done with CH_4 product, which was recompressed and

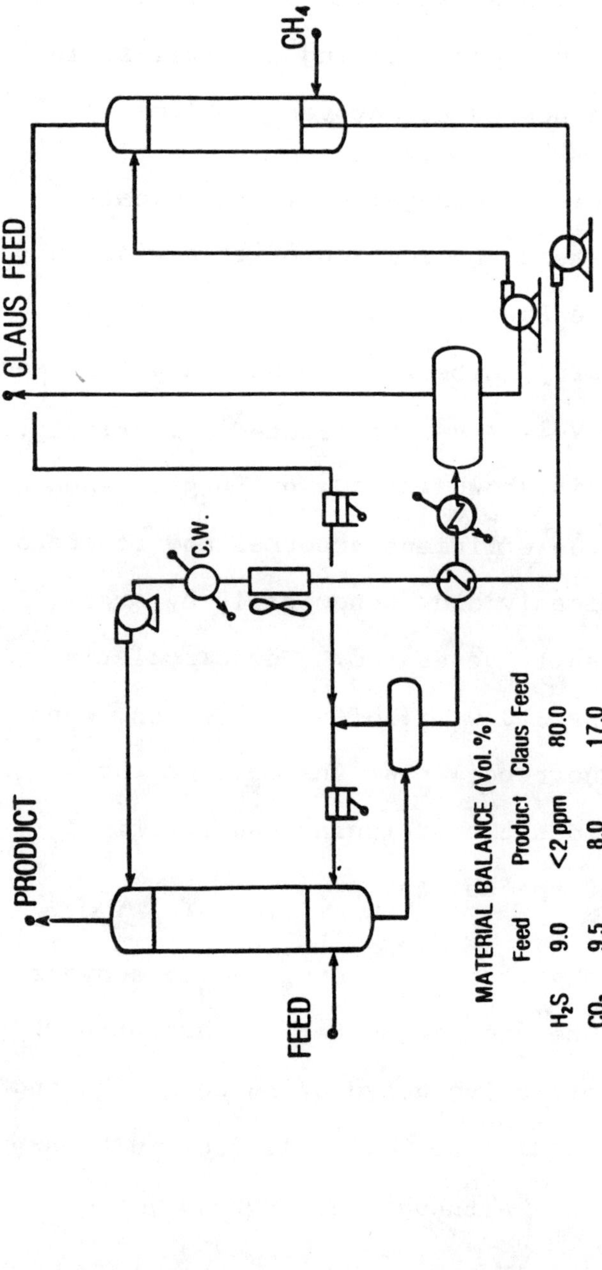

MATERIAL BALANCE (Vol. %)

	Feed	Product	Claus Feed
H_2S	9.0	<2 ppm	80.0
CO_2	9.5	8.0	17.0
C_1	76.8	86.6	3.0
C_2+	0.2	0.2	—
N_2	4.5	5.2	—
COS	130 ppm	70 ppm	—
RSH	100 ppm	<1 ppm	—

Figure 5. Norddeutsche Erdgas—Aufbereitungsgesellschaft mbH, Neag II Plant.

recycled back to the main absorber. It was known that stripping with steam would be more economical and effective, but some piloting and full-scale operations were needed to uncover problems.

This experience was gained in the Dueste plants. The latest application of this kind in operation is the Pan Canadian Morley plant, located near Calgary, Alberta, Canada, in which feed containing 5.0 vol. % H_2S is treated selectively. The flowscheme is shown in Fig. 6. H_2S is removed selectively in an efficient absorber and regeneration is done cleanly and economically by steam which is internally generated. The circulating solvent contains 3.0 wt. % of water, which, when vaporized in the reboiler at the base of the stripper, provides the stripping gas for final regeneration of the solvent.

SELEXOL solvent is also expected to dehydrate the raw gas. One can see in Fig. 1 that SELEXOL solvent is an effective dehydrating agent, having a solubility relative to CH_4 of 11,000, much more than H_2S and CO_2. Although the dryness

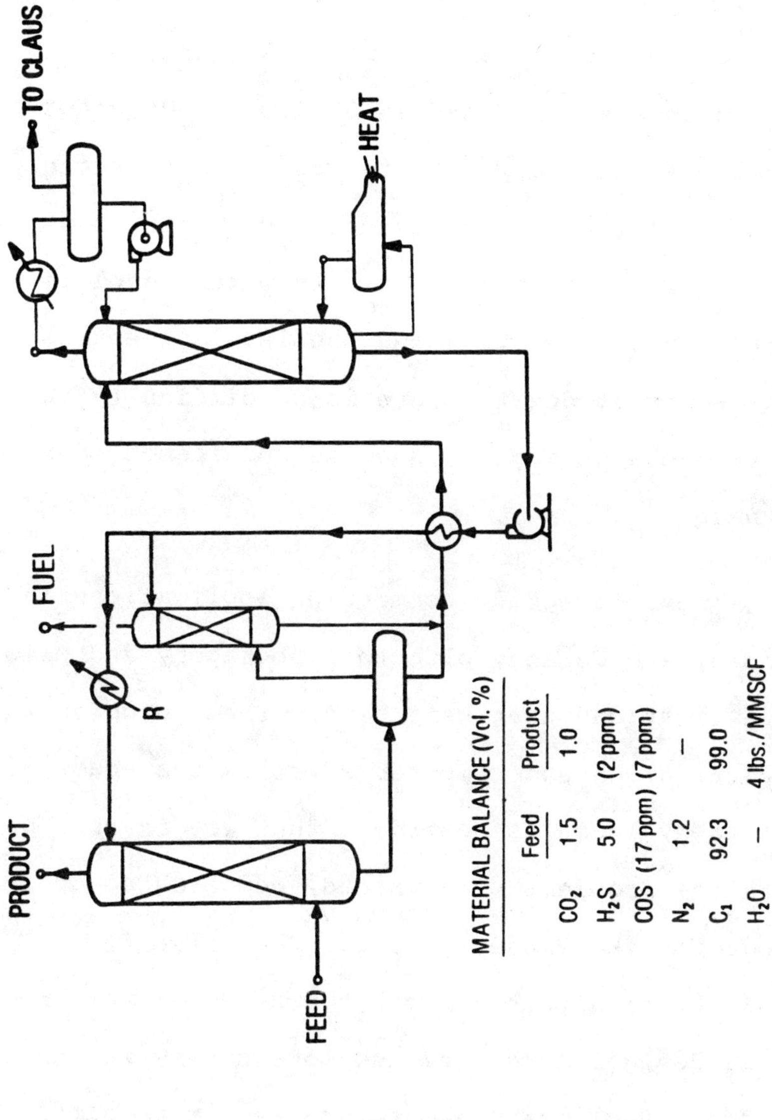

MATERIAL BALANCE (Vol. %)

	Feed	Product
CO_2	1.5	1.0
H_2S	5.0	(2 ppm)
COS	(17 ppm)	(7 ppm)
N_2	1.2	—
C_1	92.3	99.0
H_2O	—	4 lbs./MMSCF

Figure 6. Pan Canadian Petroleum, Ltd., Morley Plant.

requirement in Canada is more stringent than that for the U.S. (4 lbs./MMSCF vs. 7 lbs./MMSC F), SELEXOL solvent does accomplish satisfactory drying in the Pan Canadian plant, and does not require further drying with TEG after treatment. This is so even with a quantity of water, 3 wt. %, in the circulating solvent. If water level is kept down to 3.0 wt. % and the lean solvent is chilled to 35 deg.F, there is no difficulty in meeting the stringent Canadian requirement for dryness.

These foregoing commercial applications in Germany and Canada, plus the laboratory and pilot investigations that went before them, plus further investigation and computer advances that came after them, leave us confident that we can cope well with the problems of advanced, complicated projects like Project No. 1. The solvent, SELEXOL, remains the same, but as more is learned about designing for it, the more surely we can extend our applications into the ever more difficult treating challenges now arising.

Selective Removal of Sulfur: Project No. 2

Project No. 2 is a pilot project intended to demonstrate the effectiveness of CO_2 injection in increasing production from an oil field which has tight sands. A suitable source of natural CO_2 has been found only 90 km away, but it contains 4000 ppmv of H_2S, which must be removed for safe transportation of the gas over the 90 km distance.

The CO_2 is intended to so reduce the viscosity of the oil that production rates can be increased significantly over present levels. If the pilot project succeeds, the SELEXOL treating facility now under construction will be considerably expanded.

The raw gas contains 90.6 vol. % CO_2, 9.0 vol. % CH_4-N_2, and 0.4 vol. % H_2S. All that is required is removal of H_2S to 50 ppmv, since the CO_2 will be sufficiently pure for the purpose without removal of N_2. This requires a difficult selective absorption of H_2S from a background of almost pure CO_2.

The flowscheme for the plant, shown in Fig. 7, is not very different from the Pan Canadian plant of Fig. 6, but, since CO_2 is a very non-ideal gas being handled in this project near critical conditions, new VLE data were required to insure that the design was valid and the objectives would be achieved.

This project is significant in the development of EOR by CO_2 injection, since success in this area will depend on having adequately large supplies of CO_2, which in turn will depend on the ability to remove H_2S from near-pure CO_2 to permit safe pipeline transport of injection gas.

Removal of small amounts of H_2S with CO_2: Project No. 3

Project No. 3 involves removal of a large amount of CO_2 - 36% - and a relatively small amount of H_2S - 90 ppmv. Regeneration for CO_2 is very economical, requiring only flashing; regeneration for H_2S is less economical, generally requiring heat. It is desirable to use a flowscheme which allows separate regenerations,

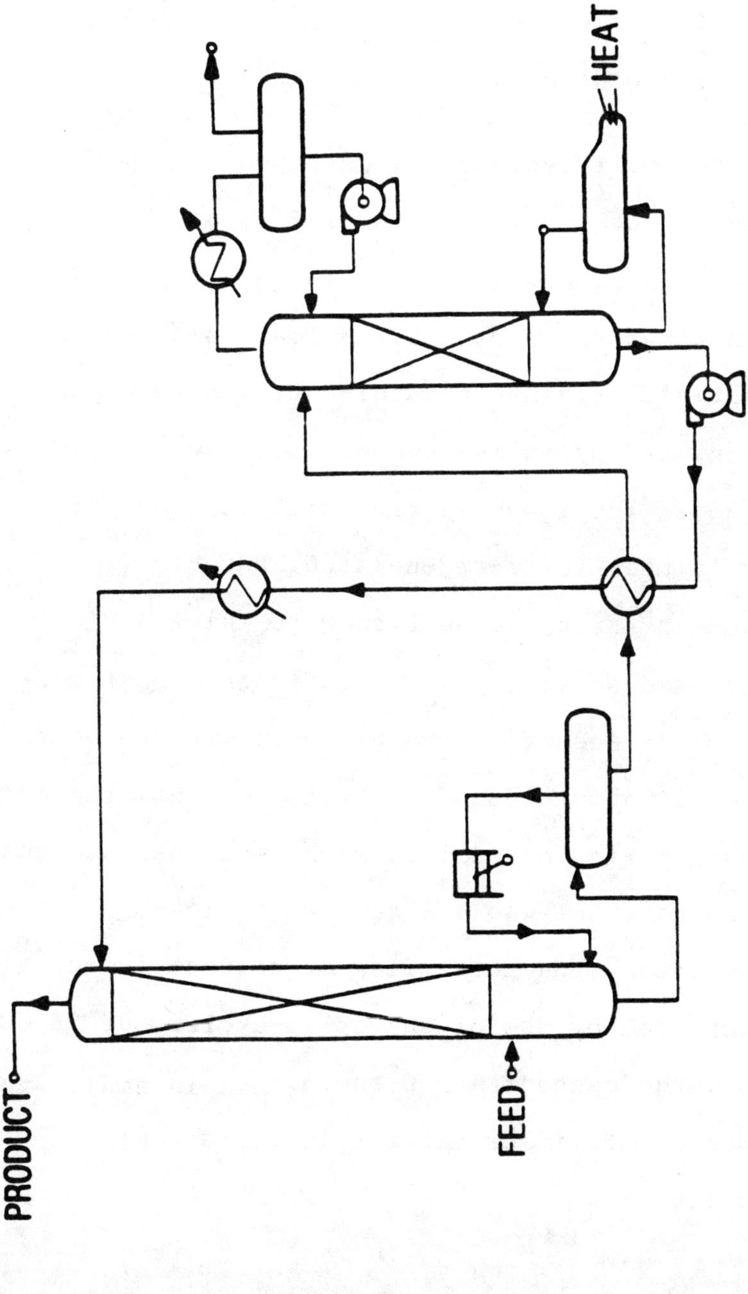

Figure 7. Project No. 2.

even if the two components are removed together. This is the split cycle, first used in the Oates plant of the Northern Natural Gas Co., located near Ft. Stockton, Texas.

The split cycle is illustrated in the flowscheme of the Oates plant, Fig. 8. Both H_2S and CO_2 are removed in a single absorber, but the absorber has two sections: a lower section, fed by partially regenerated solvent, and an upper section, fed by fully-regenerated solvent. CO_2 is removed in the lower section with solvent which has only been flash-regenerated, and H_2S is removed in the upper section with fully regenerated solvent. The flow in the upper section is far less than in the lower; in the Oates case, only about 10% of the total. Thus, the cost of regenerating overall is much less than it would have been if the whole stream were fine-regenerated. The total flow of solvent is actually increased by use of the split-cycle, but if CO_2 is in large concentration and H_2S is in small, the savings in utilities makes this flowscheme attractive.

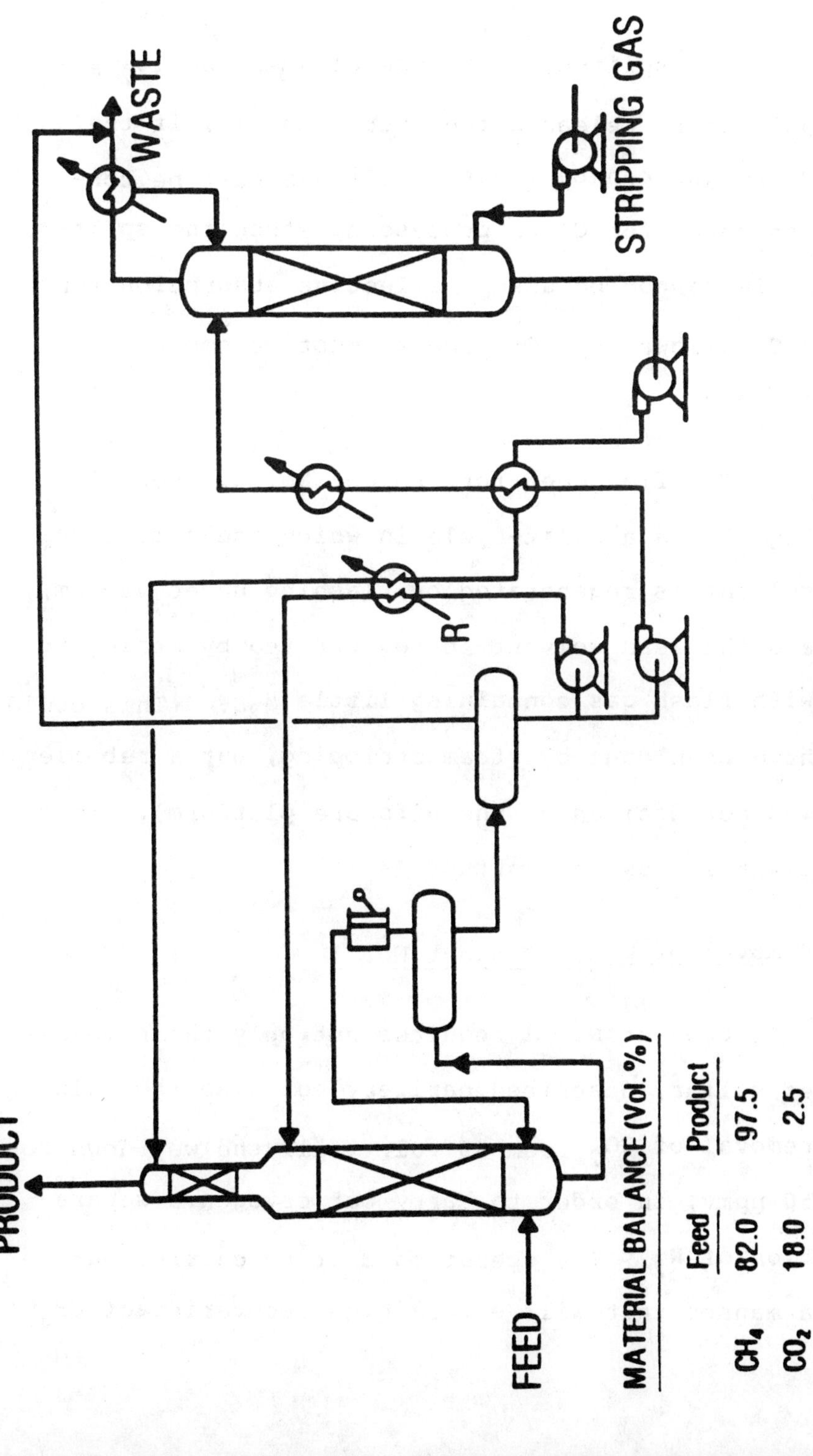

Figure 8. Northern Natural Gas Co., Oates Plant.

The split-cycle is useful when the CO_2 and H_2S can be released together; that is, if the waste gas can be flared. (If H_2S must be concentrated for Claus processing, then the split-cycle cannot be used - selective absorption for H_2S followed by CO_2 removal must be employed instead.)

The flowsheet for Project No. 3, shown in Fig. 9, is a split-cycle in which the semi-lean solvent is regenerated by flashing under vacuum, and the lean solvent is regenerated by stripping with flash gas containing little H_2S. (This could have been done by steam stripping, but a reboiler was not desired on the offshore platform). This plant is now in startup.

Removal of CO_2: Project No. 1

Project No. 1 requires not only the removal of sulfur, described earlier, but also the bulk removal of CO_2 from 65 vol. % all the way down to 50 ppmv, in order to carry out cryogenic separation of N_2. The operation must be carried out in a manner that will secure high recoveries of CH_4

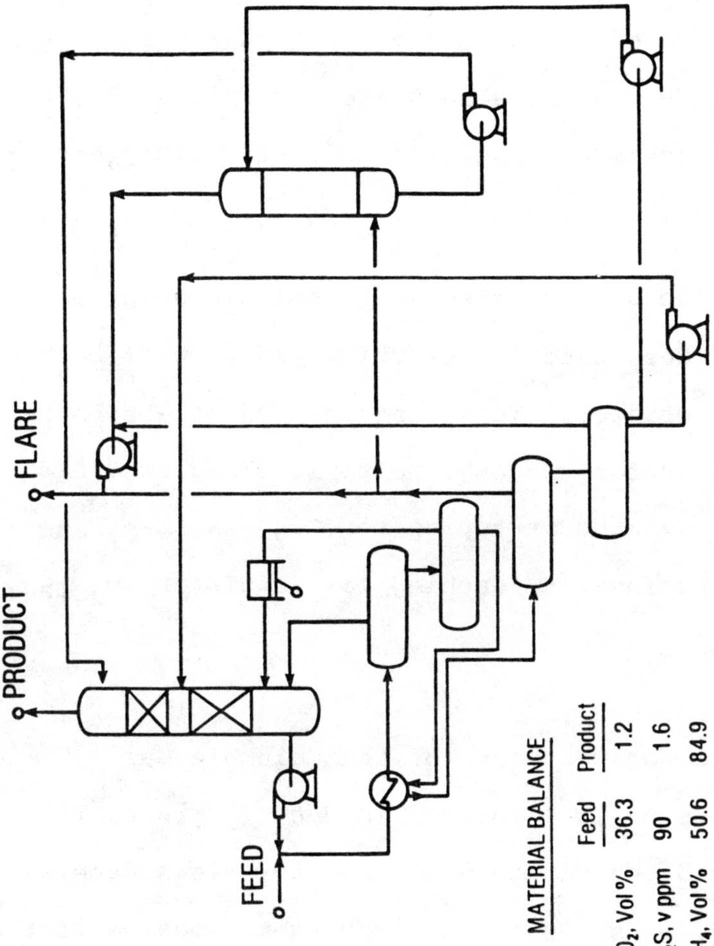

MATERIAL BALANCE

	Feed	Product
CO_2, Vol %	36.3	1.2
H_2S, v ppm	90	1.6
CH_4, Vol %	50.6	84.9
C_2+, Vol %	11.1	10.5
N_2, Vol %	2.0	3.4

Figure 9. Project No. 3.

(99%) and CO_2 (98%) of the main products, and produce CO_2 of at least 97% purity and 7 lbs./mmscf dryness. (Sulfur has already been taken care of in the H_2S Removal Section). This must be done with as much economy as possible to control costs of what will be a relatively expensive operation overall.

To secure these ends, techniques which have already been used in operating plants will be used in this project. These are the use of absorber-exchangers for economy, multiple flashing for economy, vacuum stripping for CO_2 recovery, and gas stripping for reaching low levels of CO_2 in CH_4 product.

The Kanawha plant of the Columbia Gas Transmission Co., located in West Virginia, illustrates the first three. On the flowscheme, Fig. 10, is shown a heat exchanger, located in the drawing near the absorber. This is actually part of the absorption train, operating in series with the absorber proper. Raw feed gas makes its first contact with solvent in the absorber-exchanger, in

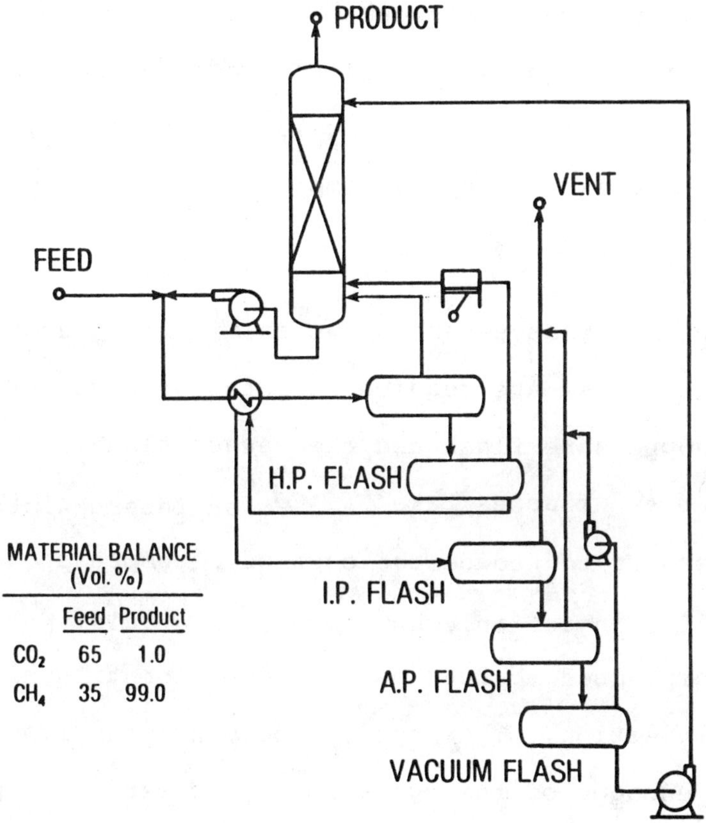

Figure 10. Columbia Gas Transmission Corp., Kanawha Plant.

which about 40% of the CO_2 is removed, leaving the remaining 60% for the absorber proper. Temperature in the exchanger is controlled at a low value, which reduces the required solvent circulation. (Conditions at the last, or bottom, contact determine the circulation required.) The absorber-exchanger thus improves econcomy by reducing circulation.

Four flashes are shown in Fig. 10, but only two of these are required for the process: The high-pressure flash and the vacuum flash. The other two flashes take off CO_2 at intermediate pressures to reduce the cost of recompression of CO_2 for injection, which must, as a rule, be recompressed to 2000 psig or more for transmission or injection. The physical solvent thus improves the economy of the operation by conserving a portion of the original pressure of the CO_2 in the feed.

The Kanawha plant employs vacuum regeneration in order to reach 1.0 vol. % CO_2 in the product without having to use gas or heat regeneration.

This technique, also proved out in four other DMPEG plants, will be used in Project No. 1 to help bring recovery of CO_2 to the high level of 98%.

Removal of CO_2 by the SELEXOL process must be quite complete - down to 50 ppmv - in order to permit downstream cryogenic separation of N_2 from CH_4. Levels such as this are unusual in natural gas purification, but not in synthesis gas purification, in which the SELEXOL process has been extensively employed[4]. CO_2 can be removed down to any level desired, if the means of regeneration is adequate. This was demonstrated several years ago in a commercial ammonia synthesis gas purification plant employing the SELEXOL process, which in regular operation reduced CO_2 from 33 vol. % down to 0.3 vol., % or 3000 ppmv. In tests, however, the CO_2 level was brought down progressively to only 12 ppmv as the regeneration by stripping gas was made more complete. This plant, owned by Ammoniaque Synthetique et Derives (ASED) of Willebroek, Belgium, provided the actual experience to convince us and our clients that such low levels as 50 ppmv can be reached with gas stripping of DMPEG.

The flowscheme for the CO_2 section of Project No. 1 is shown in Fig. 11A-B. This design employs an absorber-exchanger, multiple flashing, vacuum regeneration, and gas stripping - all of which have been used in commercial plants before. This prior experience gives us confidence that the exacting demands of this important project can be met.

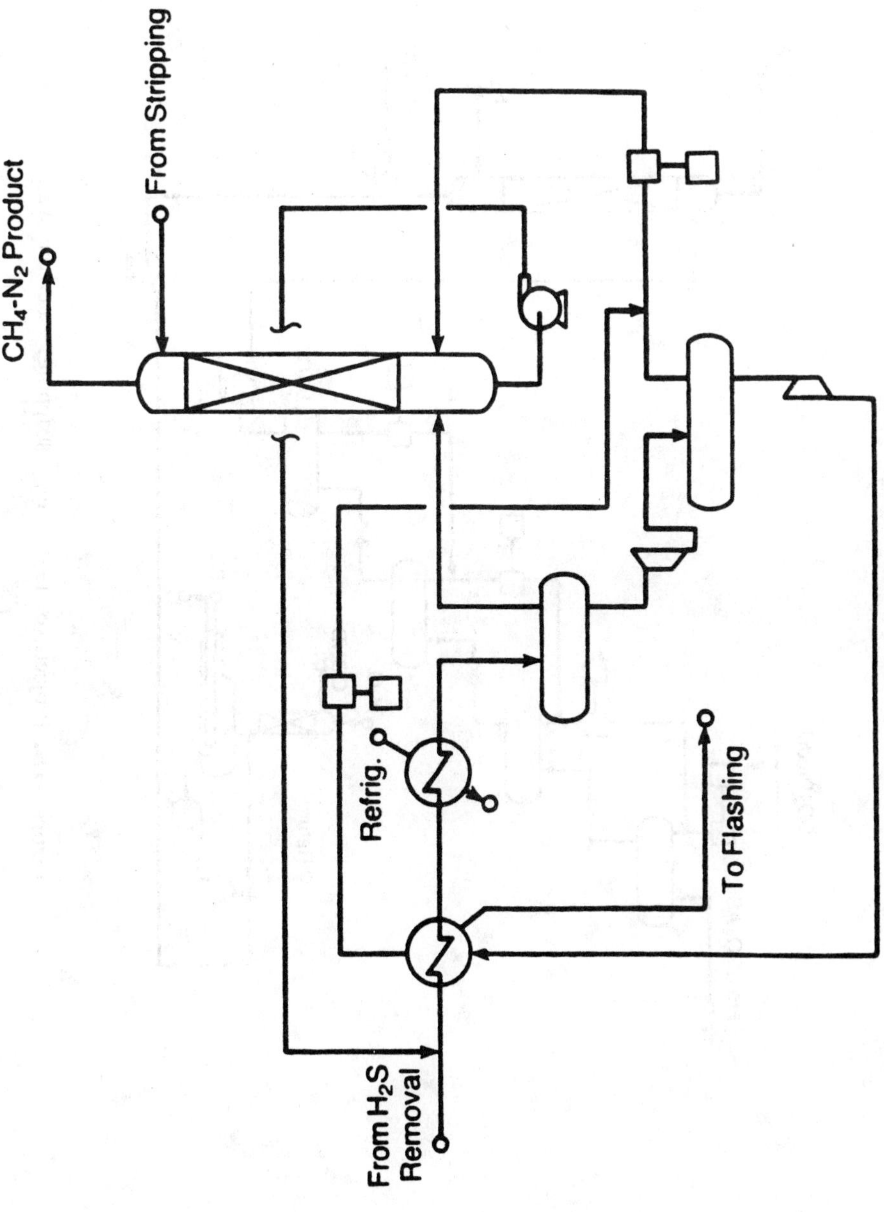

Figure 11A. Project No. 1—CO_2 absorption.

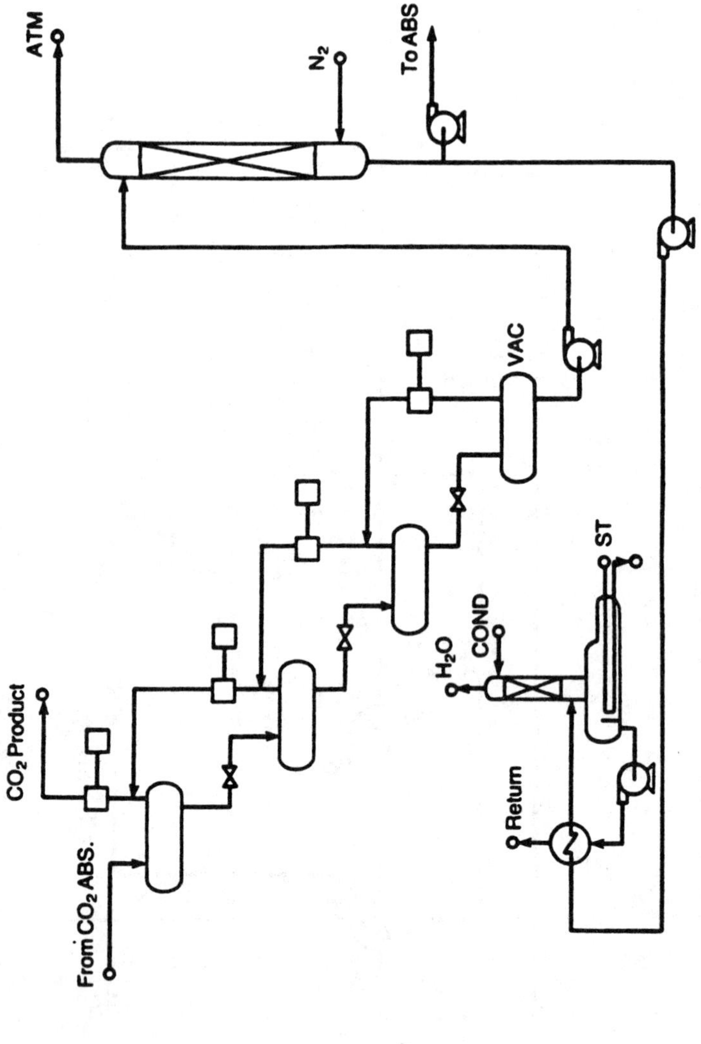

Figure 11B. Project No. 1—flashing, stripping, water removal.

Carbonyl sulfide removal: Project No. 1

Natural gases are not supposed to contain COS, but they often do if they contain an appreciable amount of H_2S. Project No. 1 raw gas will contain some COS, somewhere in the 100-200 ppmv range. This must be considered in process design because in DMPEG processing, much of the COS will appear in the CO_2 product, if nothing is done about it. This is not a problem with CO_2 for injection, but if the CO_2 is released to the atmosphere, as it may be on occasion, the COS needs to be removed.

COS was encountered early in SELEXOL process experience in the NEAG II plant. In the testing of NEAG II after its conversion to the SELEXOL process, it was determined that COS could be removed as completely as any other component, provided that absorption and regeneration are adequate.

COS can then be removed effectively with the SELEXOL process, but it may not be advisable to take this route if selective absorption for H_2S is involved. Fig. 1 shows that the relative solubilities of the 3 components involved in selective absorption are: CO_2, 15.0; COS, 35.; and H_2S, 134. As described earlier, the difference in solubility between H_2S and CO_2 is sufficient to give a good basis for selective removal of H_2S, but the figures above show that CO_2 and COS have not so favorable a difference: if COS must be removed to a high degree, more CO_2 will be carried with it, other things being equal. If COS is removed along with H_2S, the feed to the Claus unit will be diluted with more CO_2 than it would be with H_2S alone.

This problem becomes acute with synthesis gases from coal or oil, which typically will contain 30-35 vol. % CO_2, 0.5 to 0.7 vol. % H_2S (5000-7000 ppmv), and 200-600 ppmv COS. The COS must be removed to protect catalyst, but there is not enough H_2S to make an acceptable feed for the Claus plant if COS must be removed too.

This problem was met in the Tennessee Valley Authority's coal-to- ammonia plant by use of

Haldor-Topsoe COS hydrolysis. High conversion of COS to H_2S can be accomplished in a vapor-phase catalytic hydrolysis, such that COS virtually disappears as a component, leaving only H_2S and CO_2 as keys. These have relative concentrations and solubilities suitable in SELEXOL processing to produce a Claus feed containing 20 vol. % H_2S.

In Project No. 1, COS hydrolysis will be used not so much for the sake of Claus gas richness but for CO_2 recovery; less CO_2 to the Claus plant means more CO_2 for the pipeline. Fig. 3 shows how the hydrolysis will be carried out at a convenient point after removal of most of the H_2S. With H_2S removed, the hydrolysis reaction can be more completely to the right, since H_2S is a product:

$$COS + H_2O = H_2S + CO_2$$

Process improvement:

Development of our process goes on. We are getting (1) from the laboratory: new vapor-liquid equilibrium data to cover new system conditions not previously considered commercial, new thermal data (heats of absorption) to make closer designs possible, and testing of elastomers which may have advantages over those now used; (2) from the office: improved VLE data correlation for better interpolation and extrapolation of experimental data, and extension of computer use further into the rigorous and integrated calculations not previously possible; (3) from the plant: plant performance data to confirm designs already built; pilot and plant data on performance of packings, investigating both new and old packings under the more extreme conditions now of interest; and testing of new devices for improvement of power recovery and economy.

We are confident that growth in this fashion, improvements confirmed by practice, will lead to wider adoption of adaptable, easily-operated physical solvent processing.

REFERENCES

1. Johnson, J.; Homme, A.: "SELEXOL® Solvent Process Reduces Lean High-CO_2 Natural Gas Treating Costs," Am. Inst. Chem. Engrs. 1983 Summer National Meeting, Gas Processing Symposium.

2. Michaeli, W.; Stein, W.' Unland, H.:" "Das SELEXOL® Verfahren - Ein Beitrag zur selektinen Aufbereitung von sauren Erdgas," ("The Selexol Process - a contribution to selective treatment of sour gas"), Erdoel Erdgas Zeitschrift, Oct. 1975.

3. Woelfer, W.: "Construction and Operation of a SELEXOL® Natural Gas Purification Plant," 27th Canadian Chemical Engineering Conference, Calgary, Alberta, Oct. 27, 1977.

4. Swanson, Carl G., Jr.: "Carbon Dioxide Removal in Ammonia Synthesis Gas by SELEXOL®", 71st Annual Meeting, Am. Inst. of Chem. Engrs., Miami Beach, Florida, Nov. 12-16, 1978.

2

Comparison of Physical Solvents Used for Gas Processing

R. W. Bucklin
Fluor Engineers, Inc.

R. L. Schendel
Fluor Engineers, Inc.

INTRODUCTION

With recent emphasis on energy and capital savings, physical solvent processes have received considerable attention. El Paso's Terrell County Treating Plant, built in the late 1950's, was a first of a kind using a physical solvent (propylene carbonate) at mild sub-ambient temperatures for carbon dioxide removal. Fluor's process objectives of lowest possible capital and operating costs were achieved. Neither external heat nor alloy steel equipment were required to satisfy the process conditions. The process was unique in that the only significant energy consumers were the solvent circulation pumps and the recycle gas compressor. The process design was very simple, involving only a high pressure contactor where the CO_2 was removed from the methane, followed by a series of flash tanks at successively lower pressures to achieve solvent regeneration. A compressor was used to recycle the flash gases from an intermediate pressure flash tank to minimize methane losses. Hydraulic turbines were used to recover about half of the required pumping energy. The same type of mechanical arrangement and flow scheme have since been used in several gas treating plants with competitive physical solvents developed since that time.

The chief criterion for selection of propylene carbonate for Fluor Solvent was its high CO_2 solubility concurrent with a relatively low methane solubility. Even today, propylene carbonate ranks somewhat better than other physical solvents for bulk CO_2 removal with minimum hydrocarbon loss as the only contingent requirement.

As developments in physical solvent processing matured, other solvents, better suited to meet specific process requirements, have been developed.

COMPARISON OF COMMERCIAL PHYSICAL SOLVENTS

Today the commercially proven physical solvent processes and their solvents are:

Estasolvan	-	tributyl phosphate or TBP
Fluor Solvent	-	propylene carbonate or PC
Purisol	-	normal methyl pyrrolidone or NMP
Rectisol	-	methanol
Selexol	-	dimethyl ether of polyethylene glycol or Selexol
Sepasolv-MPE	-	mixture of polyethylene glycol dialkyl ethers or Sepasolv

Of these solvents, methanol is relatively high in vapor pressure at normal process conditions and therefore requires deep refrigeration or special recovery methods to prevent high solvent losses. This paper will omit methanol from comparisons of solvents since the processing conditions and equipment are so unlike the others.

Most of the equilibrium data are proprietary to the process licensors. Therefore, definitive comparative information about solvent performance cannot be published without violating existing secrecy agreements. It is possible, however, to use public information to indicate relative circulation rates, relative recycle stream volumes, and stream compositions for identical process configurations and conditions in order to illustrate how some solvents differ in basic character and performance. Selexol and propylene carbonate are compared in this manner in the case study later in this paper.

Table 1 (3,4,5,6,7) is a comparison of miscellaneous solvent data. All of the solvents are noncorrosive, nontoxic and require only carbon steel construction for a simple cycle process scheme.

All the solvents have a pleasant or no odor except NMP which is quite noxious to personnel.

The relatively poor CO_2 solubility of TBP may explain why no commercial plants using the Estasolvan process have been built.

Table 2 (4,7,8,9,10) compares each solvent's affinity for various gases relative to carbon dioxide.

There is a wide variation in molecular weights and significant differences in densities of the various solvents. Therefore, the comparison of mol fraction or weight portion of solute in the solvent at saturation

Table 1A (3,4,5,6,7)
Miscellaneous Comparative Data of Solvents

Process Name	Selexol	Fluor Solvent
Solvent Name	Selexol	PC
Solvent Cost \$/lb FOB Fact.	1.32	.74
Licensor	Norton	Fluor
Viscosity @ 25°C, cp.	5.8	3.0
Specific Gravity @ 25°C, KG/M^3	1030	1195
Mol Weight	280	102
Vapor Pressure @ 25°C, MM Hg	7.3×10^{-4}	8.5×10^{-2}
Freezing Point, °C	-28	-48
Boiling Point, °C @ 760 MM Hg		240

Table 1A (3,4,5,6,7) (Continued)
Miscellaneous Comparative Data of Solvents

Process Name	Selexol	Fluor Solvent
Thermal Conductivity Btu/Hr/Ft^2/(°F/Ft)	0.11	0.12
Maximum Operating Temp., °C	175	65
Specific Heat @ 25°F	0.49	0.339
Water Solubility @ 25°C	∞	94 gm/l
Solvent Solubility in Water @ 25°C	∞	236 gm/l
Ft^3 CO_2 Solubility/ U.S. Gal @ 25°C	0.485	0.455
Number of Commercial Plants	32	13

Table 1A (3,4,5,6,7) (Continued)
Miscellaneous Comparative Data of Solvents

Process Name	Selexol	Fluor Solvent
Bulk CO_2 Removal		
Synthesis Gas	6	3
Natural Gas	6	10
Landfill Gas	3	0
Selective H_2S Removal		
Synthesis Gas	9	0
Natural Gas	8	0

Table 1B (3,4,5,6,7)
Miscellaneous Comparative Data of Solvents

Process Name	Purisol	Sepasolv MPE	Estasolvan
Solvent Name	NMP	Sepasolv	TBP
Solvent Cost $/lb FOB Fact.	1.34		
Licensor	Lurgi	B.A.S.F.	Uhde & IFP
Viscosity @ 25°C, cp.	1.65	-	2.9
Specific Gravity @ 25°C, KG/M^3	1027	-	973
Mol Weight	99	320	266
Vapor Pressure @ 25°C, MM Hg	4.0×10^{-1}	3.7×10^{-4}	$<1.0 \times 10^{-2}$
Freezing Point, °C	-24		-80
Boiling Point, °C @ 760 MM Hg	202	320	(180* @ 30 MM Hg)

Table 1B (3,4,5,6,7) (Continued)
Miscellaneous Comparative Data of Solvents

Process Name	Purisol	Sepasolv MPE	Estasolvan
Thermal Conductivity Btu/Hr/Ft^2/(°F/Ft)	0.095		
Maximum Operating Temp., °C	-	175	-
Specific Heat @ 25°F	0.40		
Water Solubility @ 25°C			65 gm/l
Solvent Solubility in Water @ 25°C			0.42 gm/l
Ft^3 CO_2 Solubility/ U.S. Gal @ 25°C	0.477	0.455	0.329
Number of Commercial Plants	5	4	0

Table 1B (3,4,5,6,7) (Continued)
Miscellaneous Comparative Data of Solvents

Process Name	Purisol	Sepasolv MPE	Estasolvan
Bulk CO_2 Removal			
Synthesis Gas	2	0	
Natural Gas	1	0	
Landfill Gas	0	0	
Selective H_2S Removal			
Synthesis Gas	1	0	
Natural Gas	1	4	

Table 2A (4,7,8,9,10)
Solubilities of Various Gases in Solvents Relative to Carbon Dioxide at 25°C

Comp.	Selexol	PC
H_2	1.3×10^{-2}	7.8×10^{-3}
N_2	-	8.4×10^{-3}
O_2	-	2.6×10^{-2}
CO	2.8×10^{-2}	2.1×10^{-2}
C_1	6.7×10^{-2}	3.8×10^{-2}
C_2	4.2×10^{-1}	1.7×10^{-1}
C_2H_4	4.9×10^{-1}	3.5×10^{-1}
CO_2	1.0	1.0
C_3	1.02	0.51
iC_4	1.87	1.13
nC_4	2.33	1.75
COS	2.33	1.88
iC_5	4.47	3.50
C_2H_2	4.53	2.87
NH_3	4.87	-
nC_5	5.53	5.0
H_2S	8.93	3.29
NO_2	-	17.1
nC_6	11.0	13.5
2,4 DMP	-	17.5

Table 2A (4,7,8,9,10) (Continued)
Solubilities of Various Gases in Solvents
Relative to Carbon Dioxide at 25°C

Comp.	Selexol	PC
CH_3SH	22.7	27.2
nC_7	24.0	29.2
CS_2	24.0	30.9
CYCLO-C_6	-	46.7
nC_8	-	65.6
C_2H_5SH	-	-
SO_2	93.3	68.6
$(CH_3)_2S$	-	-
C_6H_6	253	200
nC_{10}	-	284
C_4H_4S	540	-
H_2O	733	300
HCN	1200	-

Table 2B (4,7,8,9,10)
Solubilities of Various Gases in Solvents Relative to Carbon Dioxide at 25°C

Comp.	NMP	Sepasolv	TBP
H_2	6.4×10^{-3}	5.0×10^{-3}	
N_2	-		
O_2	3.5×10^{-2}		
CO	2.1×10^{-2}		
C_1	7.2×10^{-2}	6.6×10^{-2}	4.0×10^{-2}
C_2	3.8×10^{-1}		
C_2H_4	5.5×10^{-1}		
CO_2	1.0	1.0	1.0
C_3	1.07		
iC_4	2.21		
nC_4	3.48		
COS	2.72	2.54	
iC_5	-		
C_2H_2	7.37		
NH_3	-		
nC_5	-		
H_2S	10.2	6.86	5.6
NO_2			
nC_6	42.7		
2,4 DMP	-		

Table 2B (4,7,8,9,10) (Continued)
Solubilities of Various Gases in Solvents
Relative to Carbon Dioxide at 25°C

Comp.	NMP	Sepasolv	TBP
CH_3SH	34.0	23.1	
nC_7	50.0		
CS_2	-		
CYCLO-C_6	-		
nC_8	-		
C_2H_5SH	78.8		
SO_2	-		
$(CH_3)_2S$	91.9		
C_6H_6	-		
nC_{10}	-		
C_4H_4S	-		
H_2O	4,000		
HCN	-		

would not be good indicators of the solvent's relative effectiveness in removing the solute. The volume of solute expressed as vapor at the reference condition per unit volume of solvent at the reference condition is a useful comparative value for solvents to be used in similar processing schemes. This is true because the size of process equipment and piping as well as power requirements for the process are largely determined by the required volume of solvent circulation.

All of the physical solvent processes being compared are concerned primarily with acid gas removal from either hydrocarbon gases (natural and landfill gas) or synthesis gases (hydrogen and carbon monoxide). In cases wherein either bulk or essentially complete CO_2 removal is desired the most significant comparative solvent data is solubility data on volume of carbon dioxide per volume of solvent at a suitable reference condition. Operating process temperature ranges from 30°C to -20°C cover most of the commercial applications, so 25°C is a suitable reference temperature for this comparison. Solubility data at 25°C and one atmosphere partial pressure of solute is in the public domain for all of the commercial solvents we are comparing except Sepasolv. The value for Sepasolv was extrapolated from published data (7) at 0°C using the same slope of a similar ether on a log log plot of Henry's constant vs 1/T.

In all cases, the solution is relatively dilute with respect to the solute. The polar compounds CO_2 and H_2S tend to deviate significantly from ideal, so that Henry's Law is applicable only in dilute solutions. Even though there can be sizeable interaction effects between the solutes in multicomponent mixtures, it is assumed for this comparison that the relative selectivities will not vary significantly in relation to each other in the majority of actual process conditions.

PROCESS SELECTION

The selection of a physical solvent process depends on process objectives and characteristics of the solvents, such as selectivity for H_2S, COS, HCN, etc., ease of handling water content in feed gas, ease of controlling water content of circulating solvent, concurrent hydrocarbon loss or removal with acid gas removal, solvent cost, solvent supply, chemical inertness, royalty cost, thermal stability and proven plant performance for various processing techniques.

Solvent Loss

All of the solvents have low vapor pressures. Although propylene carbonate has a vapor pressure much higher than the high molecular weight solvents, solvent losses have generally been very low. NMP has a vapor pressure about five times higher than PC. The licensor recommends water washing of both the treated gas and the rejected acid gases for solvent recovery (11).

Effect of Viscosity

While the higher molecular weight solvents have a lower vapor pressure, they also have a higher viscosity. This reduces mass transfer rates and tray efficiencies and increases packing or tray requirements. This effect becomes especially noticeable at reduced temperatures. Since it is usually desirable to reduce temperature to increase the solubility of acid gases and reduce the circulation rate, this can be a distinct disadvantage for the higher molecular weight solvents.

Selective H_2S Removal

The data indicate that Selexol, NMP and Sepasolv are superior to PC if selective H_2S removal from gas containing carbon dioxide is required. Actual experience confirms this prediction. The authors know of no cases where propylene carbonate would be recommended for selective H_2S removal. In fact, it is difficult to find situations where PC would be recommended if H_2S is present in more than trace concentrations. This is so because the low concentration of H_2S usually permitted in the treated gas (1/4 grain per 100 SCF for natural gas) means H_2S removal is controlling.

Effect of Water in Feed Gas

Differences in water handling flexibility can also be important. As shown in Table 1, NMP, Selexol and Sepasolv have infinite water solubility and are thermally stable at temperatures required to reject water at atmospheric pressure. Slipstreams of these solvents can be processed to control the water content of the circulating solvent stream.

PC and TBP have limited water solubility and therefore require a different solvent for hydrate control during feed gas chilldown. TBP is thermally stable for water removal by atmospheric distillation. PC slowly reacts irreversibly with water and carbon dioxide at temperatures around 90°C and is therefore unsuitable for water control by atmospheric distillation (12).

The reported design water content of the various solvents has a range from one to six percent by weight. At these levels, solvent capacity for CO_2 and H_2S is not greatly impaired for any of the solvents. The most significant penalty of water content is the cost of pumping the extra water.

Physical solvents may be used to simultaneously dehydrate the gas and meet very low treated gas specifications for CO_2 and H_2S. This is accomplished by use of a solvent regenerator using inert gas and/or heat to strip the lean solvent as required. NMP cannot be used for simultaneous gas dehydration if a water wash is used to limit solvent loss.

Both PC and Selexol tend to get slushy with water at temperatures below -18°C (0°F), so process conditions mus be held warmer than the slush temperature.

Effect of Heavy Hydrocarbons

In natural gas treating, loss of heavy hydrocarbon is a concern. NMP, Selexol and Sepasolv are miscible w water, and water may be used to reject these hydrocarbons. As shown in Figure 1, a slipstream of the circulating solvent from the lean solvent pump may be mixed with the feed gas and fed through the feed gas cooldown using water absorbed from the feed gas to separate a hydrocarbon liquid phase and then distilling off the water. Water can actually be added to this stream to reduce hydrocarbon solubility further. In this case the size of the slip-stream might be set by the desired maximum hydrocarbon content of the lean solvent rather than by the maximum water buildup in the lean solvent. The water content of

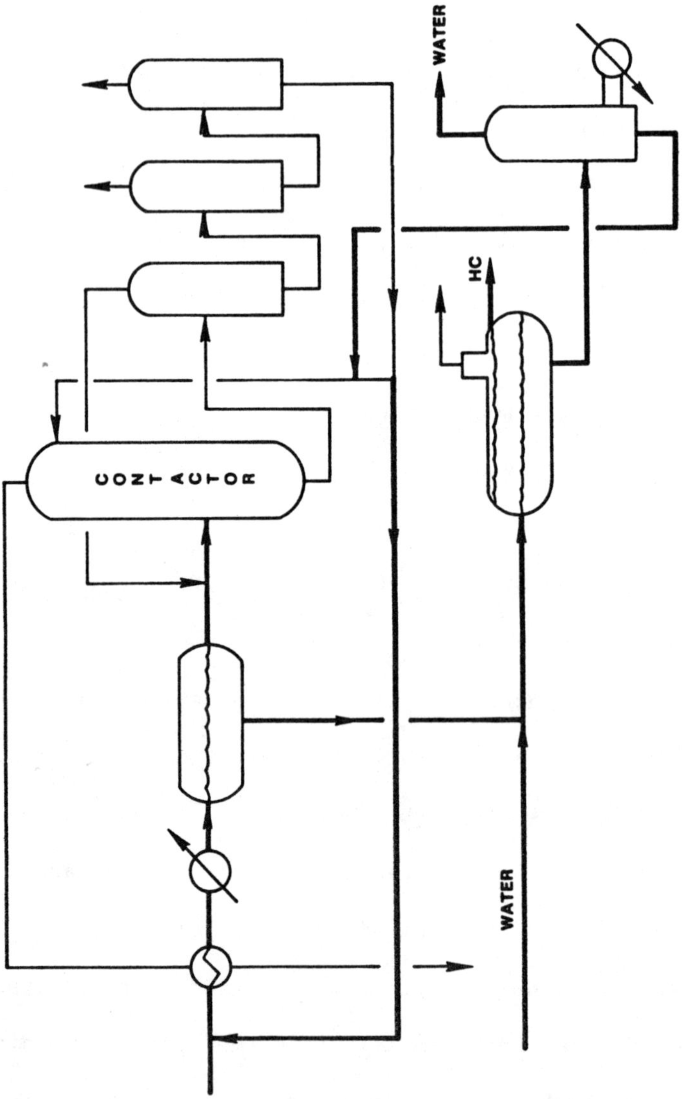

Figure 1. Example of how water may be used to reject heavy hydrocarbons.

the return solvent slipstream can be controlled by the slipstream regenerator bottoms temperature and pressure conditions.

Effect of Recycle Compressor

A major energy user in physical solvent processes is compression for the recycle of flash gas to limit methane losses. The relative compression horsepower required to recycle these intermediate flash tank gases to the high pressure contactor can be predicted from the solubility of methane in the various solvents as shown in Table 3. The higher the solubility of the methane, the higher the recycle compressor horsepower for the same amount of methane product in the treated gas.

Solvent Reactivity

Propylene carbonate reacts with amines and ammonia at all conditions and with water and carbon dioxide at elevated temperatures. PC has been used for bulk CO_2 removal followed by a downstream MEA treater in two

Table 3
Comparative Hydrocarbon Losses**
Relative to Propylene Carbnate

	Ratio of Table 2 Data				
Comp.	PC	Seloxol	NMP	Sepasolv	TBP
C_1	1.0	1.76	1.89	1.74	1.05
C_2	1.0	2.47	2.24	-	-
C_3	1.0	2.00	2.10	-	-
iC_4	1.0	1.65	1.96	-	-
nC_4	1.0	1.33	1.99	-	-
iC_5	1.0	-	-	-	-
nC_5	1.0	1.11	-	-	-
nC_6	1.0	0.81	3.16	-	-
nC_7	1.0	0.82	-	-	-

*Losses could be termed as recovery if it is desirable to either remove hydrocarbons with the carbon dioxide in order to reduce treated gas heating value or to recover the propane and heavier hydrocarbons from the CO_2 in downstream processing.

successful installations. Careful design of the Fluor Solvent Treated Gas scrubber is required to avoid the possibility of destroying both the amine and the propylene carbonate upon mixing. None of the other solvents being compared are chemically reactive with the components normally found in either natural gas or synthesis gas.

Process Configuration

Good thermal stability, chemical inertness, and thermal conductivity are also necessary to permit flexibility in process schemes. For example, selective H_2S removal can be benefitted by the use of heat. This can be particularly important in designs to produce a high concentration H_2S feedstock to a downstream Claus plant. Reboiling a solvent in a regenerator may be necessary to meet treated gas purity requirements for CO_2, H_2S or COS.

Reboiled absorbers, refrigerated solvent presaturators, absorber side chillers and absorber bottoms to feed gas pumparound chillers are examples of absorption process techniques used to minimize circulation rates which are applicable to physical solvent processes.

SYNTHESIS GAS TREATING COMPARISONS

Table 4 is a comparison of relative solubilities for some of the gases formed in steam reforming processes, by partial oxidation of heavy hydrocarbons or by coal gasification.

The differences between the solvents are not as significant as the ratios would indicate since the quantity of hydrogen and carbon monoxide absorbed concurrent with the CO_2 is relatively small. There might be a moderately significant difference in compression power requirements for intermediate pressure gases recycled to minimize product loss.

Product purity requirements might be more important in process selection. If the desired purity can be obtained with solvent regeneration by atmospheric regeneration of the solvent using inert gas stripping or by vacuum flashing, the processes are essentially equal for CO_2 removal. If H_2S is present in more than trace amounts, NMP, Selexol and Sepasolv would be favored. If control of water concentration by solvent distillation were required, PC would require an alternative independent

Table 4
Comparative Performance for Synthesis Gases

Relative to Propylene Carbonate
Ratio of Table 2 Data

Comp.	PC	Seloxol	Sepasolv	NMP
H_2	1.0	1.67	0.64	0.82
CO	1.0	1.33		1.0
C_1	1.0	1.76	1.74	1.89
CO_2	1.0	1.0	1.0	1.0

water removal step on the feed gas. The optimum choice in most of the synthesis gas applications may be a result of designer ingenuity or client preference rather than basic solvent capabilities.

CASE STUDY

As mentioned earlier, propylene carbonate has temperature limitations which prevent use of heat for solvent regeneration. This limits options for process configurations and may be a significant detriment when processing gases with hydrogen sulfide.

In applications where CO_2 removal only is required, the process selection is less clear. Hydrocarbons are more soluble when compared to CO_2 in Selexol than Fluor Solvent (see Table 3). Therefore, applications where CO_2 removal only is required and where hydrocarbons are to remain in the methane stream for downstream recovery or increased heat value, Fluor Solvent should show an advantage.

D. K. Judd has described the successful conversion of Northern Natural Gas Company's Mitchell plant from high load DEA to Selexol.[13] Since this is a simple flash regeneration scheme (see Figure 2), it is an appropriate example to examine.

Feed gas enters the Selexol plant inlet scrubber at about 120°F and 895 psig. The gas is cooled by exchange against the residue gas. The residue gas contains about 3.5% CO_2 and enters the pipeline at about 90°F. The cool inlet gas next flows through the feed gas scrubber where condensed water is removed. The gas then enters the absorption towers where the gas contacts Selexol solvent in countercurrent flow.

The rich solvent goes to a sump tank where entrained gas (methane) is allowed to separate and returned to the contactor. The solvent next is chilled with a packaged mechanical refrigeration unit to overcome heat of pumping and heat leaks. The solvent next passes through a hydraulic turbine to recover power and then to a high-pressure flash tank operating at 252 PSIG. The high-pressure vapors are compressed and recycled to the absorber feed.

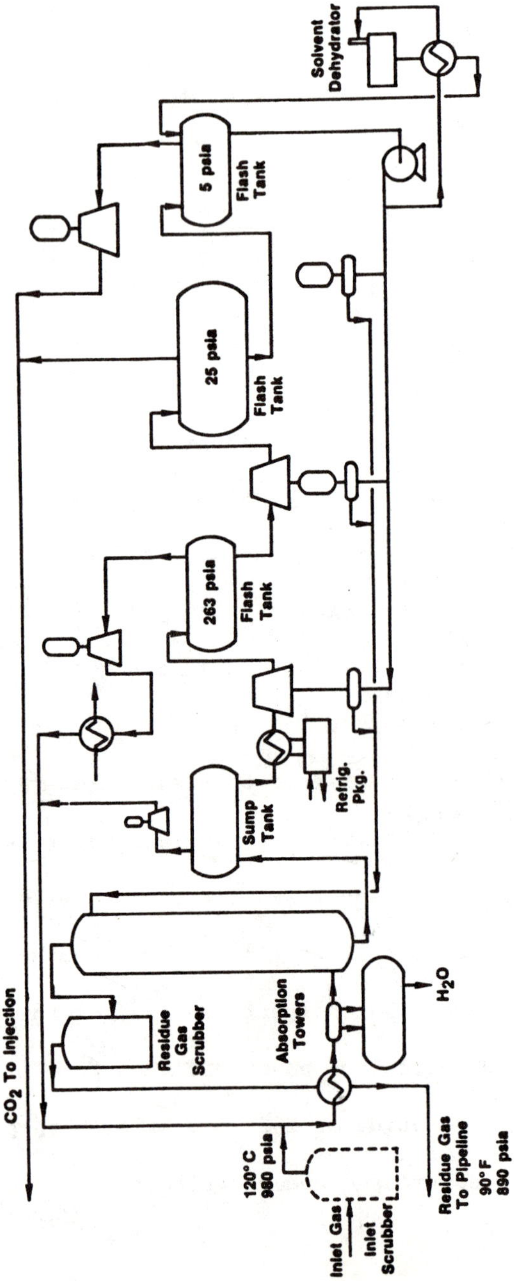

Figure 2. Mitchell Plant Selexol® Unit.

Table 5
Mitchell Plant Design

	Percent of Hydrocarbon Loss		Ratio of Hydrocarbon Loss
	Selexol	PC	Selexol/PC
C_1	3.50	2.17	1.61
C_2	14.34	5.72	2.51
C_3	98.57	35.54	2.72
iC_4	97.47	40.51	2.41
nC_4	96.85	71.65	1.35
iC_5	93.75	79.17	1.18
nC_5	91.67	79.17	1.16
nC_6	87.43	73.15	1.12

Table 6
Comparative Design Material Balance mols/hr for Selexol and Fluor Solvent

Stream No.:	1	2	
Stream Name:	Feed Gas	Residue Gas	
Solvent:	Selexol and Fluor Solvent	Selexol	Fluor Solvent
Comp.			
N_2	71.1	71.0	71.0
CO_2	4,398.3	395.7	407.1
H_2S	0.2	--	--
C_1	11,110.2	10,721.4	10,993.6
C_2	131.1	112.3	123.6
C_3	34.9	0.5	22.5
iC_4	7.9	0.2	4.7
nC_4	12.7	0.4	3.6
iC_5	4.8	0.3	1.0
nC_5	4.8	0.4	1.0
C_6*	17.5	2.2	4.7
H_2O	37.6	0.2	0.6
	15,831.1	11,304.6	11,633.4

Table 6 (Continued)
Comparative Design Material Balance mols/hr for Selexol and Fluor Solvent

Stream No.:	3		4	
Stream Name:	Atmospheric Flash Gas		Vacuum Flash Flash	
Solvent:	Selexol	Fluor Solvent	Selexol	Fluor Solvent
Comp.				
N_2	0.1	0.1	--	--
CO_2	3,490.5	3,325.3	513.0	665.9
H_2S	0.2	0.2	--	--
C_1	388.1	115.9	0.7	0.7
C_2	16.5	7.3	2.3	0.2
C_3	30.7	11.0	3.7	1.4
iC_4	6.4	2.7	1.3	0.5
nC_4	10.0	7.1	2.3	2.0
iC_5	3.2	2.7	1.3	1.1
nC_5	3.1	2.6	1.3	1.2
C_6*	10.5	8.5	4.8	4.3
H_2O	3.9	2.8	1.7	1.8
	3,963.2	3,486.2	532.4	679.1

Table 7
Comparative Requirements of Power Users for Selexol and Fluor Solvent after Retrofit

Item No.	Description	Selexol	Fluor Solvent
P-101 A&B	Lean Solvent Pumps, BHP	2,700*	2,700
	Power Recovery Turbines, BHP	1,350*	1,350
	Net Pump Motor, BHP	1,350*	1,350
R-12	Refrigeration, Tons	600	700
C-102	Recycle Compressor Motor	2,000 BHP Motor	1,000 BHP Motor
	Estimated Shaft BHP	1,500 Shaft BHP	850 Shaft BHP
C-103	Vacuum Compressor Motor	700 BHP Motor	900 HP Motor

*Developed from Judd's description of the refrigeration requirements and assumed pump efficiencies of 70%.

The solvent next passes through a second hydraulic turbine to the intermediate pressure (25 psia) flash tank. Vapors from the flash tank are routed to compressors for pipeline transmission and injection for enhanced oil recovery. The solvent is further regenerated by flashing to 5 psia vacuum before returning to the contactor. The vapors from the 5 psia flash are compressed and join the 25 psia flash tank vapors for further compression.

A Fluor Solvent plant was designed for the same application. Table 6 presents the composition of the feed gas to the Mitchell plant and the material balance, as presented in the 1978 paper with the addition of Fluor Solvent data for comparison. The Fluor Solvent design produces an additional 148.8 MMBTU/hour in residue gas. This reflects the low solubility of hydrocarbons in Fluor Solvent, as shown in Table 5. The Fluor solvent design does require water removal at the front-end.

Table 7 shows a comparison of horsepower requirements based on information available from the Judd paper and the Fluor Solvent design. As indicated, there is a net savings in horsepower as well as improved hydrocarbon recovery. We suspect that the 2,000 BHP motor used in the

Mitchell plant retrofit is considerably larger than required. The recycle compressor in the Fluor Solvent design is 1,000 BHP. We expect the comparable Selexol requirement is about 1,500 BHP.

It should be noted, where separate recovery of natural gas liquids or heavier hydrocarbons are desirable, process configurations may be employed which deviate significantly from this simple pressure let-down scheme. The value of the hydrocarbons to be recovered may dictate the process configuration and solvent choice.

SUMMARY AND CONCLUSIONS

A preliminary screening of physical solvent characteristics can eliminate those solvents that are clearly unsuitable or noncompetitive for a particular application.

Considerable weight should be given to proven performance with the type of process configuration proposed. Innovative designs into unproven territory might lead to undesirable results. The high cost of testing and commercializing a new solvent in any process configuration is a good incentive to stick with proven processes if they can do a satisfactory job.

The demands on physical solvent processes are increasing, losses of valuable components must be minimized, removal of acid gas and trace components to lower levels must be achieved and processes must be capable of selective H_2S removal with simultaneous production of suitable Claus sulfur plant feed. This is causing a revival of performance improving processing techniques common to absorption processes such as the refrigerated oil absorption plants for natural gas liquids recovery built two or three decades ago. The process designer's ingenuity and innovations might easily outdistance small inherent advantages of one solvent over another.

Selexol has a clear experience advantage over all other solvents in all applications involving H_2S and CO_2 removal in hydrocarbon systems.

Fluor Solvent and Selexol both enjoy a clear experience advantage over the other processes in applications for CO_2 removal only.

BIBLIOGRAPHY

1. Kohl, A. L. & Buckingham, P. A. "The Fluor Solvent CO_2 Removal Process", Oil and Gas Journal, May 9, 1960. Vol. 58, No. 19.

2. Buckingham, P. A. "Fluor Solvent Plants: How They Are Working", Hydrocarbon Processing & Petroleum Refiner, April 1964.

3. Kohl, A. L. & Riesenfeld, F. C. "Gas Purification", Third Edition 1979, Gulf Publishing Company, pp. 768-793.

4. Fluor Corporation - unpublished data.

5. Volkamer, K., Wagner, E. and Schubert, F. "New Developments in Purification of Synthesis Gas", Plant/Operations Progress, Vol. 1, No. 2., April 1982, pp. 134-8.

6. Greene, P. A. & Kutsher, G. S. "Carbon Dioxide Removal and Recovery - New Polyglycol Ether Processes", Paper presented at Gas Conditioning Conference, Norman, Oklahoma. April 6-7, 1965.

7. Renon, H. Bulletin of IFP #18-330, July 1970.

8. "Sepasolv MPE for Gas Purification", Sept. 1981 Bulletin of BASF.

9. "Solubility Relationships of Paraffin Hydrocarbons and Sulfur Compounds in M-PAROL®, Technical Report #44 of General Aniline & Film Corporation.

10. Sweny, John W. "High CO_2-High H_2S Removal with Selexol Solvent", Paper presented at 59th Annual GPA Convention, Houston, Texas. March 17-19, 1980.

11. Hochgesand, G. "The Purisol Process for Acid Gas Treatment", Paper presented at Gas Conditioning Conference, Norman, Oklahoma. April 1-2, 1969.

12. Texaco (Jefferson) Chemical Company, Technical Bulletin on Propylene Carbonate. January, 1960.

13. Judd, D. K. "Northern Natural Gas Converts Mitchell Plant to Selexol", Paper presented at the Gas Conditioning Conference, Norman, Oklahoma. March 6-8, 1978.

3

The Rectisol Wash
New Developments in Acid Gas Removal from Synthesis Gas

Gerhard Ranke
Senior Consultant, Process Plant Engineering & Construction Division, Linde AG/TVT Munich (West Germany)

Volker H. Mohr
Manager, Project Development
Lotepro Corporation, New York, N.Y.
(Subsidiary of Linde AG)

INTRODUCTION

The Rectisol wash is a physical wash system using an organic solvent, and operating at temperatures below 32°F. The preferred solvent is methanol, but other solvents like toluene can be substituted for special applications.

The following text outlines the characteristics of physical washes as compared to chemical washes, and compares Rectisol with other physical wash systems. Further applications of Rectisol for the acid gas removal from synthesis gas - in particular, from partial oxidation gas are described in detail and illustrated with several flowsheets.

FUNDAMENTALS AND AREAS OF APPLICATION

Comparison of Physical and Chemical Washes

Chemical wash systems remove the acid gas by forming a chemical compound with the solvent. In the physical wash systems, each component of the feedgas is absorbed in the solvent according to its partial pressure in the gas phase and its solubility coefficient, which depends on the temperature. One and the same solvent can be used for the selective removal of various components and the recovery in separate fractions. Depending on the process requirements, a complete removal and a highly concentrated recovery of individual components can be achieved.

The process designer's main task is to provide optimum conditions for the removal of certain components from a feedgas by proper selection of the solvent as well as solvent flow rate, temperature and type of regeneration.

The primary application of physical washes is the removal of acid gas components. Usually, a selective removal and concentrated rejection of sulfur compounds and CO_2 is required. Oil washes for the recovery of C_3^+ from natural gas and for the removal and recovery of C_2H_2 from cracked gas are operating according to the same principle.

In chemical washes the solvent flow rate is approximately proportional to the amount of acid gas to be removed. Since the solvent loading capacity is nearly independent from the pressure, they are preferred in treating gases with low acid gas content and at low total pressure. Most chemical solvents are saturated at an acid gas partial pressure of approximately 1 bar.

In physical washes the solvent flow rate is proportional to the total feedgas flow rate and is inversely proportional to the total pressure and the solubility coefficient of the component to be removed. According to Henry's law, the amount of gas absorbed in the solvent should be proportional to the partial pressure of the component and to the solubility coefficient, which is a binary coefficient depending only on the temperature.

The unit of this coefficient is volume of dissolved gas per pressure and per amount of solvent. A widely used unit is normal cubic meter (M^3n) of gas per ton solvent and bar pressure. Figure 1 shows the solubility coefficients for CO_2, H_2S, COS, and CH_4 at a pressure of one bar in methanol (solid lines) and Sepasolv (interrupted lines), depending on the temperature. In Table 1 the physical properties of common solvents are listed. Table 2 shows the

Table 1
Miscellaneous Comparative Data of Solvents *

Process Name	Selexol	Fluor Solvent	Purisol	Sepasolv MPE	Estasolvan	Rectisol
Solvent Name	Selexol	PC	NMP	Sepasolv	TBP	Methanol
Solvent Cost $/lb FOB Fact	1.32	.74	1.34			
Licensor	Norton	Fluor	Lurgi	B.A.S.F.	Uhde & IFP	LINDE AG LURGI
Viscosity @ 25°C, cp.	5.8	3.0	1.65	-	2.9	0.6
Specific Gravity @ 25°C, KG/M^3	1030	1195	1027	-	973	785
Mol Weight	280	102	99	320	266	32
Vapor Pressure @ 25°C, MM Hg	7.3×10^{-4}	8.5×10^{-2}	4.0×10^{-1}	3.7×10^{-4}	$< 1.0 \times 10^{-2}$	
Freezing Point, °C	-28	-48	-24		-80	-92
Boiling Point, °C @ 760 MM Hg		240	202	320	(180 @ 30 MM Hg)	65
Thermal Conductivity Btu/Hr/Ft^2/(°F/Ft)	0.11	0.12	0.095			0.122
Maximum Operating Temp., °C	175	65	-	175	-	
Specific Heat @ 25°F	0.49	0.339	0.40			0.566
Water Solubility @ 25°C	-	94 gm/l	-	-	65 gm/l	
Solvent Solubility in Water @ 25°C	-	236 gm/l	-	-	0.42 gm/l	
Ft^3 CO_2 Solubility/ U.S. Gal @ 25°C	0.485	0.455	0.477	0.455	0.329	1.799**
Number of Commercial Plants	32	13	5	4	0	approx. 100
Bulk CO_2 Removal						
Synthesis Gas	6	3	2	0		6
Natural Gas	6	10	1	0		1
Landfill Gas	3	0	0	0		-
Selective H_2S Removal						
Synthesis Gas	9	0	1	0		Rest
Natural Gas	8	0	1	4		2

* from (1), figures for Rectisol added.

** at -25°C

Table 2
Solubilities of Various Gases in Solvents Relative to Carbon Dioxide at 25°C *

Comp.	Selexol	PC	NMP	Sepasolv	TMP	Methanol at -25°C
H_2	1.3×10^{-2}	7.8×10^{-3}	6.4×10^{-3}	5.0×10^{-3}		5.4×10^{-3}
N_2	-	8.4×10^{-3}	-			1.2×10^{-2}
O_2	-	2.6×10^{-2}	3.5×10^{-2}			2.0×10^{-2}
CO	2.8×10^{-2}	2.1×10^{-2}	2.1×10^{-2}			2.0×10^{-2}
C_1	6.7×10^{-2}	3.8×10^{-2}	7.2×10^{-2}	6.6×10^{-2}	4.0×10^{-2}	5.1×10^{-2}
C_2	4.2×10^{-1}	1.7×10^{-1}	3.8×10^{-1}			4.2×10^{-1}
C_2H_4	4.9×10^{-1}	3.5×10^{-1}	5.5×10^{-1}			4.6×10^{-1}
CO_2	1.0	1.0	1.0	1.0	1.0	1.0
C_3	1.02	0.51	1.07			2.35
iC_4	1.87	1.13	2.21			
nC_4	2.33	1.75	3.48			
COS	2.33	1.88	2.72	2.54		3.92
iC_5	4.47	3.50	-			
C_2H_2	4.53	2.87	7.37			3.33
NH_3	4.87	-	-			23.2
nC_5	5.53	5.0	-			
H_2S	8.93	3.29	10.2	6.86	5.6	7.06
NO_2	-	17.1				
nC_6	11.0	13.5	42.7			
2,4 DMP	-	17.5	-			
CH_3SH	22.7	27.2	34.0	23.1		
nC_7	24.0	29.2	50.0			
CS_2	24.0	30.9	-			
CYCLO-C_6	-	46.7	-			59.5
nC_8	-	65.6	-			
C_2H_2SH	-	-	78.8			
SO_2	93.3	68.6	-			
$(CH_3)_2S$	-	-	91.9			
C_6H_6	253	200	-			
nC_{10}	-	284	-			
C_4H_4S	540	-	-			
H_2O	733	300	4,000			
HCN	1200	-	-			

* from (1), data of methanol added (3)

relative solubilities of various gases which are important for acid gas washes.

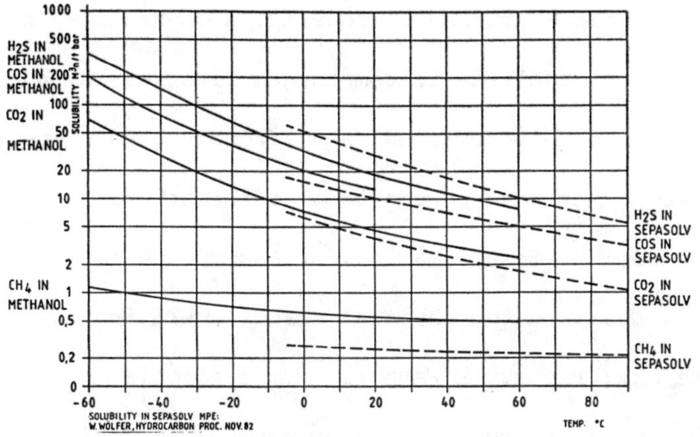

Figure 1. Solubility of gases in Sepasolv MPE and methanol.

Practical experience has shown that the solubility coefficient depends on the pressure, if the gas temperature is close to the dew point of the acid gas to be removed. Since the dew point is influenced by the presence of the other components, the amount which is absorbed depends on the gas composition and the total pressure.

Thus, the calculation of a physical wash requires the knowledge of the solubilities of all components depending on the operating temperature, composition, and pressure.

The above mentioned characteristics make the use of physical washes with gases at high total pressure and high content of acid gas economical. By lowering the temperature the solvent flow rate can be lowered. A physical wash should only be applied, if components with very high solubilities have to be removed, and if due to the properties of the components to be removed a different process, e.g. by condensation, rectification, or adsorption is not possible. Examples for physical washes operating at low pressure are: The water wash to remove ammonia or HCN, or the absorption of organic solvent vapors from tailgases.

Acid Gas Removal With Physical Washes

In the application of physical washes for acid gas removal two general areas have to be distinguished:

Bulk Removal of CO_2. The bulk removal of CO_2 from gas mixtures, preferably from natural gas, increases its heating value. Significant for a bulk removal is that a certain residual content of CO_2 in the treated gas is acceptable. This can often be achieved by simple expansion or stripping of the loaded solvent with an inert gas. Compared to chemical washes, which require a steam consumption approximately proportional to the amount of CO_2

removed, this represents a considerable energy saving.

Desulfurization and CO_2 Removal. If besides the CO_2 also sulfur compounds have to be removed, additional specifications apply. For natural gas, the product purity with respect to H_2S is typically maximum 4 ppm, according to ¼ grain sulfur per hundred standard cubic feet. For synthesis gas usually a sulfur content of less than 0.1 ppm is required. Because of its lower toxicity, COS contents of up to 100 ppm are acceptable with natural gas or fuel gas, whereas with synthesis gas mostly 0.1 ppm are specified.

Sulfur Processing

For acid gases which are vented to the atmosphere environmental regulations usually limit the total sulfur concentration to 10 ppm. The minimum H_2S content of the H_2S fraction depends on the further treatment of this fraction. Since in many geographical areas due to air pollution control sulfur compounds like H_2S, COS and SO_2 cannot be emitted to the atmosphere, the H_2S has to be converted. Widely used are the conversion to elemental sulfur, or the production of sulfuric acid.

For the conversion of H_2S to elemental sulfur, two major processes are available: The Claus process according to the reaction

$$2\ H_2S + SO_2 \longrightarrow 3/x\ Sx + 2\ H_2O \qquad (1)$$

and the direct conversion of H_2S with SO_2 to elemental sulfur and water, using a catalytically acting solvent. On the basis of those two reactions a variety of processes for H_2S conversion was developed, each of them requiring different H_2S concentrations in the feedgas.

Next to the H_2S concentration, other components influence the sulfur conversion, e.g.

- Organic sulfur compounds like COS or Mercaptans do not participate in oxidizing sulfur removal processes. Thus, no complete desulfurization is possible.
- HCN and certain other trace components act as catalyst poison in oxidizing sulfur removal processes, and have to be removed upstream.
- Higher hydrocarbons (C_3^+) decompose with the partial combustion of the H_2S in a Claus plant and form

soot. This prevents the production of yellow sulfur.

Accordingly, the specification of the sulfur recovery plant influences the selection of the acid gas removal system very strongly.

Comparison of Rectisol with Other Physical Wash Systems

R.W. Bucklin and R.L. Shendel (1) compared various physical washes operating at ambient temperature. The Rectisol wash, which operates at temperatures below 32°F, was excluded from this comparison. In Table 1, data for Rectisol at a temperature of -13°F have been added.

Table 2 shows the relative solubilities of various gases compared to the solubility of CO_2 at 77°F. It is obvious that all wash systems considered are about equal for CO_2 bulk removal. For methanol, data at a temperature of -13°F have been added.

Treatment of Natural Gas. For natural gas physical washes can only be economically applied at low contents of heavy hydrocarbons. Even C_3H_8 has a higher solubility than CO_2 and is simultaneously removed. If only H_2S has to be removed from natural gas, Selexol, Sepasolv and Purisol can be used, provided that the natural gas has a low content of heavy hydrocarbons. If besides the H_2S COS also has to be removed, those solvents are not optimal, since the solubility of COS is remarkably worse than the solubility of H_2S.

If the natural gas is rich in heavy hydrocarbons (e.g., in associated gas), the use of a physical system for the acid gas removal is usually not economical, since too many valuable components are simultaneously removed. Those acid gases can be treated economically by a combination of rectification and physical absorption.

If the solubility of methanol is compared with other solvents it becomes obvious that methanol is not applicable as a solvent for the removal of acid gas from natural gas. Even the removal of H_2S is only in special cases economical.

Treatment of Partial Oxidation Gas. If synthesis gases produced by partial oxidation of coal, oil, heavy residue or peat have to be treated, the conditions are definitely in favor of Rectisol. Some important criteria are discussed in the following in more detail.

The Rectisol wash operates at temperatures below 32°F, usually close to the dew point of CO_2 taking advantage of the increased CO_2 solubility.

Although the supply of refrigeration at low temperatures requires much power, this disadvantage is outweighed by a considerable reduction of the solvent flow rate for the CO_2 removal. The advantage is particularly pronounced in large plants due to the reduced size of heat exchangers and vessels.

Selexol, Sepasolv and NMP have a higher selectivity of H_2S/CO_2 than methanol. However, the selectivity of COS/CO_2 is higher in methanol. Another advantage is the high solubility of COS in liquid CO_2. Due to the operating conditions, the amount of CO_2 dissolved in methanol is much higher than in the other solvents. This allows a selective removal of COS and a concentration of the COS in the H_2S fraction produced by a Rectisol wash.

If H_2S has to be removed from a feedgas, but CO_2 may remain in the purified gas, Selexol, Sepasolv and NMP are more suitable than methanol. The standard version of Rectisol using pure methanol is not competitive in those cases. If however toluene or a toluene/methanol mixture is used, the Rectisol wash is more economical than the other wash processes.

A particular advantage of toluene is the fact that COS is even more soluble than H_2S, making a complete desulfurization possible without additional expenditure.

However, for a simultaneous selective removal of H_2S and CO_2 toluene is not suitable since the absolute CO_2 solubility is considerably lower than in methanol.

DIFFERENT WAYS OF PROCESSING PARTIAL OXIDATION GASES AND THEIR INFLUENCE ON THE DESIGN OF THE RECTISOL SYSTEM

For the acid gas removal from a feedgas produced by gasification of coal or oil, different versions of the Rectisol wash can be applied. The particular design depends on the overall concept of the plant and the product specifications. Figure 2 shows the most important alternatives.

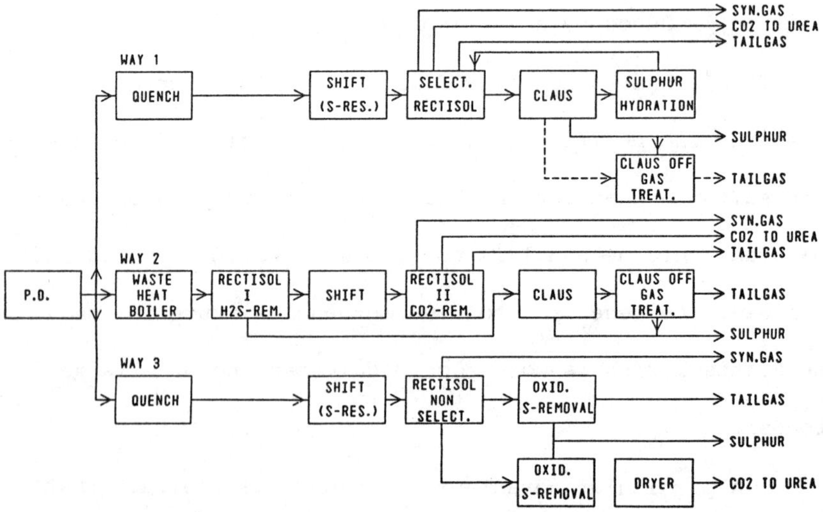

Figure 2. Different ways of processing partial oxidation gases.

If the heat from the partial oxidation gas is removed by a quench cooler, which is followed by a sulfur resistant CO shift conversion unit, either a selective (way 1) or a nonselective (way 3) Rectisol system can be applied. The term "selective" is applied, when the wash system produces at least two separate tailgases: a sulfur free CO_2 fraction, and a sulfur fraction which is rich enough in sulfur to be used as feed for a Claus plant. The term "nonselective" is applied, when only one tailgas is produced, in which the ratio CO_2/H_2S corresponds to the ratio of those components in the feedgas.

In cases where a waste heat boiler is used for heat removal from the partial oxidation gas and a sweet shift for the CO conversion, Rectisol is applied in a two-stage version (way 2), but using the same solvent and a common regeneration system. In Figure 2, the two stages are marked with I (for the desulfurization) and II (for the removal of CO_2). Since the CO_2/H_2S ratio is low in front of the conversion, it is relatively easy to recover a sulfur fraction suitable as feed for a Claus plant, and the two-stage wash is typically of the selective type.

One and two-stage system have been compared in detail elsewhere (2). If a selective acid gas removal is desired, the one-stage system is more economical than the two-stage system.

In comparing selective and nonselective systems, it is important to include the effect on the subsequent sulfur recovery plant. The flowsheet of a nonselective system (Figure 3) is simpler, but heavy restrictions apply to the further handling of the tailgas. Since the CO_2/H_2S ratio in the feedgas is usually in the range of 30/1...300/1, only an oxidizing wash system like Stretford, Unisulf or Sulfolin is applicable.

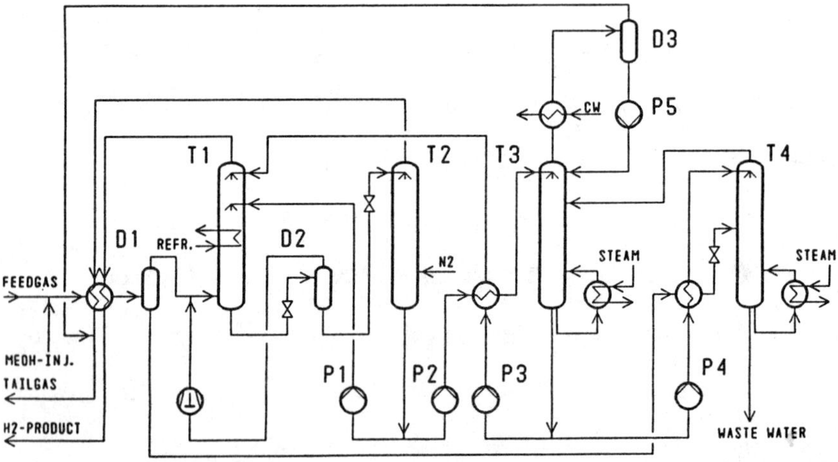

Figure 3. Rectisol wash for the non-selective removal of H_2S and CO_2.

Process Description of a Nonselective Wash (Figure 3)

After injection of methanol the feedgas is cooled against cold separation products. The condensed methanol/water mixture is separated, and in wash column T-1 the acid gas is absorbed by methanol, using lean methanol as reflux to the top of the column, and semilean methanol (containing CO_2 and H_2S) from stripping column T-2 at the middle of the column. The majority of the heat of solution is compensated by warming up of cold methanol, and the residual cold is supplied by external refrigeration.

In order to recover coabsorbed hydrogen, the rich solvent is flashed to an intermediate pressure in separator D-2. Afterwards the solvent is expanded to the top of the stripping column T-2, thus relieving a considerable part of the CO_2 and H_2S. After cold stripping with nitrogen a part of the methanol is recycled to the wash column by pump P-1 as semilean solvent, the residue is regenerated in the hot regeneration column T-3 by methanol vapors completely. The gases removed in the stripping column and the regeneration column contain H_2S in low concentration (0.3 to 3% H_2S, depending on the sulfur content in the feedgas), and have to be treated in a subsequent oxidizing wash for the recovery of elemental sulfur.

If CO_2 for urea or enhanced oil recovery has to be recovered, an intermediate flash step has to be installed upstream of the stripping column T-2, and the recovered gas has to be desulfurized in a parallel oxidizing wash. Since COS cannot be removed by such a wash system, the residual sulfur content of the CO_2 fraction depends primarily on the COS content of the feedgas to the Rectisol wash.

Table 3 shows a material balance and utilities for the nonselective system.

Table 3
Material Balance and Utilities of a Nonselective Rectisol Wash

	Feedgas mol %	Product mol %	Tailgas mol %	Stripping Gas mol %
H_2	54.7	94.7	0.2	
N_2	0.2	0.4	8.1	100
Ar + CO + CH_4	2.8	4.5	0.4	
CO_2	42.2	0.4	91.0	
H_2S + COS	0.1	0.5 ppm	0.3	
Flow/kmol/hr	10,000	5,766	4,614	380
Pressure/bar	32	30	1.1	3
Steam	14,300 lbs/hr			
Electr. Energy	1,100 kW			
Refrigeration	20×10^6 BTU/hr			

The solvent flow rate, which is required in a non-selective wash for the removal of CO_2 consists of two parts:

1. hot regenerated methanol which is fed to the top of the column, and
2. solvent which is regenerated only by stripping, and which is received cold from the stripping column.

In the following, the utilities of the nonselective system are compared to a selective system.

Since the cold methanol is preloaded with CO_2, the total flow rate has to be increased as compared to a selective wash. On the other hand the solvent flow rate which is hot regenerated can be reduced, since the gas flow rate is lower and the temperature is lower in the fine purification section. Due to their high solubilities, H_2S and COS are simultaneously removed to the specified level.

According to the reduced methanol flow rate, the steam consumption in the regeneration column is lowered, whereas the steam consumption in the methanol water separation column is independent from the type of the wash system. The refrigeration requirements are lower in the nonselective system, since the exchange losses of the methanol in a selective system are larger than the cold losses by cold pump P-1 in a nonselective system.

The electrical energy consumption is similar in both systems. In a nonselective wash, the disadvantage of the large methanol flow rate is compensated by the elimination of some cold intermediate pumps. A disadvantage is the use of two high pressure pumps in the nonselective system.

An important point of view is the difficulty to recover sulfur free CO_2. The desulfurization plant has either to be split in two parallel trains, one of them purifying the CO_2 product, and the other purifying the CO_2 nitrogen mixture, or the inert gas stripping is replaced by a vacuum compressor. However, all consumption figures increase considerably with vacuum stripping, since the recovered CO_2 cannot be warmed up against feedgas in order to save pressure drop, and the methanol to be treated in the hot regeneration is increased due to a higher CO_2 loading of the cold regenerated methanol. Up to date nonselective Rectisol wash systems have been preferably used for feedgases with a high content of hydrocarbons, e.g., after a coal gasification unit,where the use of a Claus plant is prevented. The nonselective wash can also be economical at low H_2S contents in the feedgas or at small feedgas flow rate in combination with an oxidizing sulfur recovery unit.

SPECIAL AREAS OF APPLICATION

The following examples illustrate applications of the Rectisol wash process in various areas of synthesis gas purification. Wherever sulfur is present in the feedgas, it is removed selectively. Some recent developments are also included.

Production of Ammonia Synthesis Gas and Recovery of CO_2 for Urea

In fertilizer plants, the NH_3 is frequently converted to urea. For the urea production, CO_2 is needed which has to have low contents of sulfur, hydrogen and CO. The CO_2 can be recovered from the synthesis gas by modifying the selective one-stage Rectisol process as shown in Figure 4.

Shifted feedgas is injected with methanol to prevent icing and cooled by cold product streams in heat exchanger E-1. The condensed methanol/water mixture is withdrawn from separator D-1 and distilled in the methanol/water separation column T-5. The wash column T-1 has two sections; in the upper section the CO_2 is absorbed by lean, cold methanol;

and in the lower section H_2S and COS are absorbed by a part of the CO_2 loaded methanol coming from the upper section. The purified synthesis gas can be processed in a subsequent nitrogen wash unit (not shown on this flowsheet) to produce ammonia synthesis gas.

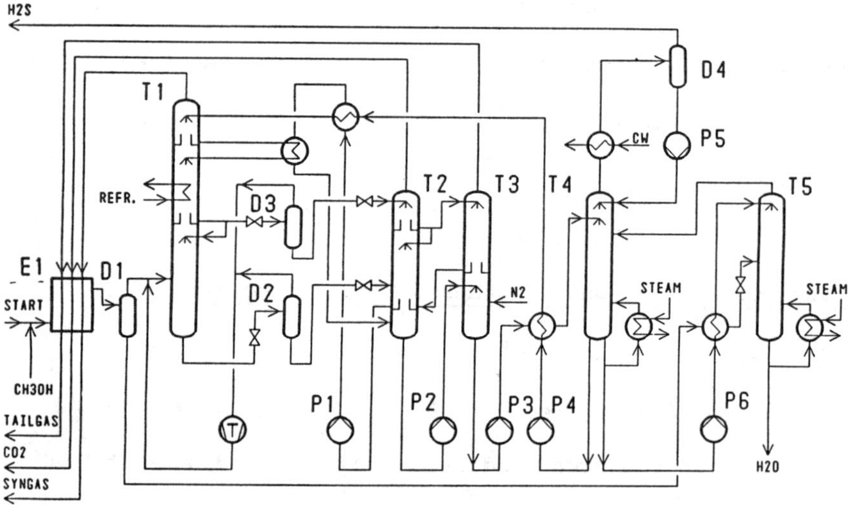

Figure 4. One-stage selective Rectisol wash for production of synthesis gas and CO_2.

Coabsorbed hydrogen is recovered by flashing in the drums D-2 and D-3, and recycled to the feedgas line. Further flashing of the loaded mehtanol streams releases the CO_2 product in column T-2. Additional CO_2 recovery is achieved by warming up the methanol against lean and loaded methanol streams. The bulk of the residual CO_2 is stripped off the solvent by nitrogen in the cold stripper T-3. In order to keep the CO_2 fraction and the tail gas sulfur free, H_2S and COS are reabsorbed in the upper sections of the columns T-2 and T-3 by sulfur-free, CO_2 loaded solvent streams.

The solvent leaving the bottom of T-3 is warmed up against lean methanol and fed to the hot stripper T-4, where it is completely regenerated by reboiling with steam. The methanol vapors are condensed in separator D-4. The gas phase out of D-4 is the sulfur fraction, which can be delivered to a Claus unit. The plant is refrigerated typically using cold at -40°F. Depending on the amounts of waste heat or low pressure steam available, it is worthwhile to consider the use of an ammonia refrigeration system.

Maximum CO_2 recovery can be achieved, if vacuum stripping is applied, thus eliminating the need of nitrogen and column T-3. After compression, the recovered CO_2 is treated in the CO_2 column T-2 for final purification. This allows the recovery of about 95% of the CO_2 at the expense of increased power consumption.

Table 4 shows a material balance around the Rectisol system of a plant producing 1,000 metric tons per day of ammonia, and a summary of the utility consumptions.

Table 4
Material Balance and Utilities for a 1,000 mtpd NH_3 Plant in mol %

	Feed	Purif. Gas	CO_2	Tail Gas	H_2S Fraction
H_2	62.47	95.27	0.84	0.07	0.31
N_2 + Ar	0.51	0.76	0.08	25.32	4.36
CO + CH_4	2.67	3.97	0.33	0.04	0.08
CO_2	34.10	20 ppm	98.75	74.57	68.01
H_2S + COS	0.25	0.1 ppm	2 ppm	5 ppm	27.24
Flow/kmol/hr	6,021.1	3,936.1	1,283.9	1,003.6	54.2
Press./bar	78	76	1.8	1.1	2.5

Power (shaft, without power recovery)	1,100 kW
Steam (75 psig)	14,300 lbs/hr
Refrigeration	6.75×10^6 BTU/hr
Cooling water (delta t = 18°F)	1,320 gpm
Stripping gas at 45 psig	256.7 kmol/hr

Reduction of the CO Emission

In many applications, the CO content of the CO_2 product stream or the tail gas is limited due to process specifications or environmental regulations. In ammonia plants, Rectisol is usually applied after the shift, and the amount of CO absorbed in the methanol is relatively low due to its low partial pressure in the feedgas. In methanol plants, the CO content in the feedgas is remarkably higher, especially in cases, where excess CO is produced in a cryogenic separation unit.

The recovery of coabsorbed gases with low solubility like hydrogen and CO in an intermediate flash step is only economical, if the flash pressure is not lowered below the partial pressure of the CO_2 in the feedgas, otherwise too much CO_2 is outgassing and the recompression of the flash-gas becomes too expensive. On the other hand, the flash-gas cannot be recovered as separate CO-rich fraction because of its high H_2S content.

Figure 5 shows a modification of the Rectisol process, which provides for a sulfur free CO-rich tailgas, thus lowering the CO content in the CO_2-rich tailgas considerably. A small additional wash column T-3 is installed, in which the sulfur containing methanol from D-2 is expanded, and the sulfur is reabsorbed from the flash gas by CO_2 loaded methanol coming from D-4. This modification can also be installed in existing plants by adding a few components.

The CO-rich tailgas can be incinerated, thus lowering the CO emission considerably. Even in large plants, the CO emission does not exceed 25 short tons per year.

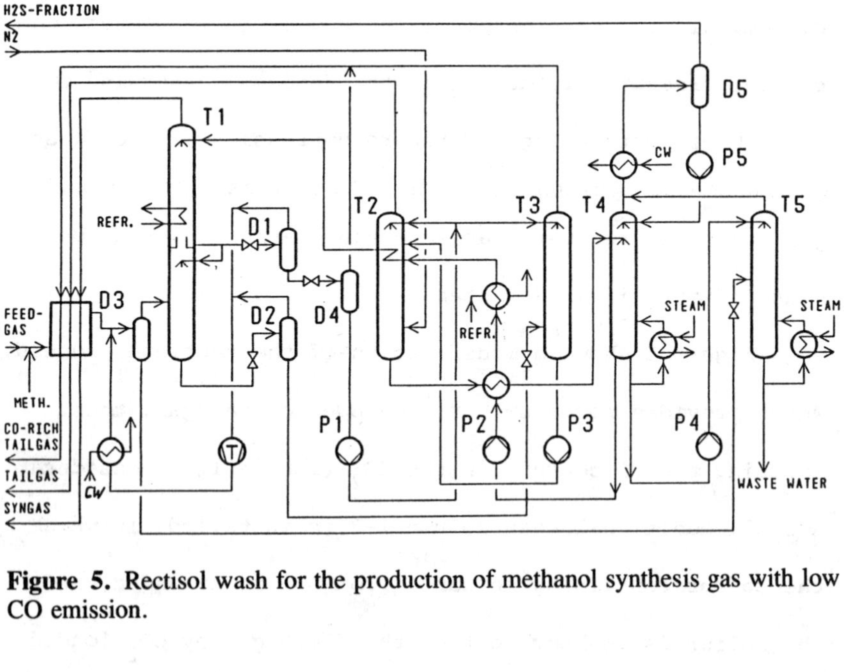

Figure 5. Rectisol wash for the production of methanol synthesis gas with low CO emission.

Combining Rectisol Wash and Claus Plant

A Claus plant treating the sulfur rich tailgas out of the Rectisol unit converts typically 90...95% of the H_2S to elemental sulfur. The residual sulfur has to be removed from the Claus plant tailgas in a relatively expensive system. If a Scott process is used, organic sulfur compounds and SO_2 are first hydrogenated to form H_2S, CO_2 and water. Then, the H_2S is removed by a selective chemical wash and recycled to the Claus plant feed stream. Investment cost and energy requirements are high, since the H_2S has to be removed from a low pressure stream selectively.

A combination of selective Rectisol wash and Claus plant offers an attractive, cost saving alternative. The sulfur rich Claus tailgas is hydrogenated as described above, but then recycled to the cold stripping column (e.g. column T-2 in Figure 5). There the CO_2 is added to the tail-gas, and the H_2S is absorbed in the cold methanol. Thus, an additional wash system is saved, and only the hydrogenation unit and a recycle blower are required.

The utility requirements for the wash system are slightly increased, since additional refrigeration is needed in order to compensate the heat of solution of the recycled H_2S, and additional steam is required for the H_2S stripping. But those effects are minor and do not require additional equipment.

Combining Rectisol Wash and Cryogenic Separation Plant

The production of CO is usually accomplished in a cryogenic separation plant. In many cases it is advantageous to use Rectisol for the purification of the feedgas to the cryogenic unit.

Figure 6 shows an application for a sulfur free feedgas coming from a steam reformer. Instead of nitrogen, the tailgas from the cryogenic unit is used as stripping gas in the cold stripper T-2. Due to the absence of sulfur, the hot stripper is eliminated. For precooling in E-1, product gases from the cryogenic unit are used. Additional refrigeration is supplied via heat exchanger E-2.

Figure 7 shows a selective Rectisol system treating a sulfur containing feedgas coming from a partial oxidation unit. Again, the tailgas from the cryogenic unit is used as stripping gas in column T-2. The H_2S containing methanol from drum D-2 is warmed up in exchanger E-4 and hot regenerated in column T-3. Since the CO_2 content in the unconverted feedgas is low, an H_2S enrichment step is usually not required. The recycle compressor prevents the enrichment of sulfur and CO in the vent gas. This compressor can also be used to recompress the CO-rich tailgas from the cryogenic unit simultaneously in order to improve the CO yield.

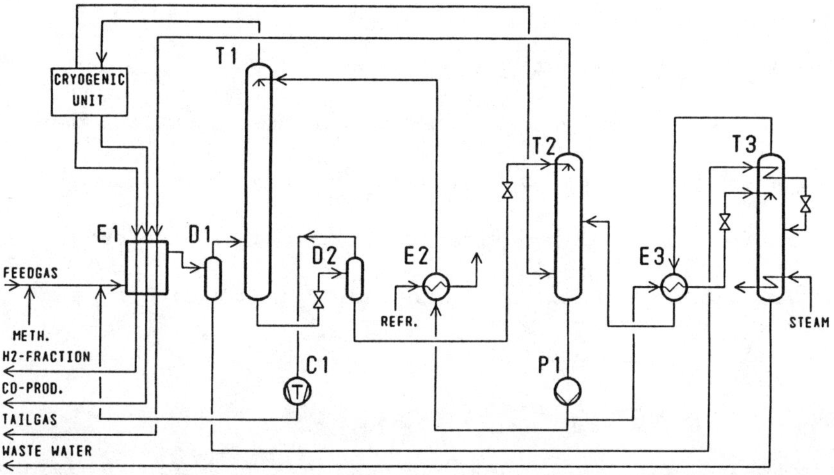

Figure 6. Rectisol wash for treating sulfur-free feedgas to a CO production unit.

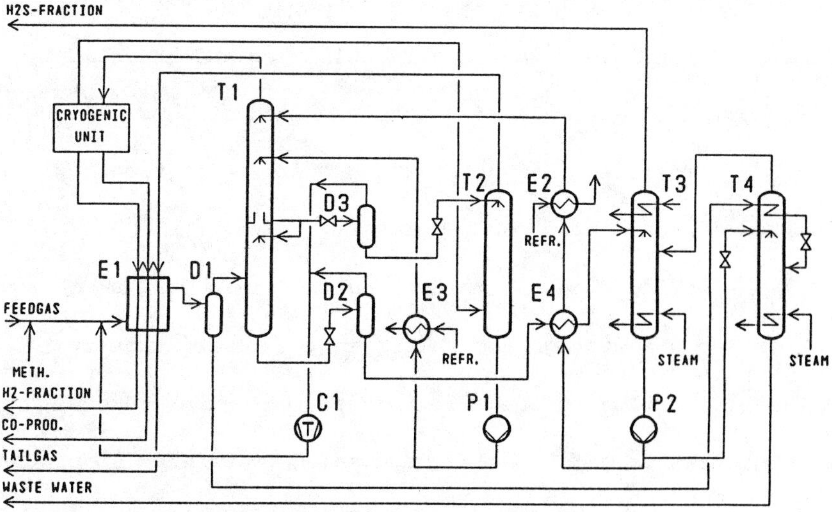

Figure 7. Rectisol wash for treating sulfur containing feedgas to a CO production unit.

Other elements of the flowsheets shown in Figures 6 and 7 like the methanol injection, methanol/water separation and distillation etc. are similar to the preceding flowsheets.

Production of Methanol Synthesis Gas

To produce methanol synthesis gas from a raw gas generated by partial oxidation, a splitstream of the raw gas is shifted and combined with the unshifted part before entering the synthesis unit. The acid gas can either be removed from the combined streams (in which case a flowsheet like Figure 5 could be employed), or shifted and unshifted streams are treated in separate wash columns, using a common train of regeneration columns. Depending on the concentrations and flowrates of the two feedgas streams, the separate treatment can lead to savings in utilities and investment, if compared to a treatment of the combined streams. Figure 8 shows a simplified flowsheet of a Rectisol unit treating shifted and unshifted gas separately.

The shifted feedgas is treated in column T-1, removing CO_2 in the upper stage and sulfur in the lower stage by methanol. CO_2 loaded methanol is used for sulfur removal, and the heat of solution is compensated by warming up of the solvent and by two cooling stages E-2 and E-3. The sulfur content in the purified gas is less than 0.1 ppm, and the CO_2 content is adjusted according to the specifications for the methanol synthesis. Due to the excellent sulfur removal, no guard adsorber is required for the synthesis gas.

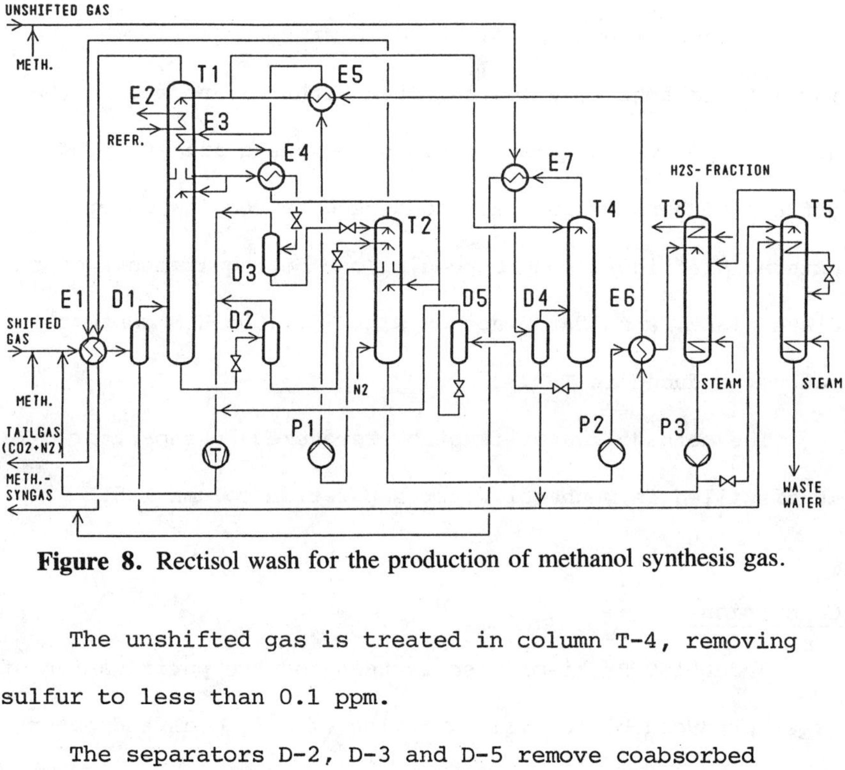

Figure 8. Rectisol wash for the production of methanol synthesis gas.

The unshifted gas is treated in column T-4, removing sulfur to less than 0.1 ppm.

The separators D-2, D-3 and D-5 remove coabsorbed hydrogen and CO, which is recycled to the shifted feedgas by a compressor.

In the cold stripper T-2, the bulk of the CO_2 is stripped off the solvent by nitrogen. Sulfur compounds are reabsorbed in the upper column section by sulfur free methanol. With a sulfur content in the range of a few ppm, the tailgas can be released to the atmosphere without further treatment. CO_2 recovery is possible by installing an additional separator (not shown). The stripping is enhanced by warming up of the methanol in the exchangers E-5, E-3 and E-4.

The bottom product of the cold stripper is warmed up in heat exchanger E-6 against lean methanol and fed to the hot stripper T-3, where the dissolved gases are stripped off and methanol vapors are condensed by cooling water and external refrigeration. Depending on the requirements of the Claus plant, a sulfur fraction with 25...60% H_2S content can be produced in T-3.

The methanol/water fraction recovered in separator D-1 is distilled in methanol/water separation column T-5.

Conclusion

About 100 Rectisol wash systems for the purification of gases produced by partial oxidation of oil, coal and other carbon containing feedstocks have been built since 1954 by Linde AG and Lurgi GmbH, the patent owners and licensors of the Rectisol process. Considerable experience has been accumulated, and the process has been continuously improved and adapted to changing conditions with respect to feedgas and product gas specifications, as well as more stringent environmental regulations.

REFERENCES

1. Bucklin, R.W. and R.L. Schendel, "Comparison of Fluor solvent and Selexol process," AIChE Symposium, Atlanta, March 11-14, 1984.

2. Ranke, G. and V.H. Mohr, "Comparison of various processes for acid gas removal," AIChE Symposium, Atlanta, March 11-14, 1984.

3. Landolt and Boernstein, "Zahlenwerte und Funktionen aus Physik, Chemie, Astronomie, Geophysie une Technik," 4. Band Technik, 4. Teil, Bestanteil C1.

4. US Patent 4,425,317 (1983)

4

Improved AMISOL Process for Gas Purification

Manfred Kriebel
Assist. Vice President, Gas Treating and Sulfur Recovery
LURGI GmbH, Frankfurt a.M.

The AMISOL Process was developed already in the 1960s.

The solvent consists of an amine as a chemical component in conjunction with a physical solvent.

According to its absorption characteristics, the AMISOL Process may be classified among the physical/-chemical gas purification processes.

Amines used in the past have been the well known alkanol amines, i.e., mono or diethanolamines mixed with methanol. This solvent mix was successfully used in the first commercial AMISOL Plants at the beginning of the 70s.

The main advantage of this solvent lies in the good physical absorption of the physical solvent component in combination with the chemical reaction of the amine. Good mass transfer between gas and liquid ensures that the number of trays in the absorption column can be kept low.

The combination of a chemically active amine with a low-boiling polar physical solvent such as methanol offers major advantages in the absorption of carbon dioxide and sulfur components:

1. Very low clean gas sulphur contents of less than 0.1 ppmv, as they are required for syngases.
2. Very low CO_2 contents in the purified gas.
3. Good absorption of trace components, such as HCN, COS, mercaptans and higher hydrocarbons.
4. Low regeneration temperature owing to the fact that the solvent methanol has a boiling point approximately 35°C below that of water.
5. This means that the absorption and regeneration temperatures are not far apart so that the process can do without the usual heat exchangers for lean/rich solvent.

6. The solvent is non-corrosive so that carbon steel equipment can be used.

As the conventional alkanolamines such as monoethanolamine (MEA) and diethanolamine (DEA) which have been used in the past are non-selective, the sulfur components H_2S and COS are absorbed together with the CO_2 and occur jointly in the offgas.

This is a disadvantage wherever sulfur rich offgases have to be treated, for instance in Claus Plants.

Nor are these conventional amines sufficiently stable chemically and thermally so that a reclaimer is often required to eliminate the products of chemical degradation.

Efforts have therefore been made to find new amines by which these disadvantages can be eliminated in a mixture with methanol.

Extensive research led to the discovery that aliphatic alkylamines will meet these requirements.

These amines are:

1. Diisopropylamine - abbreviated DIPAM
2. Diethylamine - abbreviated DETA

Such aliphatic alkylamines are secondary amines characterised by a number of very desirable properties.

Table 1 summarises some of the physical properties. They are contrasted with the corresponding parameters for diethanolamine, methyl diethanolamine and methanol.

These new DIPAM and DETA amines offer the following additional advantages for the AMISOL solvent:

1. High thermal and chemical stability so that no reclaimer is needed.
2. Higher effective CO_2 and H_2S loadings.
3. High selectivity between H_2S and CO_2.
4. High selectivity of DETA AMISOL also between COS and CO_2.
5. Good industrial availability of these amines at moderate cost.
6. Smaller difference between absorption and desorption temperatures.
7. The amines allow for the same boiling range as the solvent methanol.
8. No foaming tendency due to low surface tension.

Table 1
Properties of Amines and Methanol

	DIPAM	DETA	DEA	MDEA	METHANOL
CHEM. FORMULA	$[(CH_3)_2CH]_2NH$	$(C_2H_5)_2NH$	$(HOC_2H_4)_2NH$	$[CH_3(HOC_2H_4)_2]N$	CH_3OH
MOL. WEIGHT	101,2	73,14	105,14	119,47	32,04
BOILING POINT °C 760 mm Hg	84	56	269	247,2	65
VAPOUR PRESSURE mbar at 20°C	100	253	0,1	0,01	128
MELTING POINT °C	- 61	- 50	28	- 21	- 98
SPEC. GRAVITY kg/m^3_{20}	716	704	1092	1042	792
THERM. CONDUCTIVITY $\frac{W}{m \cdot K_{20}}$	0,117	0,175	0,220	~ 0,2	0,202
SPECIFIC HEAT $\frac{kcal}{kg \cdot °C}$	0,63 (20°C)	0,512 (25°C)	0,531 (30°C)	~ 0,63	0,60 (20°C)
SOLVENT COST DM/kg	2,70	3,30	3,00	12,00	0,45

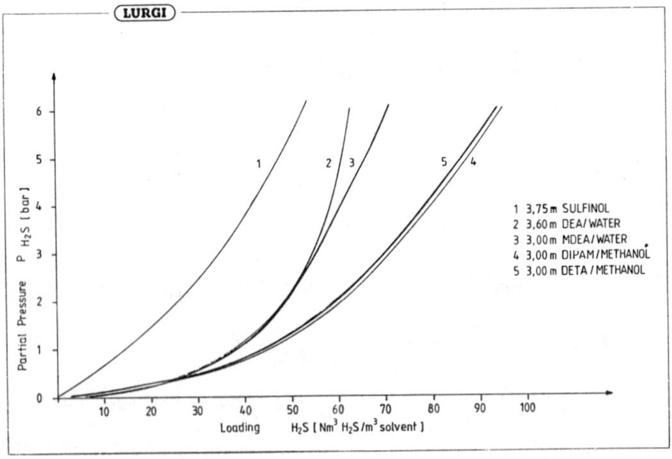

Figure 1. H_2S loading capacity of amine solvents, temperature 75°C.

9. No corrosion problems.
10. Considerable reduction in vapor pressure of the amine neutralized by sour gases.

The relatively high effective H_2S and CO_2 loadings are shown in the Figures 1 and 2 as a function of partial pressures.

For comparison, diethanolamine which has so far been used in the AMISOL solvent as well as familiar solvents like MDEA/water and SULFINOL (DIPA, SULFOLAN and water) are shown as well. The pictures show the markedly higher H_2S and CO_2 pick-up of the AMISOL solvent with the new amine.

A particular advantage of the aliphatic alkyl amines, as already mentioned above, is the high selectivity of the solvent between sulfur components and CO_2, along with the very low residual sulfur contents in the clean gas.

Even with very low sulfur and high CO_2 contents in the feed gas, it is unproblematic to produce a high sulfur offgas containing more than 25% H_2S, which is suitable for Claus Plants.

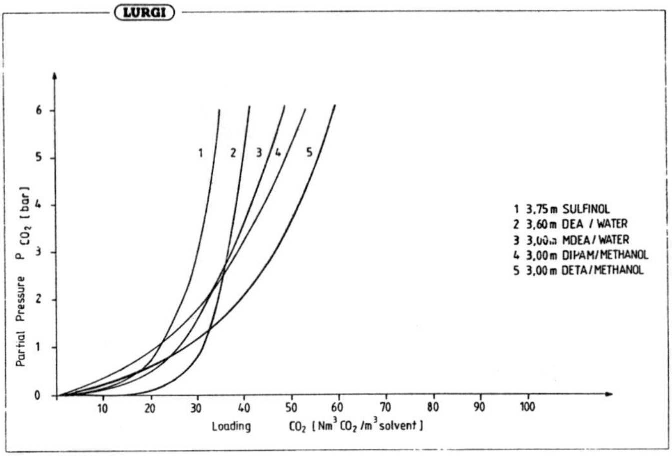

Figure 2. CO_2 loading capacity of amine solvents, temperature 75°C.

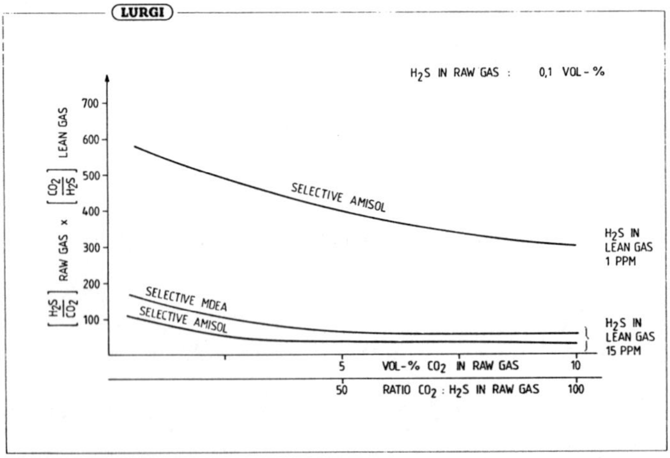

Figure 3. Selectivity of DETA-AMISOL

A remarkable feature in this context is also the high COS selectivity of the DETA/methanol mix.

This enables not only the COS to be removed from the feed gas without difficulty, but the COS also occurs together with the H_2S in the high-sulfur offgas to be fed to the Claus Plant.

The CO_2 which is then removed from the desulfurised gas is free from COS so that, even with today's very stringent environmental pollution requirements (which have recently been introduced also for COS), it can be freely discharged to the atmosphere. Wherever pure CO_2 is required, for instance for urea synthesis, it can be produced in the same manner, using the selective DETA AMISOL solvent.

In order to demonstrate the selectivity, the Figure 3 shows the product of the ratios of H_2S and CO_2 contents in the raw gas and clean gas plotted against the CO_2 content in the raw gas. This method is particularly suitable to demonstrate the selectivity of chemical solvents.

It takes account of the effect that increased selectivity of chemical solvents is mostly associated with only moderate gas purity. Therefore, the selectivity of the AMISOL solvent has been compared in Figure 3 with that of the gaseous solvent methyl diethanolamine (MDEA). The selectivity is of the same order for a gas purity of 15 ppm H_2S. If only 1 ppm H_2S is allowed in the gas, it can clearly be seen that the AMISOL solvent is superior and that this purity cannot be reached with MDEA.

Taking advantage of the varied properties of aliphatic alkylamines, the AMISOL Process lends itself mainly for the following applications:

1. Joint absorption of H_2S, COS and CO_2, as well as trace contaminants, such as HCN, NH_3, mercaptans, thiophenes etc. from raw gases to produce highly pure syngases.
2. Selective absorption of all sulfur components from raw gases to obtain highly pure product gas and a high sulfur Claus gas. CO_2 is removed in a second stage and can be recovered in pure form.

In this case too, all trace contaminants are removed.

3. Selective desulfurization of natural gases, simultaneously removing all organic sulfur compounds, such as COS and mercaptans, and production of high-sulfur Claus gases.

This opens up a wide field of commercial applications for the AMISOL process to treat various raw gases which may be produced through gasification of coal or oil. It must be pointed out here that, owing to the predominantly chemical characteristic of the AMISOL solvent, the AMISOL process is particularly cost-effective for all gases with low to medium partial pressures of the sour gas. If sour gas partial pressures are high, physical gas purification processes should in most cases still take preference.

The first commercial AMISOL Plant was successfully used in an ammonia plant as early as 1970 to purify a raw gas produced through partial oxidation of oil.

Meanwhile, the process has been put to a wide variety of practical applications.

An interesting case in point, for instance, is the purification of a coal gas produced through hydrogenating gasification, which after purification is separated in a cryogenic unit into hydrogen, methane, and a CO/H_2 fraction - a process which requires highly pure gases.

Figure 4 shows the simplified process arrangement. On the left hand we have the absorber with a benzene prewash unit, in the centre is the regenerator in a split-flow connection, while the right-hand side shows a necessary third column to separate methanol and water in the bottom section and to wash out the remaining methanol from the acid gas in the top section.

This AMISOL Plant, which was constructed already towards the end of the 70s, is still using conventional DEA/methanol mixtures as a solvent. It can, however, be operated also with the new alkylamines. Selectivity is not required here.

This AMISOL plant is part of a Demonstration plant for hydrogenating coal gasification.

The merits of combining an AMISOL Gas Purification Plant with a cryogenic separation plant have been excellently proven by this arrangement. The highly pure gas from the AMISOL Plant is directly fed to the separation plant, where the methanol vapors in the pure gas are condensed in the cold box and recycled to the top of the AMISOL absorber.

An example of another commercial AMISOL Plant, which already uses the new alkylamine, however, is the purification of a coal gas produced by an entrained bed gasifier.

The concept was developed for a clean fuel power station. Although, in this case, high syngas sulfur purity is not required, the object is to produce a high sulfur Claus gas containing at least 30 vol.% H_2S from a raw gas with a $H_2S:CO_2$ ratio of 1:20. In addition, a number of trace contaminants had to be eliminated.

As both the sour gas partial pressures and the total raw gas pressure were low, conditions were ideal to use the AMISOL Process with a DETA/methanol solvent mix.

The process arrangement is shown on the Figure 5.

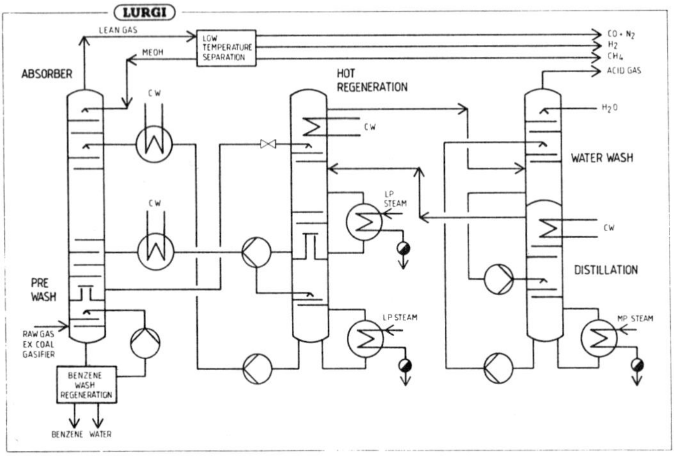

Figure 4. AMISOL plant for the purification of coal gas (RBW-Hydrogasifier).

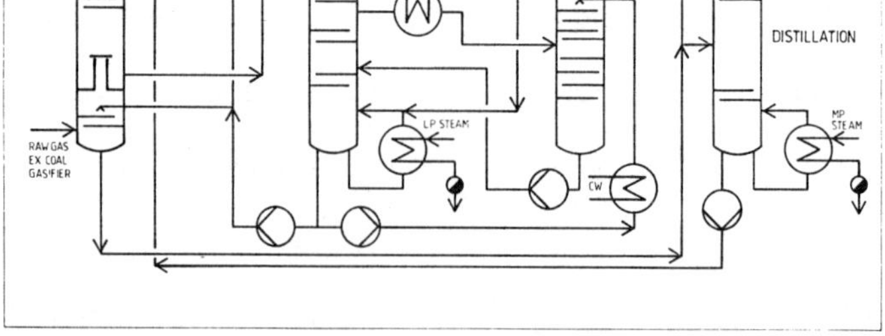

Figure 5. AMISOL process for selective desulfurization (DETA/Meoh) for coal gas (VEW entrained gasifier).

The absorber/regenerator loop on the left is the same as in the case of conventional amine solvents.

The absorber bottom again contains a prewash stage to remove traces of HCN, NH_3 etc. Water is injected in the top section of the absorber to remove the methanol from the clean gas together with the volatile amine.

Likewise, water is used to recover evaporated solvent also from the acid and offgases. The distillation column on the right is used to separate the alkylamine/-methanol mix from the water. Owing to its low boiling point, the alkylamine boils off together with the methanol overhead, while the bottoms consist of pure water which is again recycled to the wash stage.

This arrangement features an additional reabsorber shown in the centre. It serves to increase the Claus gas concentration by extracting an H_2S/CO_2 mixture from the regenerator and reabsorbing the H_2S by means of regenerated solvent.

The H_2S-rich solvent is recycled to the central portion of the regenerator, and a high-sulfur Claus gas is extracted overhead. Via this additional reabsorber loop, the Claus gas concentration can, for all practical purposes, be increased arbitrarily. The only limit is set by utility consumption rising beyond economic proportion.

This relationship between Claus gas concentration and utility consumption is shown for this DETA/AMISOL Wash Unit in the Figure 6. The Figure is based on an CO_2/H_2S ratio of approximately 20 in the raw gas as it is normal for entrained bed gasification.

Such a Claus gas concentration by means of a reabsorber has in the past been reached practically only for physical wash processes. This is the first instance where it is used with an amine solvent, as it is possible only through the special characteristics of the alkylamines.

As a last example for an interesting application of the AMISOL Process, the purification of coal gas produced by a slagging gasifier (Lurgi/British Gas) will be described. Here is a design example:

While CO_2 contents are low, the raw gas contains, as a rule, relatively high amounts of sulfur with CO_2:H_2S ratios between 8 and 10, so that conditions are favorable for the AMISOL Process, allowing a straight forward process arrangement.

On the right-hand side, Figure 7 shows the AMISOL Process consisting of the absorber/regenerator and the distillation column as already described. Selectivity is not required as a high sulfur Claus gas will be produced also with joint sour gas removal.

The gas purities may be conditioned to suit the specific applications of the clean gas, ranging from fuel gas quality to syngas purity.

As the slagger gas passes a carbonization zone in the gasifier, it contains a number of saturated and unsaturated hydrocarbons, gum formers and trace contaminants, such as HCN, NH_3 and other organic sulfur compounds.

An appropriate precleaning stage is therefore required.

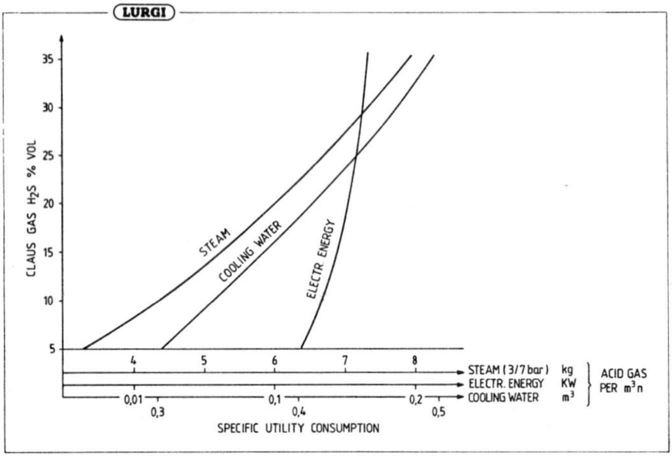

Figure 6. Selective removal of H_2S and COS, dependency of the utilities on the H_2S concentration in the acid gas, based on the conditions for VEW raw gas.

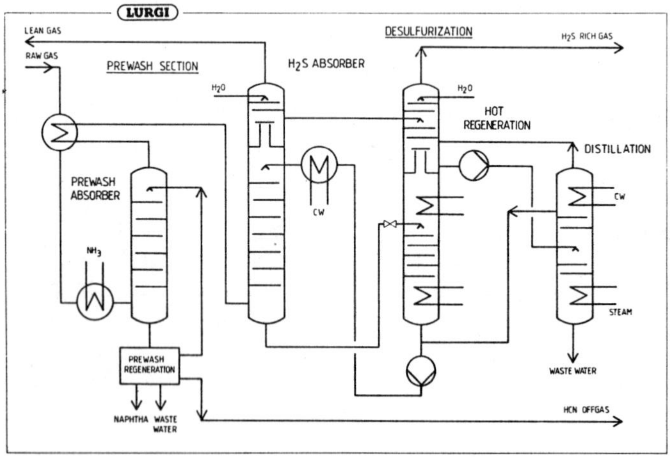

Figure 7. AMISOL plant for the purification of slagger gas (BGL gasifier).

To this end, the AMISOL Process is preceded by a prewash stage using pure methanol as shown on the left side of the picture. This prewash stage is well known from the RECTISOL Process used to purify gases from the Lurgi pressure gasification units. It is a process stage which has been successfully used in a large number of plants. In this case, a similar version is combined with the AMISOL Process - an obvious arrangement, as methanol is used in both wash stages. For this application, too, the prewash methanol is cooled to moderate temperatures and regenerated separately through extraction and distillation. All trace contaminants are absorbed and desorbed along with it, producing a sulfur-laden naphtha and an offgas containing HCN. The HCN offgas is fed to the Claus Unit to be incinerated in a separate burner muffle.

As the prewash stage is at the same time used to dry the raw gas, the waste water discharged from it should best be routed to the waste water treatment unit handling the gas liquor from the coal gasification unit, particularly since the flow rate is relatively small.

If the slagger gas is shift-converted for synthesis, a CO_2 removal stage will as a rule be required as well; this duty can also be handled by an AMISOL stage.

Typical raw gas compositions for the three examples dealt with here are compiled in Table 2.

The bottom section of the table lists a variety of trace contaminants, although only the most important ones are mentioned here which can all be removed without problems by the improved AMISOL Process.

Table 2
Gas Composition for Amisol Plant Downstream of Coal Gasifiers

	RBW HYDRO GASIFIER	VEW ENTR.BED GASIFIER	BGL SLAGGER GASIFIER
CO_2 VOL %	1,8	4,17	3,96
CO	7,2	22,36	60,65
H_2	53,1	11,46	25,15
N_2	0,8	61,48	0,87
CH_4	36,3	} 0,2	8,11
C_2H_6	0,6		0,33
C_2H_4	-	-	0,14
C_3-HYDROCARB.	-	-	0,16
C_4-HYDROCARB.	-	-	0,04
C_5-HYDROCARB.	-	-	0,01
C_6H_6	0,05	-	-
H_2S	0,1	0,21	0,49
TRACES (PPMV)			
COS	5	244	200
RSH	134	-	-
CS	-	14	200
HCN	9	66	600
NH_3	9	34	100
THIOPHEN	-	-	10
NAPHTHALENE	-	-	7
HCl, HF, NH_4, Cl, C_5	-	< 10	-
PRESSURE BAR	51	19	55
TEMPERATURE °C	40	60	40

RBW: Rheinische Braunkohlenwerke A.G., Köln
VEW: Vereinigte Elektrizitätswerke Westfalen, Dortmund
BGL: British Gas Corporation, London

5

Regeneration of Physical Solvents in Conditioning Gases from Coal

J. S. Staton
R. W. Rousseau*
J. K. Ferrell
Department of Chemical Engineering
North Carolina State University

It has become apparent that improved technology is needed to meet current environmental standards on processes utilizing coal for the production of synthetic fuels and chemical feedstocks. Coal gasification produces a crude coal gas containing a number of sulfur, hydrocarbon and other compounds that must be separated from hydrogen and carbon monoxide in order that these latter two compounds may be used safely and economically. Carbon dioxide is the primary diluent and hydrogen sulfide the primary sulfur compound to be removed from crude coal gas. Because of their acidic nature, the unit in which CO_2 and H_2S are separated from the crude coal gas is referred to as an acid

----

* author to whom correspondence should be addressed

gas removal system (AGRS). An AGRS used in a coal gasification process, however, must be able to handle a range of compounds that includes carbonyl sulfide, mercaptans, thiophene, aromatic hydrocarbons and others.

The composition of gases obtained from a fluidized bed gasifier and fed to an acid gas removal system was discussed in a recent report by Rousseau and Ferrell (1). That work considered the effects of feeding various kinds of fuels to the gasifier on the operation of the acid gas removal system using refrigerated methyl alcohol as solvent. Fuels used included a devolatilized char, a subbituminous coal, a peat and a lignite. The acid gas removal system was operated with essentially one set of regeneration conditions in one configuration.

The effect of absorber operating conditions on the performance of the AGRS was evaluated by Kelly (2). Methyl alcohol was used as the solvent in that work. It was found that absorption of acid gases is governed primarily by conditions acheived in solvent regeneration. In other words, the ability of the absorber to reduce the concentration of acid gases in the sweet gas to extremely low levels is set by the regeneration of the solvent and, hence, the levels of the acid gases in the clean solvent being returned to the absorption column.

As a consequence of the observation noted in the preceding paragraph, this study was conducted to evaluated various schemes for solvent regeneration. In particular, the AGRS was operated with and without multiple flash tanks between the absorber and stripper, and with stripping by nitrogen and/or heat supplied to a reboiler at the base of the stripper. The study was conducted using Texas lignite as the fuel fed to the gasifier and refrigerated methyl alcohol in the acid gas removal system. The objective of this study was to evaluate the effect of system configuration on the quality of the conditioned gas leaving the absorber of the AGRS.

It should be mentioned that this work was coupled with a study involving modeling of the packed stripping column used in the acid gas removal system (3); this model is an extension of that developed for the description of an adiabatic packed absorber (4). Although gas compositions predicted by these models were in close agreement with measured values, quantitative information on constituents present in concentrations below approximately 100 ppm (0.01 mole %) could not be obtained from model simulations. This experimental work was necessary, therefore, to determine

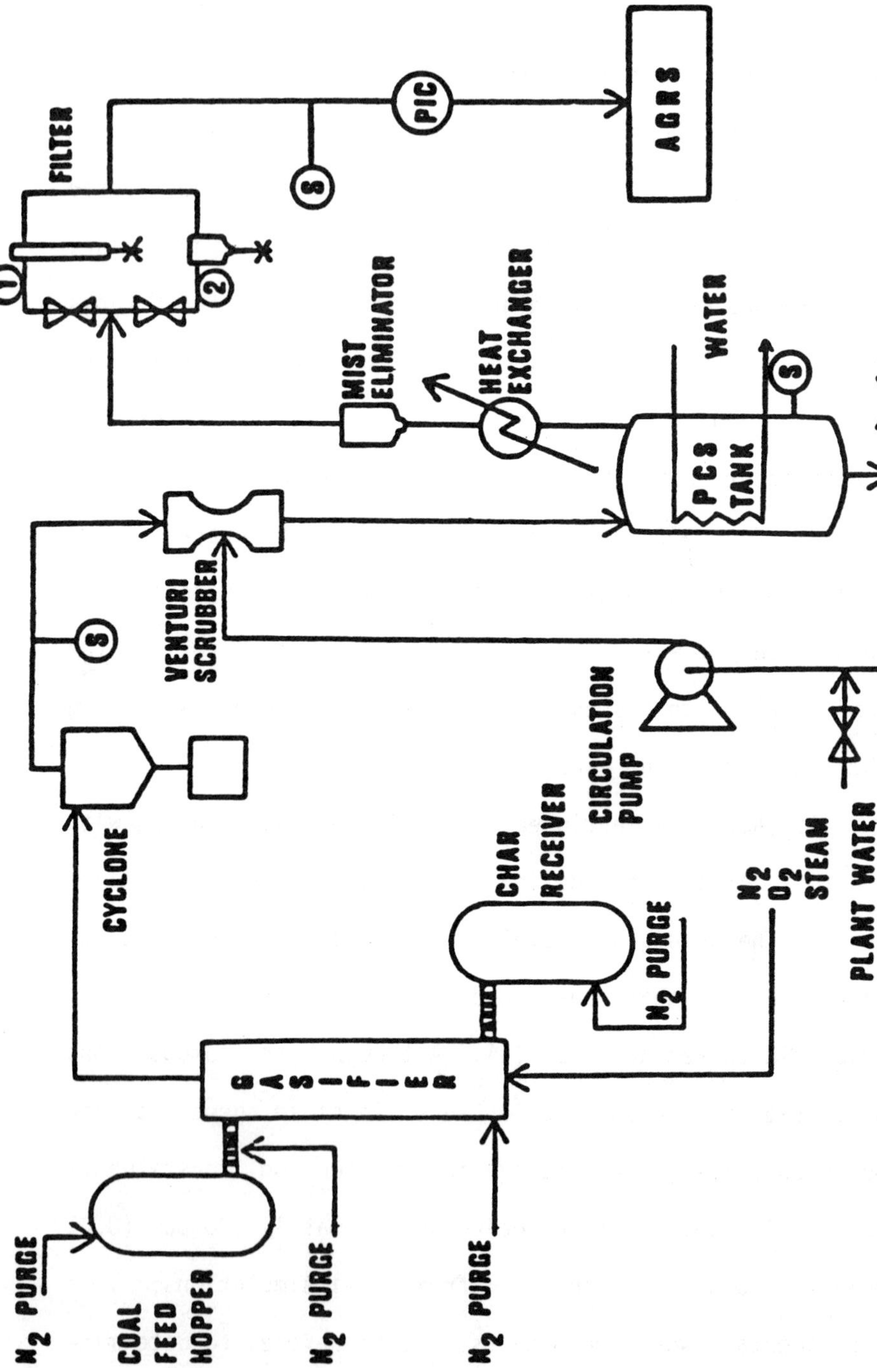

Figure 1. Gasifier/PCS system.

relationships between operating conditions and the ability of the acid gas removal system to reduce the concentrations of CO_2, H_2S and COS in the gas exiting the absorber to extremely low levels.

DESCRIPTION OF FACILITY

The pilot plant used in this study consisted of two major subsystems: one in which the fuel was gasified and the resulting gas quenched by direct contact with water in a Venturi scrubber for removal of particulates, condensables and water-solubles (referred to as the Gasifier-PCS system) and the acid gas removal system (AGRS). Each subsystem could be operated independently. A detailed schematic of the gasifier-PCS system is given in Figure 1.

In the operation of the gasifier, crushed coal was fed to the gasifier from the feed hopper while steam, oxygen, and nitrogen flow into the bottom to form a fluidized bed. The oxygen and steam reacted with the coal to form a large number of compounds, and the product gas left the top of the gasifier while spent char was removed from the bottom. The reaction bed normally was operated at about 7.8 atm absolute (100 psig) and temperatures ranging from 870 to

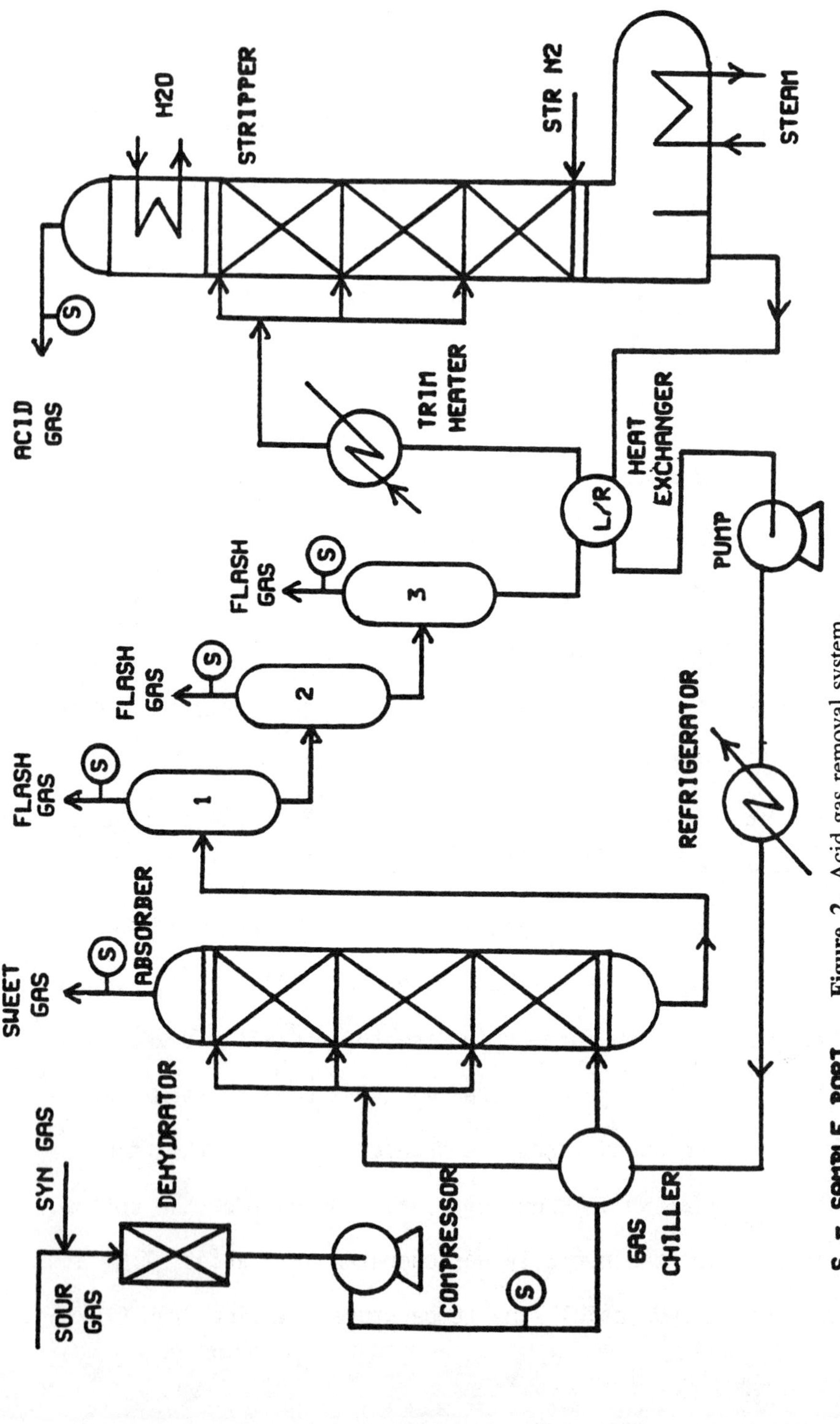

Figure 2. Acid gas removal system.

1040°C. The gas stream exiting the gasifier went through a cyclone separator that collected particulates, and was quenched with water in a Venturi scrubber. This step in the process separated entrained particles, tars and heavy oils, and water-soluble compounds from the gas, which also was cooled to near ambient temperature. Condensable and water soluble species were deposited in the water circulation tank (referred to a PCS tank in Figure 1). The gas stream was then passed through a mist eliminator and filter, and fed to the AGRS.

A detailed schematic of the AGRS is shown in Figure 2. Gas entering the AGRS was dried in a molecular sieve dehydrating tower, compressed to about 35 atm absolute (500 psig), cooled to approximately 15°C and fed to the absorber. The typical absorber operating pressure during this study was 32 atm absolute (450 psig). Acid gases (CO_2, H_2S, COS, etc.) were removed from the coal gas by contact with methyl alcohol in the absorber. Methanol, containing absorbed acid gases, left the bottom of the absorber and was fed to the regeneration units. During regeneration, acid gases were removed from the methanol and the resulting clean solvent was cooled and returned to the absorber. The conditioned gas from the absorber, and the gases leaving the flash tank(s) and stripping column were sent to a flare.

The absorber and stripper columns consisted of three sections of packing, each of which contained 2.2 m (7.1 ft) of 0.0064-m (1/4-in) ceramic Intalox saddles. One, two or three sections of packing could be used in either the absorber or stripper. Each section of packing contained two redistributers of liquid flow to prevent channeling. The columns, flash tanks and all pipes were insulated so that heat absorbed from the surroundings was negligible.

The entire process was controlled with a Honeywell TDC 2000 process control computer that regulated 16 control loops. Approximately 100 process sensors monitored temperatures, pressures, flow rates, and liquid levels throughout the plant and sent collected information to a data aquisition system linked to a DIGITAL PDP 11/23 computer.

EXPERIMENTAL METHODS

The system was considered to be at steady-state when measured flow rates, temperatures, pressures and gas compositions reached constant values. Operation was continued at these conditions for a two to three hour period, during which five-minute average values of flow rates, temperatures, pressures, etc. were stored by the PDP 11/23 computer.

Gas and liquid samples were taken periodically to determine steady-state compositions throughout the process. Gas samples were taken in one-liter stainless steel or one-liter glass bombs; samples to be analyzed for trace components were taken in stainless steel bombs internally coated with Teflon. Although information obtained from liquid samples was used in other research activities, quantitative information about the concentration of the acid gases (CO_2, H_2S, COS) in the methanol could not be obtained. Gas chromatography was used to analyze gas samples; major gas concentrations were determined using a thermal conductivity detector (TCD), sulfur compounds using a dual flame photometric detector (FPD), and hydrocarbon species using a flame ionization detector (FID).

Closures of mass balances were used as an indication of the accuracy of measured flow rates and compositions for each run. Values of process variables used in the balances were obtained by averaging measured values over a fixed period of time (typically one hour). If the overall mass balance closure for a given run was not within 10%, data

Table 1
A Typical Mass Balance

STREAM COMPOSITION (MOL %)

	SOUR GAS	SWEET GAS	FLASH GAS	STRIP N2	ACID GAS
CO2	27.510	0.010	31.960	0.000	54.730
H2S	0.304	0.002	0.086	0.000	0.600
COS	0.004	0.000	0.002	0.000	0.008
MEOH	0.000	0.000	0.343	0.000	5.840
H2	36.710	52.330	18.600	0.000	0.000
CO	14.210	19.420	19.310	0.000	1.549
N2	15.590	22.780	15.220	100.000	33.350
CH4	5.350	5.460	13.540	0.000	3.030
C2H4	0.297	0.000	0.666	0.000	0.451
C2H6	0.424	0.000	0.997	0.000	0.657

MASS BALANCE (LB-MOLES/HR)

	SOUR GAS	STRIP N2	SWEET GAS	FLASH GAS	ACID GAS	PERCENT RECOVERY
CO2	0.654	0.000	0.000	0.054	0.585	97.59
H2S	0.007	0.000	0.000	0.000	0.006	85.71
COS	0.000	0.000	0.000	0.000	0.000	97.80
MEOH	0.000	0.000	0.000	0.001	0.116	0.00
H2	0.873	0.000	0.840	0.031	0.000	99.78
CO	0.338	0.000	0.312	0.033	0.017	106.75
N2	0.371	0.373	0.366	0.026	0.356	100.54
CH4	0.127	0.000	0.088	0.023	0.032	112.26
C2H4	0.007	0.000	0.000	0.001	0.005	84.02
C2H6	0.010	0.000	0.000	0.002	0.007	86.28
TOTAL (LB-MOLES/HR)	2.379	0.373	1.605	0.169	1.004	100.9

TOTAL METHANOL LOSS = 0.116 LB-MOLES/HR

Table 2
Regeneration Schemes

Description	Runs
1 flash tank and stripping with N_2	AMIL-1,2,3,7
2 flash tanks and stripping with N_2	AMIL-11
3 flash tanks	AMIL-12,16
1 flash tank and stripping with reboiled methanol	AMIL-14,15

Table 3
Acid Gas Removal System Operating Conditions

Run number	AMIL-1	AMIL-2	AMIL3	AMIL-7
date	7-3-82	8-3-82	8-12-82	10-12-82
solvent circulation rate (l/min)	4.8	4.8	4.8	3.3
absorber				
pressure (atm)	31.6	31.5	31.5	31.5
packing height (m)	4.33	4.33	4.33	4.33
sour gas flow (scm/min)	0.35	0.35	0.40	0.34
liquid inlet temp (C)	-34.7	-34.2	-33.3	-36.7
gas inlet temp (C)	14.1	16.9	17.0	13.4
flash tank 1				
pressure (atm)	6.01	6.04	6.05	6.04
liquid inlet temp (C)	-16.7	-17.6	-16.0	-15.7
stripper				
pressure (atm)	1.34	1.34	1.41	1.34
packing height (m)	6.48	6.48	6.48	6.48
nitrogen flow (scm/min)	0.03	0.03	0.06	0.06
liquid inlet temp (C)	20.7	26.6	31.9	31.9
gas inlet temp (C)	32.9	33.1	30.1	26.2

Table 3 (Continued)

Run	AMIL-11	AMIL-12	AMIL-16
date	12-2-82	1-18-83	3-31-83
solvent circulation rate (l/min)	3.3	3.3	3.3
absorber			
pressure (atm)	31.5	31.5	31.4
packing height (m)	4.33	4.33	4.33
sour gas flow (scm/min)	0.34	0.33	0.28
liquid inlet temp (C)	-33.9	-40.7	-38.2
gas inlet temp (C)	14.8	7.49	8.11
flash tank 1			
pressure (atm)	6.02	7.69	7.78
liquid inlet temp (C)	-14.5	-23.6	-20.4
flash tank 2			
pressure (atm)	2.41	5.98	6.11
liquid inlet temp (C)	27.5	25.9	24.9
flash tank 3			
pressure (atm)	-	1.13	1.13
liquid inlet temp (C)	-	19.4	17.8
stripper			
pressure (atm)	1.34	-	-
packing height (m)	6.48	-	-
nitrogen flow (scm/min)	0.03	-	-
liquid inlet temp (C)	23.7	-	-
gas inlet temp (C)	27.2	-	-

Table 3 (Continued)

Run	AMIL-14	AMIL-15
date	2-24-83	3-7-83
solvent circulation rate (l/min)	3.3	3.3
absorber		
pressure (atm)	31.1	31.1
packing height (m)	4.33	4.33
sour gas flow (scm/min)	0.27	0.27
liquid inlet temp (C)	-37.0	-36.1
gas inlet temp (C)	14.43	14.83
flash tank 1		
pressure (atm)	5.59	5.59
liquid inlet temp (C)	-19.1	-17.7
stripper		
pressure (atm)	1.32	1.32
packing height (m)	6.48	6.48
liquid inlet temp (C)	70.0	70.0
steam flow (kg/hr)	7.70	7.60
steam temp (C)	157	158

Table 4
Gas Analysis Summary For AMIL-1
(13:00 Sample Time)

Species	Sour Gas	Sweet Gas	Flash Gas	Acid Gas
H_2	34.09	47.63	18.33	0.0000
CO	16.02	21.24	22.17	1.6600
CH_4	4.59	4.53	11.68	2.47
CO_2	27.73	0.00	31.13	64.50
N_2	17.47	24.72	17.06	21.28
H_2S	0.2428	0.0048	0.1007	0.6147
COS	0.0058	0.0000	0.0033	0.0138
Thiophene	0.0022	0.0000	0.0000	0.0020
CH_3SH	0.0391	0.0000	0.0010	0.0138
C_2H_5SH	0.0000	0.0000	0.0000	0.0000
CS_2	0.0001	0.0000	0.0000	0.0002
Ethylene	0.3729	0.0000	0.7627	0.7589
Ethane	0.4890	0.0000	1.0699	0.9954
Propylene	0.1225	0.0000	0.0711	0.2747
Propane	0.0426	0.0000	0.0333	0.0966
Butylene	0.0533	0.0000	0.0201	0.1164
Butane	0.0142	0.0000	0.0074	0.0314
Benzene	0.0341	0.0000	0.0015	0.0270
Toluene	0.0055	0.0000	0.0001	0.0024
Ethylbenzene	0.0002	0.0000	0.0000	0.0000
P-Xylene	0.0001	0.0000	0.0000	0.0000
M-Xylene	0.0007	0.0000	0.0000	0.0001
O-Xylene	0.0005	0.0000	0.0000	0.0000
Methanol	0.7478	0.0000	0.2888	5.79

Table 5
Gas Analysis Summary For AMIL-2
(13:00 Sample Time)

Species	Sour Gas	Sweet Gas	Flash Gas	Acid Gas
H_2	34.38	46.89	18.49	0.0000
CO	19.99	25.25	26.57	1.4917
CH_4	4.88	4.83	12.38	2.21
CO_2	25.47	0.00	28.38	59.79
N_2	15.58	21.57	15.10	18.91
H_2S	0.2577	0.0024	0.0878	0.5989
COS	0.0078	0.0000	0.0043	0.0222
Thiophene	0.0019	0.0000	0.0000	0.0019
CH_3SH	0.0037	0.0000	0.0005	0.0085
C_2H_5SH	0.0002	0.0000	0.0000	0.0004
CS_2	0.0000	0.0000	0.0000	0.0002
Ethylene	0.3621	0.0000	0.8080	0.8224
Ethane	0.5581	0.0000	1.3021	1.2492
Propylene	0.0921	0.0000	0.0561	0.2468
Propane	0.0342	0.0000	0.0275	0.0903
Butylene	0.0385	0.0000	0.0107	0.1163
Butane	0.0106	0.0000	0.0039	0.0314
Benzene	0.0546	0.0000	0.0014	0.0538
Toluene	0.0108	0.0000	0.0001	0.0051
Ethylbenzene	0.0006	0.0000	0.0000	0.0000
P-Xylene	0.0003	0.0000	0.0000	0.0000
M-Xylene	0.0015	0.0000	0.0000	0.0000
O-Xylene	0.0009	0.0000	0.0000	0.0000
Methanol	0.0073	0.0000	0.3292	9.98

Table 6
Gas Analysis Summary For AMIL-3
(12:40 Sample Time)

Species	Sour Gas	Sweet Gas	Flash Gas	Acid Gas
H_2	36.71	52.33	18.60	0.0000
CO	14.21	19.42	19.31	1.5489
CH_4	5.35	5.46	13.54	3.03
CO_2	27.51	0.00	31.96	54.73
N_2	15.59	22.78	15.22	39.35
H_2S	0.3042	0.0016	0.0863	0.4707
COS	0.0039	0.0000	0.0025	0.0081
Thiophene	0.0012	0.0000	0.0000	0.0020
CH_3SH	0.0022	0.0000	0.0004	0.0037
C_2H_5SH	0.0000	0.0000	0.0000	0.0003
CS_2	0.0000	0.0000	0.0000	0.0001
Ethylene	0.2971	0.0000	0.6660	0.4506
Ethane	0.4241	0.0000	0.9971	0.6573
Propylene	0.1290	0.0000	0.0710	0.2264
Propane	0.0530	0.0000	0.0411	0.0913
Butylene	0.0550	0.0000	0.0128	0.1011
Butane	0.0167	0.0000	0.0048	0.0291
Benzene	0.0371	0.0000	0.0012	0.0572
Toluene	0.0110	0.0000	0.0001	0.0060
Ethylbenzene	0.0008	0.0000	0.0000	0.0000
P-Xylene	0.0004	0.0000	0.0000	0.0000
M-Xylene	0.0018	0.0000	0.0000	0.0003
O-Xylene	0.0009	0.0000	0.0000	0.0000
Methanol	0.0000	0.0000	0.3428	10.84

Table 7
Gas Analysis Summary For AMIL-7
(12:00 Sample Time)

Species	Sour Gas	Sweet Gas	Flash Gas	Acid Gas
H_2	35.87	50.43	15.84	0.0000
CO	12.05	16.47	13.77	0.8019
CH_4	4.76	5.39	10.58	1.72
CO_2	28.67	0.00	41.41	49.67
N_2	17.84	27.71	16.58	33.37
H_2S	0.2141	0.0000	0.0846	0.3716
COS	0.0029	0.0000	0.0022	0.0054
Thiophene	0.0021	0.0000	0.0000	0.0023
CH_3SH	0.0033	0.0000	0.0004	0.0053
C_2H_5SH	0.0000	0.0000	0.0000	0.0000
CS_2	0.0000	0.0000	0.0000	0.0000
Ethylene	0.2701	0.0000	0.7222	0.4390
Ethane	0.4047	0.0000	1.1174	0.6489
Propylene	0.0709	0.0000	0.0501	0.1339
Propane	0.0260	0.0000	0.0253	0.0477
Butylene	0.0309	0.0000	0.0076	0.0677
Butane	0.0122	0.0000	0.0027	0.0281
Benzene	0.0408	0.0000	0.0012	0.0465
Toluene	0.0048	0.0000	0.0001	0.0052
Ethylbenzene	0.0002	0.0000	0.0000	0.0000
P-Xylene	0.0001	0.0000	0.0000	0.0000
M-Xylene	0.0006	0.0000	0.0000	0.0003
O-Xylene	0.0003	0.0000	0.0000	0.0000
Methanol	1.0498	0.0000	0.4780	15.10

Table 8
Gas Analysis Summary For AMIL-11
(11:45 Sample Time)

Species	Sour Gas	Sweet Gas	Flash 1Gas	Flash 2Gas	Acid Gas
H_2	34.07	47.35	17.27	0.59	0.00
CO	14.42	18.82	18.09	2.14	0.00
CH_4	5.25	5.37	14.74	3.64	0.55
CO_2	26.46	0.00	34.45	89.23	44.33
N_2	19.14	28.46	18.64	1.59	44.51
H_2S	0.1516	0.0016	0.0638	0.2706	0.3647
COS	0.0038	0.0000	0.0026	0.0099	0.0106
Thiophene	0.0014	0.0000	0.0000	0.0010	0.0019
CH_3SH	0.0031	0.0000	0.0005	0.0031	0.0056
C_2H_5SH	0.0000	0.0000	0.0000	0.0000	0.0000
CS_2	0.0000	0.0000	0.0000	0.0005	0.0010
Ethylene	0.1734	0.0000	0.4397	0.5135	0.1742
Ethane	0.3091	0.0000	0.7970	0.8756	0.2832
Propylene	0.0827	0.0000	0.0591	0.1662	0.1771
Propane	0.0414	0.0000	0.0387	0.0884	0.0773
Butylene	0.0372	0.0000	0.0113	0.0526	0.0832
Butane	0.0108	0.0000	0.0039	0.0198	0.0323
Benzene	0.5345	0.0000	0.0010	0.0201	0.0384
Toluene	0.1200	0.0000	0.0001	0.0036	0.0040
Ethylben	0.0028	0.0000	0.0000	0.0000	0.0000
P-Xylene	0.0008	0.0000	0.0000	0.0000	0.0000
M-Xylene	0.0089	0.0000	0.0000	0.0000	0.0000
O-Xylene	0.0014	0.0000	0.0000	0.0000	0.0000
Methanol	0.0623	0.0000	0.5670	7.8877	12.54

Table 9
Gas Analysis Summary For AMIL-14
(15:30 Sample Time)

Species	Sour Gas	Sweet Gas	Flash Gas	Acid Gas
H_2	35.00	48.25	17.57	0.0000
CO	12.27	16.77	16.75	0.9533
CH_4	4.46	4.60	13.88	1.89
CO_2	27.38	0.00	31.95	77.33
N_2	20.41	28.40	18.65	0.94
H_2S	0.2579	0.0000	0.0601	0.6979
COS	0.0034	0.0000	0.0022	0.0101
Thiophene	0.0019	0.0000	0.0000	0.0099
CH_3SH	0.0034	0.0000	0.0000	0.0081
C_2H_5SH	0.0000	0.0000	0.0000	0.0009
CS_2	0.0000	0.0000	0.0000	0.0011
Ethylene	0.1590	0.0000	0.3834	0.3860
Ethane	0.2520	0.0000	0.6550	0.6179
Propylene	0.0496	0.0000	0.0273	0.1422
Propane	0.0176	0.0000	0.0140	0.0489
Butylene	0.0206	0.0000	0.0027	0.0781
Butane	0.0048	0.0000	0.0006	0.0181
Benzene	0.0781	0.0000	0.0019	0.9529
Toluene	0.0218	0.0000	0.0005	0.0743
Ethylbenzene	0.0020	0.0000	0.0000	0.0010
P-Xylene	0.0010	0.0000	0.0000	0.0000
M-Xylene	0.0020	0.0000	0.0000	0.0006
O-Xylene	0.0012	0.0000	0.0000	0.0000
Methanol	0.0000	0.0000	0.2180	15.00

Table 10
Gas Analysis Summary For AMIL-16
(15:00 Sample Time)

Species	Sour Gas	Sweet Gas	Flash 1Gas	Flash 2Gas	Flash 3Gas
H_2	32.75	45.90	19.26	0.00	0.00
CO	11.84	16.32	17.64	5.40	0.4094
CH_4	4.97	5.38	12.58	8.39	1.41
CO_2	27.10	0.63	25.38	78.12	80.02
N_2	22.90	31.77	24.29	6.30	1.92
H_2S	0.2468	0.0029	0.0618	0.2268	0.8067
COS	0.0041	0.0000	0.0025	0.0096	0.0165
Thiophene	0.0018	0.0000	0.0000	0.0007	0.0037
CH_3SH	0.0034	0.0000	0.0004	0.0018	0.0060
C_2H_5SH	0.0000	0.0000	0.0000	0.0000	0.0000
CS_2	0.0000	0.0000	0.0000	0.0000	0.0012
Ethylene	0.2097	0.0049	0.3962	0.7510	0.5468
Ethane	0.3236	0.0073	0.6345	1.1908	0.8213
Propylene	0.0676	0.0022	0.0381	0.1336	0.2261
Propane	0.0249	0.0009	0.0191	0.0573	0.0812
Butylene	0.0299	0.0007	0.0064	0.0442	0.1020
Butane	0.0077	0.0000	0.0024	0.0180	0.0269
Benzene	0.0422	0.0000	0.0005	0.0070	0.0415
Toluene	0.0092	0.0000	0.0001	0.0016	0.0161
Ethylben	0.0006	0.0000	0.0000	0.0000	0.0000
P-Xylene	0.0003	0.0000	0.0000	0.0000	0.0000
M-Xylene	0.0005	0.0000	0.0000	0.0000	0.0000
O-Xylene	0.0004	0.0000	0.0000	0.0000	0.0000
Methanol	0.0000	0.0000	0.2472	4.5339	14.00

obtained from the run were consided invalid and were discarded. Exceptions to this rule were made for individual component mass balances on species present in very small amounts (such as H_2S, COS and hydrocarbons). Table 1 shows a typical mass balance.

SOLVENT REGENERATION

As previously stated, the results of previous work indicated that the absorption of acid gases is governed by equilibrium conditions at the top of the absorber. Thus, solvent regeneration determines absorber performance, as well as energy requirements, process control problems and distribution of recovered acid gases.

Regeneration schemes for acid gas removal solvents typically employ one or more flash tanks and/or a stripping column operated with an inert gas or reboiler. Four regeneration schemes were used in this study. Table 2 presents a description of each scheme and the number of the experimental runs made with each. Table 3 lists the AGRS conditions for each run.

Results of gas analyses obtained from each run are given in Tables 4 through 10. Although this study was concerned primarily with the removal of CO_2, H_2S and COS, complete analyses are given. Conditions for Runs AMIL-14 and AMIL-15 were identical as were those for Runs AMIL-12 and AMIL-16; only the analyses of one run at each set of conditions are presented.

The two sets of paired runs, Runs AMIL-14 and AMIL-15 and Runs AMIL-12 and AMIL-16 were used to examine the reproducibility of system operation. The sweet gas samples from Runs AMIL-12 and AMIL-16 contained 0.51% and 0.63% CO_2, 0.0032% and 0.0029% H_2S, and no detectable COS. These data correspond to a 19% deviation in CO_2 composition and a 9% deviation in H_2S composition. Unfortuately, there were no detectable levels of CO_2, H_2S or COS in samples of sweet gas from Runs AMIL-14 or AMIL-15. These results and duplicate samples taken within a given run indicate reasonable reproducibility.

The objective of the AGRS is to separate acid gases from the product gas. It is reasonable then to evaluate the performance of the AGRS by examining the composition of the sweet gas obtained from each configuration. In particular, the acid gas content of this stream was considered to be the best indication of system performance. Table 11 gives percentages of CO_2, H_2S and COS in the sweet gas stream from each regeneration scheme. As the composition of the sour gas fed to the AGRS varied, CO_2, H_2S, and COS percentages in this stream are also given.

Specifications on pipeline quality gas call for a maximum CO_2 content of between 1 and 3 volume percent (5). The CO_2 content of the sour gas stream produced by the gasifier and fed to the AGRS was approximately the same for each run (Table 11). Regardless of the method used for regeneration, the gas obtained from the absorber in all runs met the allowable CO_2 levels given above.

No COS was detected in the sweet gas from any run. This may be attributed to the low levels of COS in the sour gas stream.

Pipeline quality gas must contain less than 5 to 20 ppm H_2S (6). As can be seen in Table 11, the regeneration conditions used in run AMIL-1 resulted in a sweet gas H_2S level of 0.0048 mole percent (48 ppm). This level was typical of that obtained in an earlier study (2) where absorber conditions were varied while stripper conditions were maintained at those of AMIL-1. In AMIL-2, the temperature of the liquid entering the stripper was increased to 27°C from 21°C used in Run AMIL-1. Other AGRS conditions and the concentration of H_2S in the sour gas remained approximately the same in these two runs. This

Table 11
Acid Gas Removal Summary

Run	Composition (mole %)					
	Sour gas			Sweet gas		
	CO_2	H_2S	COS	CO_2	H_2S	COS
AMIL-1	27.73	.2428	.0058	N.D.*	.0048	N.D.
AMIL-2	25.47	.2577	.0078	N.D.	.0024	N.D.
AMIL-3	27.51	.3042	.0039	N.D.	.0016	N.D.
AMIL-7	28.67	.2141	.0029	N.D.	N.D.	N.D.
AMIL-11	26.46	.1516	.0038	N.D.	.0016	N.D.
AMIL-14	27.38	.2579	.0034	N.D.	N.D.	N.D.
AMIL-16	27.10	.2468	.0041	0.63	.0029	N.D.

* N.D. (none detected) Considered to be <0.5% for CO_2 and <5 PPM (0.0005%) for H_2S and COS.

5^oC change in the temperature of the liquid fed to the stripper reduced the H_2S content of the sweet gas to 24 ppm.

In run AMIL-3, the temperature of the liquid entering the stripper was increased further to 32^oC and the flow rate of stripping N_2 was twice that of the two previous runs. The H_2S content of the sweet gas obtained from the absorber was reduced to 16 ppm, even though the concentration of H_2S in the sour gas was slightly higher for run AMIL-3 than for the previous two runs.

At a given set of operating conditions, small changes in the amount of H_2S in the gas fed to the AGRS result in corresponding changes in the amount of H_2S in the sweet gas. For example, the first set of samples on Run AMIL-1 showed that feed and sweet gases contained 0.2428% H_2S and 0.0048% H_2S, respectively; the second set of samples, taken under identical operating conditions, contained 0.2528% H_2S and 0.0059% H_2S, respectively. A four percent increase in H_2S content of the feed gas resulted in a 23 percent increase in the H_2S content of the sweet gas. As the H_2S concentration in the sweet gas depends on the concentration of this component in the regenerated solvent, these results indicate that desorption of the H_2S is more difficult than absorption. This conclusion is supported by the observation that all runs using N_2 stripping only resulted in a sweet gas that did not meet H_2S specifications.

In Run AMIL-7, the AGRS conditions were the same as for AMIL-3 except that the solvent circulation rate was decreased to 3.3 liters/min (0.87 gal/min). Although this change in liquid flow rate decreased the liquid-to-gas ratio in the absorber, it also increased the gas-to-liquid ratio in the stripper. As can be seen in Table 11, no H_2S was detected in the sweet gas for this run.

In Run AMIL-11, the conditions used were the same as those in Run AMIL-2, except two flash tanks were used instead of one. This run resulted in the gas leaving the absorber having an H_2S concentration of 16 ppm (compared with 24 ppm for run AMIL-2). However, because the H_2S concentration in the gas fed to the absorber was much lower than for AMIL-2 (Table 11), the basis of the improved performance of the AGRS is not clear.

Run AMIL-14 employed one flash tank and the stripping column operated with the reboiler instead of using a stripping N_2 stream. No H_2S was detected in the sweet gas stream from this run. This result can be compared with that of AMIL-3, which operated with a similar configuration, but with N_2 stripping. The sweet gas from AMIL-3 contained 16 ppm H_2S. It should also be noted that the solvent was heated to 70°C during operation with the reboiler, whereas its maximum temperature in N_2 stripping was 32°C.

Finally, in Run AMIL-16 regeneration was accomplished using three flash tanks and and no stripping column. Since the methanol was heated to approximately 27°C after the first flash tank, this run may be compared to Run AMIL-2 (where the methanol exited the single flash tank, was heated to approximately 27°C, and fed to the stripper). As shown in Table 11, the H_2S concentration in the the sweet gas from Run AMIL-16 was 29 ppm, whereas is was 24 ppm in Run AMIL-2.

Since the concentration of H_2S in the gas fed to the absorber in these two runs was approximately the same, regeneration using three flash tanks is nearly as effective as stripping with N_2.

DISCUSSION

All operating conditions and configurations used in this study allowed the AGRS to produce a gas from the absorber meeting CO_2 and COS specifications. However, the method of regeneration affected the ability of the AGRS to meet corresponding H_2S specifications on the product gas. Specifically, a product gas essentially free of H_2S (<5ppm) was obtained only with the use of thermal regeneration (Runs AMIL-14 and AMIL-15). No H_2S was detected in the product gas from Run AMIL-7, but this run was made using a flow rate of N_2 stripping gas thought to be unrealistic.

The results of this work indicate that the key factor in the regeneration of methanol is the maximum temperature to which the solvent is heated. A regeneration scheme that heats the methanol to moderate temperatures (for example 50°C) and employs multiple flash tanks may be adequate in situations that require a sweet gas stream essentially free from acid gases. Such a system does not require heat to be supplied to vaporize solvent, nor does it require cooling the solvent from the reboiler before it can be fed to the absorber. Consequently, it should require less energy to operate than a system utilizing a reboiled stripper.

The experimental results indicate that the absorption of acid gases in this facility is governed by equilibrium conditions at the lean end of the absorber. This means that the height of packing used in the experimental runs was sufficient to cause a pinch between equilibrium and operating line relationships for the acid gases. Therefore, the performances of the acid gas removal systems examined in this study may be considered the best that can be obtained at the given operating conditions.

The relationship between regeneration conditions and AGRS performance using a physical solvent other than methanol is thought to be similar to that observed in this study. Choosing among various regeneration schemes for these solvents will require consideration of factors whose importance varies from case to case. For example, optimizing the system to minimize energy requirements may make regeneration using multiple flash tanks desirable. In other situations, the availability of low incremental cost steam or an inert stripping gas may make those choices more attractive.

Requirements associated with specific tail gas or sulfur recovery units must also be considered in selecting a process for regeneration. If all sulfur compounds must be removed from the sweet gas (for example, to protect gas processing catalysts), regeneration in a reboiled stripper or use of a multiple flash tank scheme that includes heating the circulating methanol to high temperatures must be used.

Solvent losses vary depending on the properties of the solvent and the corresponding operating conditions in the AGRS, especially in regeneration. While the problem is made more acute with a volatile solvent such as methanol, it can be minimized by installation of condensers at points where hot gas streams exit the process. Costs of installation and operation of such condensers may make the use of a less volatile solvent more attractive.

It is often attractive to use two steps to separate CO_2 and H_2S from the gas fed to an acid gas removal system. These two steps are designed to separate two off-gas streams, one containing only CO_2 and the other containing a mixture of H_2S and the remaining CO_2. In such a facility, a regeneration scheme must be selected for recovery of each off-gas stream. The principles outlined in this study for selection of a single regeneration step can also be applied to the dual regeneration process required by an acid gas removal system operated in this fashion.

SUMMARY

An experimental acid gas removal system was operated with refrigerated methyl alcohol as the solvent to evaluate various configurations for solvent regeneration. Methods of regeneration employed single and multiple flash tanks and/or a packed stripping column. Stripping included use of an inert gas or reboiler.

Each regeneration scheme resulted in an absorber exit gas CO_2 concentration less than 1 mole percent and a COS level less than 5 ppm. Levels of H_2S in the absorber exit gas ranged from 48 ppm when stripping with an inert gas at ambient temperature to less than 5 ppm when stripping was accomplished by operation of a reboiler. Regeneration using three flash tanks was as effective as stripping with an inert gas, but not as effective as operating the stripper with a reboiler.

ACKNOWLEDGMENT

The support of the Environmental Protection Agency under Grant Number R804811 is acknowledged gratefully.

REFERENCES

1. Rousseau, R. W., and J. K. Ferrell, "Physical Absorption of CO_2 and Sulfur Gases from Coal Gasification: Pilot Plant Results," presented at the Second GRI Gas Separations Workshop, Boulder, CO, October 1982.

2. Kelly, R. M., "Use of Methanol as a Solvent in an Acid Gas Removal System Coupled with a Coal Gasifier I. Operation II. Simulation," Ph.D. Thesis, North Carolina State University, Raleigh, North Carolina, 1981.

3. Staton, J. S., "Modeling of an Adiabatic Stripping Column in a Coal Gasification Plant Acid Gas Removal System," M.S. Thesis, North Carolina State University, Raleigh, North Carolina, 1983.

4. Kelly, R. M., R. W. Rousseau and J. K. Ferrell, "Design of Packed, Adiabatic Absorbers: Physical Absorption of Acid Gases in Methanol," Ind. Eng. Chem. Process Des. Dev., 23, 102(1983).

5. Goar, B. G., "Guidelines Set for Handling Sour Gas," Oil and Gas J., p. 160,June 26, 1978.

6. Ranke, G., "The Rectisol Process for the Selective Removal of CO_2 and Sulfur Compounds from Industrial Gases,"Chemical Economy and Engineering Review, 4(5), 25(1972).

6

An Evaluation of Sources of CO_2 for EOR In Venezuela

G. Aldana
Chemical Engineer
Intevep, S. A., Caracas

R. Arai
Process Engineering Manager
Davy McKee, Houston

D. G. Elliot
Vice-president, Process Department
Davy McKee, Houston

INTRODUCTION

Around 40% of the 1.9 million BPD of oil produced in Venezuela in 1982 was associated with gas and water injection projects, 6% with EOR thermal projects, and the rest was not associated with any EOR process. Distribution of this production was 40% for heavy (21°API or less) oil, 29% for medium (21-31°API), and 31% for lighter (31°API or more) crudes. The oil industry extracted approximately 79% of 1982 oil production from the Lake Maracaibo Basin, and the remaining 21% from the Eastern Basin.

These patterns can be assumed to hold for 1983 except the total oil production decreased to 1.7 million BPD.

The number of reservoirs under gas injection was 183

Based on material presented at the 63rd Annual Gas Processors Association Convention, March 19-21, 1984, New Orleans, Louisiana.

for 1982, 29 of them in the Lake Maracaibo area, and 154 in Eastern Venezuela. The oil in place for these projects totaled more than 16 x 10^9 Bbl, with 69% corresponding to Lake projects. For the same year, 91 reservoirs under water injection were reported.

Expected additional recovery from gas injection projects was 3000 x 10^6 Bbl of oil, equivalent to 18.4% of the original oil in place, and distributed as shown in the following table:

Type of Crude	Lake Maracaibo 10^6Bbl	Lake Maracaibo % STOIP	East 10^6Bbl	East % STOIP	Total 10^6Bbl	Total % STOIP
Light	200	12.7	600	23.2	800	19.1
Medium	1900	19.6	140	15.0	2040	19.2
Heavy	--	--	160	11.3	160	11.3
	2100	18.6	900	18.0	3000	18.4

Oil in place for water injection projects was estimated at over 20.2 x 10^9 Bbl. Estimated additional recovery from these projects was 1.5 x 10^9 Bbls., 96% of which was from the Lake Maracaibo projects. The distribution of additional reserves is as follows:

Type of Crude	Lake Maracaibo 10^6Bbl	Lake Maracaibo % STOIP	East 10^6Bbl	East % STOIP	Total 10^6Bbl	Total % STOIP
Light	265	11.6	49	14.3	314	12.0
Medium	590	8.7	10	5.4	600	8.6
Heavy	585	5.5	1	7.3	586	5.5
	1440	7.3	60	11.1	1500	7.4

Gas injection has been preferred over water injection as a secondary recovery method, but the growing importance of natural gas as an energy source could lead to changes in its utilization. A comparison of oil recovery factors associated with gas and water injection projects is presented below:

WEST VENEZUELA PROJECTS

Method	Type of Crude	Primary Recovery, %	Secondary Recovery, %	Total Recovery, %
Gas Injection	Light	34.4	12.7	47.1
	Medium	32.5	19.6	52.1
	Heavy	--	--	---
Water Injection	Light	23.6	11.6	35.2
	Medium	32.4	8.7	41.1
	Heavy	19.3	5.5	24.8

EAST VENEZUELA PROJECTS

Method	Type of Crude	Primary Recovery, %	Secondary Recovery, %	Total Recovery, %
Gas Injection	Light	23.2	23.2	46.4
	Medium	22.1	15.0	37.1
	Heavy	11.8	11.3	23.1
Water Injection	Light	27.0	14.3	41.3
	Medium	18.9	5.4	24.3
	Heavy	14.0	7.3	21.3

EOR THERMAL PROCESSES

EOR thermal processes have been applied in Venezuela since 1957, when continuous steam injection was used to increase the recovery of heavy oil from a reservoir of Mene Grande field, along the East Coast of Lake Maracaibo Thermal projects in Venezuela include 56 cyclic steam injection projects, 75 continuous steam injection projects, 13 in situ combustion projects, and one hot water injection project. The estimated oil production associated with these EOR projects was 180 x 10^6 Bbls for 1981. Cyclic steam injection contributed 91% of this total, and 8% was due to continuous steam injection.

Cyclic injection has been the most successful of the thermal methods, with a recovery factor of 19.1 for 31 projects located along the East Coast of Lake Maracaibo, and 7.9 for Eastern Venezuela projects.

CO_2 INJECTION PROJECTS

The search for new and more efficient methods of recovery of light and medium oil is under way due mainly to:

1. The existence of many reservoirs containing large quantities of oil, recoverable by non-conventional methods like gas and water flooding and gas injection.
2. The expected natural gas scarcity for gas injection projects caused by an increasing internal demand.

CO_2 injection offers a promising alternative to recover additional volumes of light and medium oil from reservoirs where gas and/or water injection have been applied. A research program to study the feasibility of CO_2 injection to one reservoir of Lake Maracaibo was undertaken in 1982. The study included an analysis of the formation, fluids evaluation, PVT behavior, tube displacement tests, computer simulation, and CO_2 availability.

Studies have been initiated to evaluate CO_2 injection for two small reservoirs, where water injection was applied. Preliminary results indicate that these two reservoirs, located in Northeastern Venezuela, are good candidates for miscible CO_2 recovery. If field pilot tests are successful, new CO_2 projects will be conducted in this area.

There are two critical aspects of CO_2 projects in Venezuela: the availability and cost of CO_2. Our work is focused on this project.

CO_2 AVAILABILITY

A survey of various CO_2 sources for EOR in Venezuela was conducted in 1982. The following sources were considered:

1. Natural Sources

   1.1 CO_2 reservoirs

   1.2 Natural gas with high CO_2 content.

   A search for natural CO_2 deposits has not been conducted in Venezuela. Gas analyses from the majority of natural gas producing fields indicate that only 10 fields show a CO_2 content greater than 8% with 20 vol % as an upper limit. These fields are located in Eastern Venezuela.

2. Industrial By-products

   2.1 Ammonia plants: High purity CO_2 is produced by two plants but is totally used for urea synthesis.

2.2 Cryogenic plants: The possibility of feed gas treatment in a future cryogenic plant in the Northeast would generate 50 MMSCFD of CO_2 which could be a good source for reservoirs around this area.

2.3 Vent gases from refineries and hydrogen plants: The two largest refineries were considered in this evaluation, both located in the West, 110 miles from Lake Maracaibo oil-producing fields.

A projected hydrogen plant within the battery limits of one of the above-mentioned refineries could supply 93 MMSCFD CO_2 but the plant construction has been postponed.

2.4 Power plant flue gases: Flue gases from 17 power plants were evaluated, but after preliminary screening only two, located along the West Coast of Lake Maracaibo, were considered to be reliable sources of CO_2. They can generate between 60 to 80 MMSCFD flue gas upon combustion of natural gas and/or fuel oil. Distances to oil reservoirs from the CO_2 sources vary from

30 to 40 miles.

2.5 Flue gases from cement plants: Data from 9 cement plants with an estimated output of 500-600 MMSCFD of CO_2 in flue gases were evaluated. Only one plant was considered as a possible source of CO_2 for reservoirs in the Lake Maracaibo area. A notable disadvantage of this source is the wide fluctuation of cement production according to demand, which makes it difficult to estimate the CO_2 production from these plants.

2.6 EOR thermal projects: Steam injection projects were included in the study of CO_2 sources because the associated gas to oil production from reservoirs under effect of steam present high CO_2 contents, but also H_2S is produced simultaneously, increasing corrosion problems for gathering systems. Also, gas volumes are small because the GOR is very low.

Flue gases from steam generators used in EOR projects on the East Coast of the Lake are potential sources of CO_2 depending upon gathering costs. Total CO_2 output from steam generation plants located in this area exceeds 50 MMSCFD.

2.7 Liquid CO_2 plants: Liquid CO_2 was not taken into account in our study, but this source is limited by the cost of CO_2 due to its high purity. Reported F.O.B. prices by two plants was U.S. $13.25/MSCF for 1982. In addition, total output for each plant was less than 6 MMSCFD.

For reservoirs located in the Lake Maracaibo area, flue gases from power plants and the future hydrogen plant in the refinery in Northwest Venezuela were selected by Intevep S.A. for further technical and economic evaluation. Davy McKee in Houston, Texas conducted the study in cooperation with Intevep S.A.

PROCESSING ALTERNATIVES OF CO_2

There are many available processes for the removal of acid gases from gas streams. The selection of the appropriate process for recovery and purification of CO_2 for enhanced oil recovery is determined by the following considerations:

- o Feed gas composition & conditions
- o CO_2 product quality
- o Investment & operating costs
- o Commercial experience & reliability of process scheme

The possible processes for this type of CO_2 removal are divided into four major groups:

- o Chemical solvents
- o Physical solvents
- o Membranes
- o Cryogenic fractionation

The processes employing chemical solvents are those which chemically absorb the acid gases. The regeneration of the solvent is done by application of heat. Those processes employing physical solvents are physically absorbed by the solvent; solvent regeneration is achieved by pressure reduction. Generally, the economics of CO_2 recovery are strongly influenced by the partial pressure of CO_2 in the feed gas. At low partial pressures, physical solvents are impractical because the compression of the gas for physical absorption makes it expensive.

However, if the gas is available at high pressure, physical solvents could be a better choice. Detailed descriptions of typical processes can be found elsewhere in this paper.

The membrane process is applicable for high pressure gas containing high acid gas concentration. CO_2 recovery is accomplished by pressure-driven mass transfer through a permeable membrane. Separation is due to the differences in permeation rate of different compounds. The acid gas will be recovered at low pressure. Membranes are produced in modules which allow the producer to add membrane modules as necessary to accommodate the rising gas volume or CO_2 content in gas; however, they do not exhibit typical economies of scale. In a gas mixture, the operation of gas permeation does not yield a clear-cut split of the gases, but rather a concentration of gases. The performance of a membrane system for CO_2 separation is dependent on the following system parameters:

1. Feed gas composition and conditions (pressure and temperature).
2. Allowable CO_2 concentration in the non-permeable

residue gas stream.

3. CO_2 purity in the permeate gas.

4. CO_2 recovery requirements.

For given system conditions, there exists a definite relationship between the amount of CO_2 recovered and its purity in the permeate gas.

A high purity product containing approximately 95% CO_2 can be achieved with one or two stages, depending upon feed gas pressure and percent recovery. Increasing feed pressure and/or utilizing two stage separation with recompression of the gas from the first stage, is a matter of economic consideration. It involves an additional capital and incremental energy requirement which is related to the above mentioned parameters.

The cryogenic fractionation process is used to remove or recover CO_2 from feed gas. Generally, the process is applicable at a pressure of approximately 400-500 psig. Previous studies and our own investigation indicate the cryogenic separation is very competitive when bulk removal of CO_2 is required. In general, chemical and physical absorption systems suffer from a number of economic disadvantages in separation of feed gas containing high levels of CO_2. The conventional processes such as amine and hot potassium are too energy intensive, and have high maintenance costs due to the corrosive nature of the system and solvent degradation. Physical solvent processes have lower energy consumption but they are more attractive for CO_2 removal from relatively lean feed gas. Less energy will be required for cryogenic separation to recover CO_2 from gases with high CO_2 content. Further, CO_2 can be obtained at relatively high pressure, as opposed to other competing processes where CO_2 is obtained at low pressure. Since the CO_2 product is further compressed to 2000 psi, the capital investment and energy requirement will be much lower for the cryogenic process.

The design of a cryogenic fractionation has to consider the properties of CO_2 at cryogenic temperatures. Davy McKee has developed a cryogenic fractionation process to recover CO_2 from flue gas or offgas from hydrogen plants, where the major constituent is nitrogen.

CASE 1: 100 MMSCFD CO_2 RECOVERY FROM FLUE GAS

For recovery of CO_2 from flue gas, only a few alternatives are applicable. Two major constraints are:

1. Flue gas pressure and CO_2 content are low. The pressure is atmospheric with 11-13% volume CO_2 in the gas. As a result, CO_2 partial pressure is very low.
2. Oxygen is present in the flue gas at a concentration of 4-5% which tends to degrade the solvent at relatively high concentrations.

A physical solvent system is not economically practical due to the low pressure and CO_2 content of the gas. It will be necessary to compress the gas. The capital and operating cost of the compression and the associated physical solvent systems make the CO_2 recovery uneconomical.

Two choices are available for chemical systems: hot potassium carbonate and amine-based systems. The hot potassium process has a very low pick-up at low pressure. In order for this process to be feasible, a higher feed pressure is necessary which requires compression and makes it uneconomical. The other choice, amine-based system, can be considered.

As mentioned earlier, the presence of oxygen accelerates the degradation of an amine and also creates corrosion problems. In order to overcome these problems, one can either remove the oxygen to an acceptable level and use the conventional amine system, or use an inhibitor that could function in the oxidizing environment and provide the protection against degradation and corrosion. In this study, both options were investigated, an inhibited MEA system, and a catalytic deoxidation unit is combined with conventional MEA. The catalytic deoxidation unit uses a catalyst to promote pyrolitic reaction with the fuel gas. The product from the fuel gas combustion increases CO_2 production. The heat of combustion is recovered by generating steam which is subsequently used in the MEA unit. Figures 1 and 2 illustrate both options.

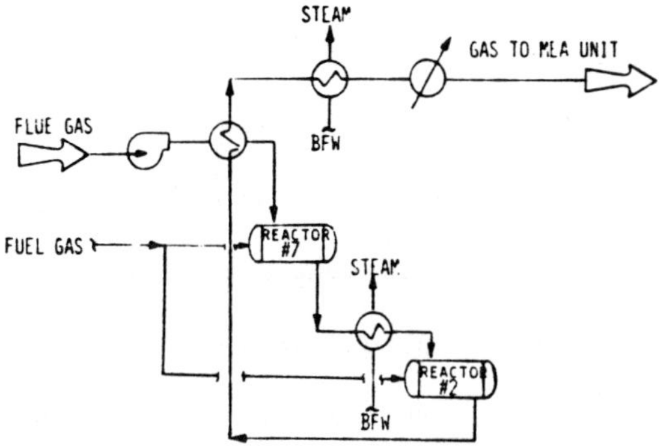

Figure 1. Catalytic Deoxo Process.

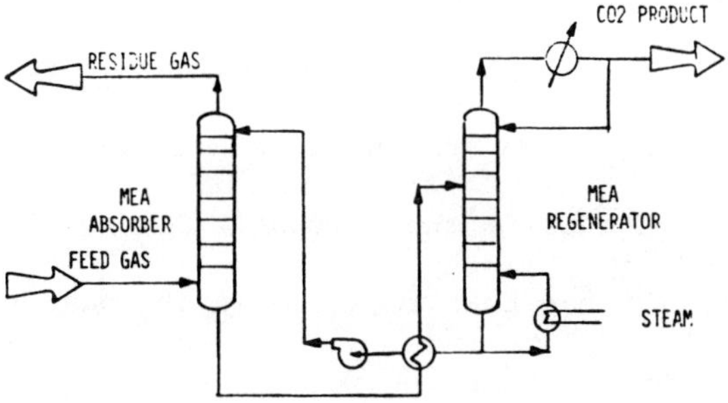

Figure 2. MEA process.

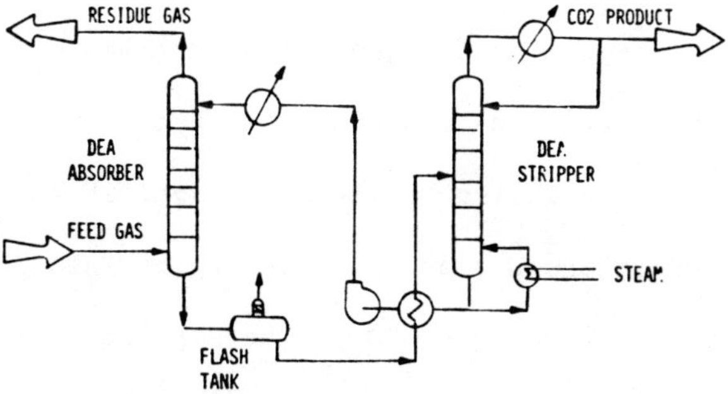

Figure 3. DEA process.

CASE 2: 94 MMSCFD CO_2 RECOVERY FROM OFFGAS HYDROGEN PLANT

There are more options for CO_2 recovery from the hydrogen plant offgas than in the flue gas case. Unlike the low pressure and low CO_2 content constraints in the flue gas, the hydrogen plant offgas contains 70% CO_2 and is available at 135 psig. The recovery of CO_2 may be accomplished by physical or chemical solvent processes, membrane or cryogenic separation. After preliminary screening, six alternatives were selected for evaluation: DEA, Benfield, Selexol, TEA, membrane and cryogenic separation.

DEA and Benfield are chemical absorbents, and heat is required to regenerate the solvent. Selexol is a physical absorbent. TEA has been regarded in the past as a chemical absorbent though it has recently developed mainly as a physical solvent. Membrane and cryogenic technologies were described above.

Figures 3 through 8 illustrate the selected schemes for CO_2 recovery. The cryogenic scheme (Figure 8) was developed by Davy McKee and consists of precompression, dehydration and chilling the feed gas against the returning CO_2 product and N_2, and introducing it into a column where N_2 is separated from CO_2. The separation is aided by the introduction of a suitable solvent which will act

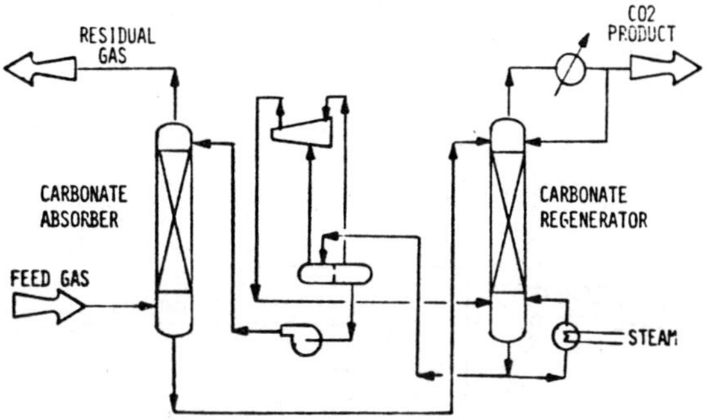

Figure 4. Benfield process.

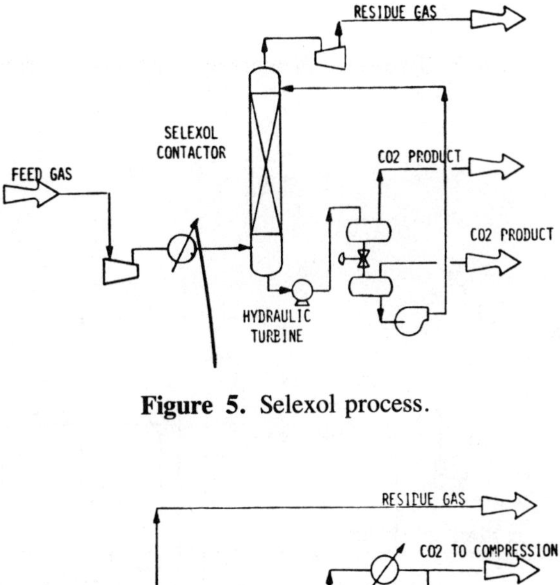

Figure 5. Selexol process.

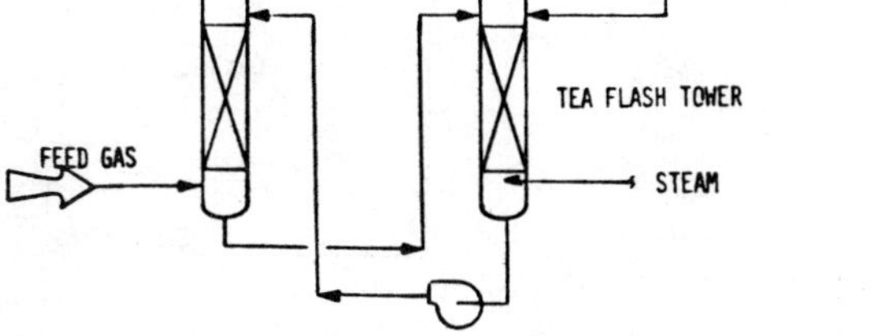

Figure 6. TEA process.

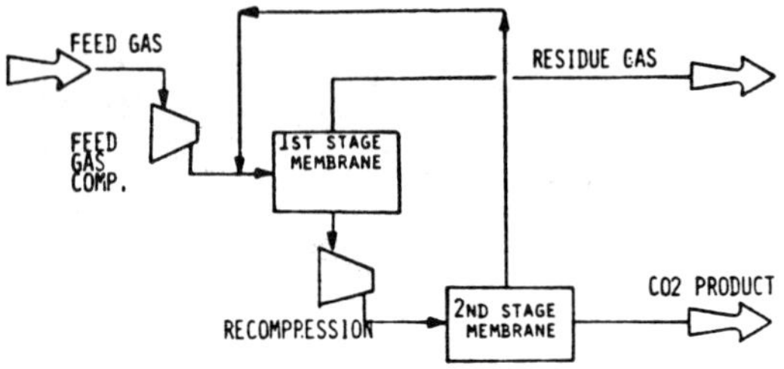

Figure 7. Two-stage membrane with recompression.

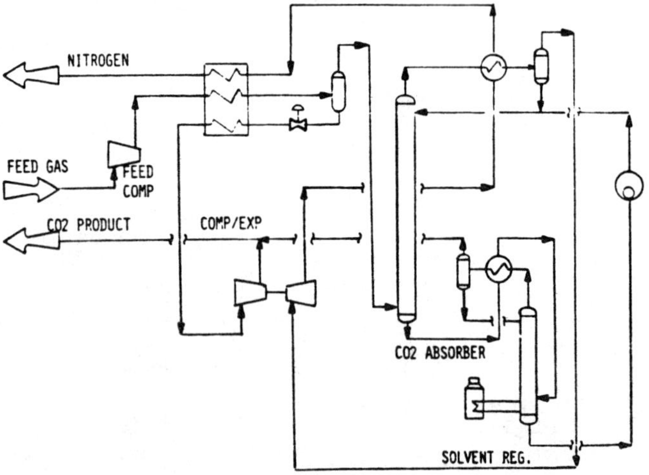

Figure 8. Cryogenic fractionation of CO_2 for EOR.

as a "lean oil" and bring down most of the CO_2 from the feed in the bottom product. The overhead nitrogen product is expanded to yield refrigeration. Supplemental propane refrigeration is required to satisfy the cooling requirements. The bottom product of the first column is introduced in the second column where it is separated into the CO_2 product and solvent, which is recycled back to the first column. CO_2 product is available at 425 psig.

BASIS FOR EVALUATION

Material and energy balances, followed by utility consumption, were prepared for each of the selected cases. Major process equipment was sized and the total investment cost was estimated, based on Davy McKee in-house procurement data and vendor budget prices for packages and specialized equipment such as membrane modules. The total installed cost was obtained by factoring the major equipments.

ECONOMIC EVALUATION

Net Present Value (NPV) analysis was used to evaluate the different process schemes. NPV combines all cash flows during the life of the project with the several assumptions applicable to Venezuela in 1983.

Pipeline Cost Data: An in-house estimation of pipeline costs was carried out. It included the material cost, grading, insulation when required, burying and labor. All costs were based on 1st quarter 1983 for U.S. Gulf Coast.

STUDY RESULTS

Summaries of investment and operating costs, utilities requirements for each of the selected schemes are given in Tables 1-2 for the flue gas from power plant, and Tables 3-4 for the offgas from hydrogen plant, respectively.

Table 1
Summary of Investment and Operating Cost, Flue Gas From Power Plant (in $1000)

	Dow/MEA (1)	UCC/MEA (1)	Deoxo/MEA (2)
Total Installed Cost, $	147,200	144,200	131,150
Operating Cost, $/Yr	33,043	29,110	25,634
Maintenance Cost, $/Yr	4,280	4,280	3,934
Total NPV, $	-189,410	-176,276	-157,973

(1) Inhibited MEA Process

(2) Conventional MEA Process

Table 2
Comparative Utilities and Compression Requirements, Flue Gas From Power Plant

	Dow/MEA	UCC/MEA	Deoxo/MEA
Total Compression, HP	64,000	64,000	64,000
Fuel Gas, MMSCFD	28.3	26.4	25.4
Electricity, KW	3,165	3,280	2,660
Cooling Water, GPM	220,000	165,000	137,000

UTILITIES COST

Cooling Water	$ 0.07/U.S. gal
Boiler Feedwater	$ 0.50/1000 lb
Electricity	$ 0.023/Kwh
Fuel Gas	$ 1.48/1000 SCF
Chemical & Catalyst	Per Vendor Quotes

Table 3
Summary of Investment and Operating Costs, Offgas From Hydrogen Plant (in $1000)

	DEA	Benfield	Selexol	TEA	Membrane	Cryogenic
Total Installed Cost, $	76,030	93,250	90,360	121,770	160,220	67,830
Operating Cost, $/Yr	21,802	8,186	6,453	7,718	6.257	5,445
Maintenance Cost, $/Yr	2,280	2,790	2,698	3,627	6,695	1,975
Total NPV, $	-111,867	-82,963	-76,053	-99,606	-125,283	-58,696

Table 4
Comparative Utilities and Compression Requirements, Offgas From Hydrogen Plant

	DEA	Benfield	Selexol	TEA	Membrane	Cryogenic
Total Compression, HP (1)	28,000	42,000	36,000	34,000	47,000	25,600
Fuel Gas, MMSCFD	26.0	9.3	7.4	7.7	8.2	8.0
Electricity, KW	2,525	3,350	6,320	6,320	1,165	550
Cooling Water, GPM	93,000	19,500	13,500	8,500	20,300	18,700
BFW, M Lb/Hr	830	180	--	60	--	--

UTILITIES COST

Cooling Water	$ 0.07/U.S. gal
Boiler Feedwater	$ 0.50/1000 lb
Electricity	$ 0.023/Kwh
Fuel Gas	$ 1.48/1000 SCF
Chemical & Catalyst	Per Vendor Quotes

CO_2 TRANSPORTATION

Various methods of transport are available for transporting CO_2. For this study, only pipeline transportation was considered. Previous studies have concluded that a high pressure supercritical pipeline is the most economical transportation mode for these volumes of CO_2 and the distance being considered in this study, i.e. 50 and 100 MMSCFD of CO_2 to a distance 100 and 300 Km. This study confirmed the same findings.

Subcritical gas phase and supercritical dense phase for pipeline transportation were evaluated. Two supercritical pressure levels were selected: 2000 and 3000 psig. For pipeline pressure drop calculations, the panhandle equation was used.

Compressor booster stations were spread along the distance as required where gas was compressed to its original pressure. Minimum downstream pressure of 400 psig was maintained in the subcritical case to eliminate condensing, while in the supercritical case, pressure was maintained above 1400 psig, which is well above CO_2 critical pressure. In all cases, CO_2 product was dehydrated at 700 psig, before its introduction into the pipeline. Table 5 summarizes the supercritical compression and transportation of

CO_2. For this analysis, the terrain over which CO_2 is transported is assumed to be flat. To account for line pressure drop losses, booster stations were equally spaced. Pressure of CO_2 at compressor suction and at injection point is atmospheric and 2500 psig, respectively.

CONCLUSIONS

It was concluded that:

1. Deoxidation followed by a conventional MEA process is the economically preferred scheme for the recovery of CO_2 from power plant flue gases. The TIC and the operating cost of this scheme are lower than the inhibited MEA system, which make it more favorable. The difference is not marginal.

2. The cryogenic process has economic advantages in capital and operating costs over conventional acid gas removal processes for CO_2 recovery from refinery hydrogen plant offgases. Membrane technology has the highest capital cost and power consumption, resulting in the highest operating cost.

3. Pipeline transportation of CO_2 is most economical at supercritical pressures of 2000-3000 psi.

Table 5
Supercritical Compression and Transportation of CO_2

FLOW RATE:	50 MMSCFD		300 MMSCFD	
DISTANCE:	50 KM	300 KM	50 KM	300 KM
Installed Cost				
Pipe Line Size In.	8	8	10	10
No. of Booster Comp. Stations	0	1	0	2
TIC of Pipe Line, $1000	11,328	68,130	14,500	85,300
TIC of Compression + Dehydration, $1000	25,418	26,575	50,837	55,836
Total TIC, $1000	36,746	94,705	65,337	141,136
Operating and Maintenance Cost, $/YR	3,454	3,972	6,366	6,766
Utilities Consumption				
Guel Gas, MMSCFD	3.30	3.40	6.60	6.80
Electricity, KW	550	560	1,000	1,130
Compression, HP	18,000	19,000	36,000	38,000

7

Process Considerations in Selecting Amines

John Polasek
Process Engineer,
Bryan Research & Engineering, Inc.

Jerry Bullin
Professor, Chemical Engineering
Department, Texas A & M University

INTRODUCTION

The technology of using alkanolamines for sweetening natural gases has been used for many decades. In the 1960's and 70's, several new amines came into general use. However, little information has been reported in the literature concerning the best amine for a particular service.

The basic flow scheme for an amine sweetening plant is shown in Figure 1. In the process, the overriding consideration is that the sweet gas meet the required purity specifications. After this consideration, the primary desire is to select the amine which gives the

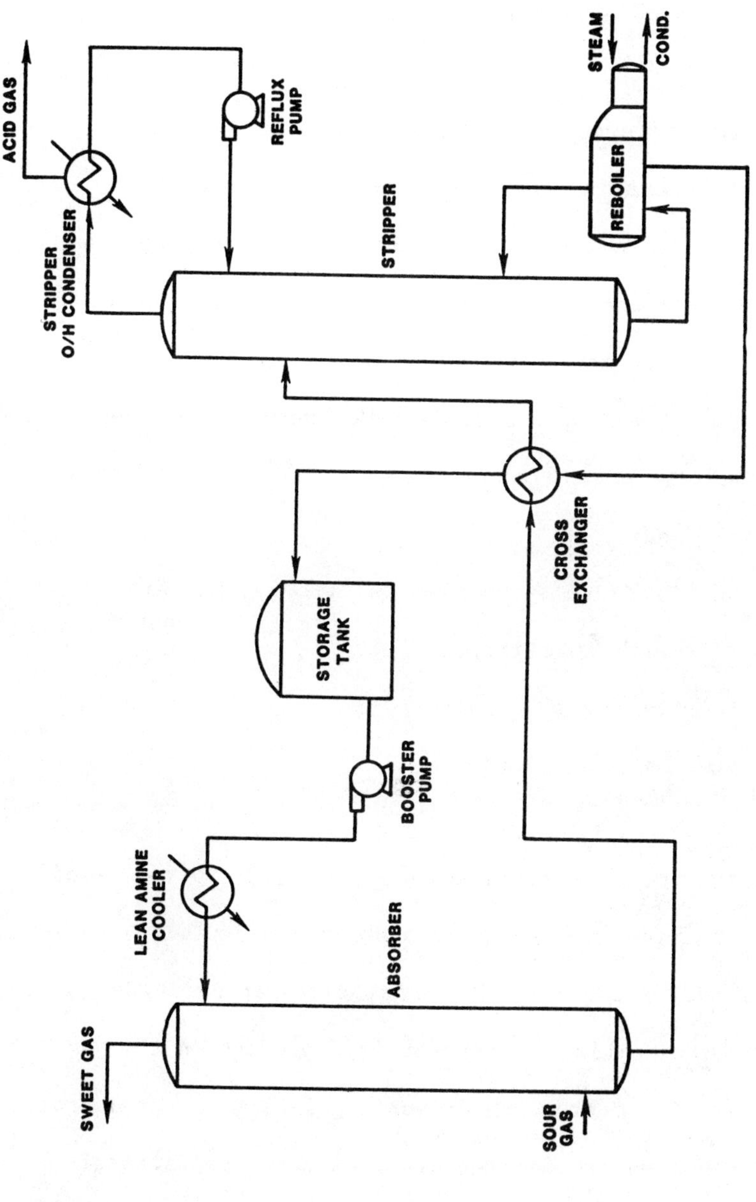

Figure 1. Process flow sheet for common sweetening plant.

optimum equipment size and operating costs. The factors to be considered include:

1.) Alternative flow schemes.
2.) Lowering the amine circulation rate by selecting another amine, by increasing the amine concentration, and/or by increasing the solution loading.
3.) Reducing the reboiler/condenser size and duty.
4.) Selectively absorbing the H_2S from the sour gas.
5.) Pressure swing plants for bulk CO_2 removal with no stripper.
6.) Corrosion problems.
7.) Solution losses.

According to Astarita, et al. (1983) 50 to 70% of the plant investment is directly associated with the magnitude of the solvent circulation rate. Another 10 to 20% of the initial investment is dependent on the regeneration energy requirement. Additional consideration must be given to the regeneration energy requirements since about 70% of the operating costs, excluding labor, are due to this energy requirement. Thus the proper selection of the amine will often have a profound influence on the equipment size and operating costs for a sweetening unit.

Alternative flow schemes for amine sweetening plants have been previously discussed in detail by Polasek, Donnelly and Bullin (1983) and will be lightly treated in the present work. Due to the lack of sufficient data, detailed analysis of corrosion problems will also not be possible in the present work.

GENERAL CONSIDERATIONS FOR SELECTING AMINES

The general considerations for selecting amines in sweetening plants have changed over the years. For example, until the 1970's monoethanolamine (MEA) was the amine first considered for any sweetening application. This fact was exemplified in papers by Beck (1975) and Butwell and Perry (1975) which reported favorable experiences of switching from MEA to diethanolamine (DEA).

As different conditions are tested and proven with a particular amine, they become accepted on an industry wide basis. Thus each amine has a currently "accepted" range of process conditions and parameters associated with it. These "accepted" conditions and parameters are discussed below. Some of the typical operating conditions and data for amines are summarized in Table 1.

Monoethanolamine (MEA)

MEA is generally used as a 10 to 20 % solution in water. The acid gas loading is usually limited to 0.3 to 0.4 moles acid gas per mole of amine for carbon steel equipment. Loadings as high as 0.7 to 0.9 mole/mole have been used in stainless steel equipment with no corrosion problems. MEA itself is not considered to be particularly corrosive. However, its degradation products are very corrosive. COS, CS_2, SO_2 and SO_3 can partially deactivate MEA but most of the deactivated amine can be recovered with a reclaimer. Since MEA is a primary amine, it has a high pH. This enables MEA solutions to produce gas containing less than 1/4 gr H_2S per 100 SCF at very low H_2S partial pressures. When MEA is used, essentially all of the CO_2 must be absorbed to produce quarter grain gas with respect to H_2S. This is particularly burdensome since the heat of reaction for CO_2 in MEA is about 825 BTU/lb CO_2. The heat of reaction for all amines is a function of loading and other conditions. It usually varies by only 50 to 60 BTU/lb up to about 0.5 mole/mole total acid gas loadings. Above this loading, the heat of

Table 1
Typical Operating Conditions and Data for Amines

Amine	MEA	DEA	DGA	MDEA
Solution Strength, wt %	15-20	25-35	40-60	30-50
Acid gas loading, mole/mole	0.3-0.4	0.3-0.4	0.3-0.4	Unlimited
ΔH_r^* for H_2S BTU/lb	550	511	674	522
ΔH_r^* for CO_2 BTU/lb	825	653	850	600
Ability to perferrentially absorb H_2S	NO	Under some Conditions	NO	Under Most Conditions

* BTU/lb and for total loadings below 0.5 moles acid gas/mole amine

reaction varies considerably and should be calculated as a function of loading.

Diethanolamine (DEA)

DEA is most commonly used in the 25 to 35 wt.% range. The loading for DEA is also limited to 0.3 to 0.4 mole/mole for carbon steel equipment. DEA can safely be loaded to equilibrium when using stainless steel equipment or when using inhibitors as in the SNPA process (Kohl and Reisenfield, 1979). The degradation products of DEA are much less corrosive than those of MEA. COS and CS_2 may react irreversibly with DEA to some extent. Since DEA is a secondary alkanolamine, it has a reduced affinity for H_2S and CO_2. As a result, for some low pressure gas streams, DEA can not produce pipeline specification gas. In general, as the gas pressure is lowered, the stripping steam must be increased or a split flow design must be used. In some cases, even these measures will not be sufficient and another solvent must be used.

Under some conditions, such as low pressure and a liquid residence time on the tray of about 2 sec, DEA will be selective toward H_2S and will permit a significant fraction of the CO_2 to remain in the sales gas. At least one case of selective absorption of H_2S with DEA has been documented by Donnelly and Henderson (1974). In this particular case, selective absorption was not desired but was encountered due to a low pressure of 11 psig and a short liquid residence time on the trays. An H_2S/CO_2 ratio of one or above would probably rule out any chance of significant CO_2 slippage while producing 1/4 gr H_2S/100 SCF sales gas. The heat of reaction for DEA and CO_2 is 653 BTU/lb which is about 25% less than for MEA.

Diglycolamine (DGA)

DGA is generally used as 40 to 60 wt.% solutions in water. The reduced corrosion problems with DGA allow mole per mole solution loadings equivalent to MEA in most applications even with these high weight percents. For gas streams with high acid gas partial pressures, absorber bottoms temperatures as high as 180°F and above can occur. Obviously, this would reduce the possible loading. DGA has a tendency to preferentially react with CO_2 over H_2S. It also has a higher pH than MEA and thus can easily achieve 1/4 gr H_2S/100 SCF except in some cases where large amounts of CO_2 are present relative to H_2S. DGA has some definite advantages over the other amines in that higher DGA concentrations in the solution result in lower freezing points. In addition, DGA is not as likely to react irreversibly with COS, CS_2, SO_2 and SO_3. In fact, DGA has a known but unquantified ability to absorb COS as well as methyl- and ethyl- mercaptans. One of the primary disadvantages of DGA is the high heats of reaction as shown in Table 1.

Methydiethanolamine (MDEA)

An "accepted" set of operating conditions has not been as firmly established for MDEA as for the previously mentioned amines. This has been due to fewer MDEA plants in operation and to the nature of MDEA. Even though the operating conditions for MDEA are still evolving at a rapid pace, the range of successful applications at the present time will be discussed below.

MDEA is most commonly used in the 30 to 50 wt.% range. Due to considerably reduced corrosion problems, acid gas loadings as high as 0.7 to 0.8 mole/mole are practical in carbon steel equipment. Higher loadings may be possible with few problems. Since MDEA is a tertiary amine, it has less affinity for H_2S and CO_2 than DEA. Thus, as in the case for DEA, MDEA can not produce pipeline specification gas for some low pressure streams. MDEA has several distinct advantages over primary and secondary amines. These include lower vapor pressure, lower heats of reaction, higher resistance to degradation, fewer corrosion problems and selectivity toward H_2S in the presence of CO_2. Most of these advantages have also been reported by Blanc et al. (1982). Depending on the application, some of the advantages have special significance. For example, due to its lower heat of reaction, MDEA can be employed in pressure swing plants for bulk CO_2 removal. In pressure swing plants, the rich amine is merely flashed at or near atmospheric pressure and little or no heat is added for stripping.

At the present time, the overwhelming advantage that MDEA possesses over the other amines is that it is readily selective toward H_2S in the presence of CO_2. At high CO_2/H_2S ratios over two thirds of the CO_2 can be slipped through the absorber and into the sales gas. The enhanced selectivity of MDEA for H_2S results from the inability of tertiary amines to form a carbamate with CO_2. MDEA does not have a hydrogen attached to the nitrogen and can not react directly with CO_2 to form carbamate. At least six different mechanisms for the CO_2-MDEA reaction have been proposed by Cornelissen (1982), Barth, et al. (1981), and Danckwerts (1979). MDEA can react with H_2S by the same proton transfer mechanism of primary and secondary amines

(Jou, et al., 1982). Selective absorption of H_2S results from optimizing absorber design and circulation rate to give residence times and reaction conditions favorable to H_2S absorption into the amine solution and to CO_2 retention in the sales gas.

PROCESS DESIGN CAPABILITY

Most of the above general considerations for selecting amines have been commonly known for several years. However, since the design procedures have historically been very approximate, engineers have had no means to accurately and systematically investigate the potential of each amine. The design capability was greatly improved with the development of a process simulation program called TSWEET for gas/liquid amine sweetening units. TSWEET was developed about six (6) years ago and has undergone constant improvement and expansion in the ensuing years.

TSWEET is an advanced process simulation program using the flexible flowsheet concept. Beginning with the process flowsheet and operating parameters, complete material and energy balances for the steady state operating conditions can be performed. Rigorous tray-by-tray calculations using the Ishii-Otto (1973) method are used to predict the design of the stripper and absorber columns. Some 24 chemical species are allowed, including MEA, DEA, DGA, MDEA, water, H_2S, CO_2, and an assortment of gases commonly present in natural and synthetic gases. Vapor pressures of H_2S and CO_2 over the amine solutions are calculated by a modified Kent and Eisenberg (1976) equilibrium model. A kinetic model was developed to predict the effect of residence time, temperature, solution concentration, pressure and type of amine on the rate of CO_2 absorption. The program has been demonstrated to be very reliable in calculating the performance of amine gas sweetening plants. The program will accommodate multiple absorbers in LPG and or gas service which feed a common stripper. The capabilities of TSWEET have been previously described by Polasek et al. (1983), Holmes and Bullin (1983), and Polasek et al. (1982).

DISCUSSION OF SPECIFIC CASES

The general considerations for selecting amines were applied to several specific cases. In these cases, an operating plant was used as a base and the sweetening program TSWEET was used to explore the potential of different amines in each plant.

The first two cases are taken from a paper presented

in the early 1960's by Estep, McBride, and West (1962). This paper discussed the performance and problems of a pair of sweetening and sulfur production facilities. One of the facilities was in the United States at Worland, Wyoming and the other was at Okotoks, Alberta, in Western Canada. Both of these facilities used MEA in their amine sweetening plants and both facilities were reported to work well. The primary problem in the sweetening units was corrosion, due primarily to the fact that the 15 weight percent MEA was being loaded at 0.6 - 0.8 mole acid gas per mole of amine.

The operating data for the Worland and Okotoks plants are shown in Tables 2 and 3, respectively. The results from two TSWEET runs are also shown in each table. The first run was made at the reported operating conditions for each plant. As can be seen from Tables 2 and 3, the calculated outlet gas composition and reboiler duty matched the operating data very closely for both plants.

After demonstrating that TSWEET accurately predicted the performance of the existing operations, TSWEET was then used to explore the use of DEA as an alternate. The results for 33 wt. % DEA are shown in the last column in Tables 2 and 3. Although the outlet gas concentation was

Table 2
Comparison of Okotoks Plant Performance with Different Amines

	MEA Operating Data	TSWEET for MEA Operating Data	TSWEET for DEA Alternative
Amine Conc, wt%	15	15	33
Circulation rate, GPM	2000	2000	1500
Gas flow, MMSCFD	29.46	same	same
H_2S inlet, %	33.25	same	same
CO_2 inlet, %	10.5	same	same
H_2S out, gr/100 SCF	0.01	0.019	0.029
CO_2 out, ppm	-	0.035	0.023
Amine loading mole/mole	0.89	0.88	0.77
Approach to Equilibrium at absorber bottom	-	69	73
Reboiler Duty MM BTU/lb	109	109	78

Table 3
Comparison of Worland Plant Performance with Different Amines

	MEA Operating Data	TSWEET for MEA Operating Data	TSWEET for DEA Alternative
Amine Conc, wt%	14.7	14.7	33
Circulation Rate, GPM	1345	1345	1179
Gas Flow MMSCFD	26.37	same	same
H_2S inlet, %	33.25	same	same
CO_2 inlet, %	10.5	same	same
H_2S out, gr/100 SCF	< 0.25	0.22	0.10
CO_2 out, ppm	-	0.035	0.023
Amine loading, mole/mole	0.53	0.53	0.44
Approach to Equilibrium at absorber bottom	-	50	50
Reboiler Duty MM BTU/hr	107	107	76

essentially the same for both amines, the differences in plant performance are rather dramatic. In each case, the reboiler duty was reduced by about one third. The large heat saving resulted from two factors. The primary saving came from the reduced circulation required with DEA. The DEA circulation rate was chosen such that the approach to equilibrium at the bottom of the absorber was the same as for the MEA case. As shown in Tables 2 and 3, this resulted in a lower rich amine loading for DEA. Obviously, the circulation rate and the reboiler duty could be further reduced by increasing the DEA loading to the MEA value. Since the feed to the stripper must be heated by 40°F and since the amine solutions have a heat capacity of 0.7 to 0.9 BTU/lb, about 25 to 30 BTU's are saved for each one pound reduction in circulation rate.

The second factor responsible for the reduced reboiler duty was the lower heat of reaction for DEA. This results in a savings of about 170 BTU per pound of acid gas recovered in the still.

The North Caroline Plant of Dome Petroleum was selected for the third case since this unit was switched from DEA to MDEA. The performance and operating data for the absorber plant were reported by Daviet, et al. (1984).

The acid gas contaminants in the feed consisted of about 3.5% CO_2 and 50 ppm H_2S. When using DEA, the amine plant was operating at its maximum capacity due primarily to reboiler limitations in the stripper. The amine unit operation was also unstable. Since the plant produced gas only marginally below the 1/4 grain H_2S/100 SCF specification, any upset would bring the plant down. The H_2S came primarily from one well which was usually around 300 ppm. However, each time this well was shut in for a few hours, the H_2S from the well increased to about 1200 ppm for a short period after startup. Since the amine unit could not stand any additional H_2S, the sales gas exceeded pipeline specification and was automatically flared. Dome was understandably not satisfied with this state of affairs and explored a number of alternatives to alleviate the problem. After running TSWEET for a variety of conditions, Dome's engineering group determined that the unit could be converted to MDEA with no equipment modifications. The calculations showed that selective absorption would occur in the absorber thus increasing its H_2S scrubbing capability.

In May of 1983, the North Caroline plant was shut down, washed, charged with 33 wt. % MDEA and started up.

No operating problems were encountered during startup and the unit has performed very smoothly since startup. The operating conditions and outlet gas concentrations for the absorber are shown in Table 4, along with the TSWEET calculations. As evidenced by the H_2S level of less than 0.1 ppm in the sales gas, the conversion from DEA to MDEA greatly improved the performance of the plant at a very low cost. The operation with MDEA is also very stable.

SUMMARY AND CONCLUSIONS

Many process factors must be considered when selecting an amine for a sweetening application. The initial selection is based on the pressure and acid gas content of the sour gas as well as the purity spec for the sales gas. As different process conditions have been tested and proven with each amine, they have become widely accepted in the gas processing industry.

Based on the currently "accepted" operating conditions, MEA is usually not the first amine considered due to its high heats of reaction and lower acid gas carrying capacity per gallon of solution. However, MEA is still used for plants where the inlet gas pressure is low and pipeline specification gas or total removal of the acid gases are desired.

Table 4
Performance of North Carolina Plant with MDEA

	MDEA Operating Data	TSWEET for MDEA Operating Data
Amine Conc, wt%	33	33
Circulation Rate, GPM	123	123
Gas Flow, MMSCFD	30.24	same
H_2S inlet, ppm	55	same
CO_2 inlet, %	10.5	same
H_2S out, gr/100 SCF	< 0.1	0.31
CO_2 out, %	1.13	1.06

As operating experience was broadened during the late 1960's and early 1970's for DEA, it became the "workhorse" of the industry due to its lower heats of reaction, higher acid gas carrying capacity and resultant lower energy requirements. DEA's potential for selective H_2S removal from streams containing CO_2 has not been fully realized.

Although DGA has high heats of reaction, its very high acid gas carrying capacity usually produces very reasonable net energy requirements. DGA also has a good potential for absorbing COS and some mercaptans.

Depending on the application, MDEA has some outstanding capabilities. Due to its low heat of reaction, it can be used in pressure swing plants for bulk acid gas removal. MDEA is currently best known for its ability to preferentially absorb H_2S.

The considerations for selecting amines were applied to several specific cases involving existing plants. Using a simulation program called TSWEET, the reported operating conditions for several plants were matched. TSWEET was then used to examine the performance of other amines in these plants. This analysis demonstrated that the proper selection of the amine can have a major impact on the performance and costs of sweetening units.

REFERENCES

1.) Astarita, G., D. W. Savage, and A. Bisio, Gas Treating with Chemical Solvents, J. Wiley and Sons, (1983).

2.) Barth, D., C. Tondre, G. Lappai, and J. J. Delpuech, "Kinetic Study of Carbon Dioxide Reaction with Tertiary Amines in Aqueous Solutions", J. Phys. Chem., 85. 3660 (1981).

3.) Beck, J. E., "Diethanolamine an Energy Conserver", Gas Conditioning Conference (1975).

4.) Blanc, C., M. Grall, and G. Demarais, "The Part Played by Degradation Products in the Corrosion of Gas Sweetening Plants Using DEA and MDEA", Proc. 1982 Gas Conditioning Conf., University of Oklahoma, p. C-1 (1982).

5.) Butwell, K. F. and C. R. Perry, "Performance of Gas Purification Systems Utilizing DEA Solutions", Gas Conditioning Conference, (1975).

6.) Cornelissen, A. E., "Simulation of Absorption of H_2S and CO_2 into Aqueous Alkanolamines", Shell Laboratory, p. 3.1-3.15 (1982).

7.) Danckwerts, P. V., "The Reaction of CO_2 with Ethanolamines", Chem. Eng. Sci., 34, 443 (1979).

8.) Daviet, G. R., S. T. Donnelly and J. A. Bullin, "Dome's North Caroline Plant Sucessfully Converted to MDEA", Presented at 63rd Annual Gas Processors Association Convention, (1984).

9.) Donnelly, S. T., and D. R. Henderson, "A New Approach to Amine Selection", Presented at C.N.G.P.A. Third Quarter Meeting (1974).

10.) Estep, J. W., G. T. McBride, Jr. and J. R. West, "The Recovery of Sulfur from Sour Natural and Refinery Gases", Advances in Petroleum Chemistry and Refining, Vol. 6, (1962).

11.) Holmes, J. W., M. L. Spears, and J. A. Bullin, "Sweetening LPG's with Amines", Presented at AIChE Summer Meeting (1983).

12.) Jou, F.Y., D. Lal, A.E. Mather, and F.P. Otto, "Solubility of H_2S and CO_2 in Aqueous Methyldiethanolamine Solutions", Ind. Eng. Chem. Proc. Des. Dev., 21, 539 (1982).

13.) Kohl, A. and F. Riesenfeld, Gas Purification, Gulf Publishing Company, Houston, Texas (1979).

14.) Polasek, J. C., J. A. Bullin and S. T. Donnelly, "Analysis of Several Flow Schemes to Reduce Costs in Amine Sweetening Units", AIChE Meeting, (1982).

8

Equilibrium Between CO_2/H_2S and Ethanolamine Solutions

R. N. Maddox
Director, Physical Properties Laboratory
Oklahoma State University

A. Bhairi
Physical Properties Laboratory
Oklahoma State University

G. J. Mains
Chemistry Department
Oklahoma State University

A. Shariat
Physical Properties Laboratory
Oklahoma State University

INTRODUCTION

Ethanolamines were first used for removal of acid gas constituents (hydrogen sulfide and/or carbon dioxide) more than 50 years ago. When first introduced, ethanolamine sweetening was done almost exclusively by monoethanolamine (MEA). Later, diethanolamine (DEA) became popular. Even more recently diglycolamine (DGA) and methyldiethanolamine (MDEA) have come into use.

Based on material presented at the 63rd Annual Gas Processors Association Convention, March 19-21, 1984, New Orleans, Louisiana.

Relatively large amounts of experimental data are available for the equilibrium between MEA and DEA and the acid gas constituents. Much less data is available on DGA and MDEA. This paper reports equilibrium measurements on these two ethanolamine solutions with carbon dioxide and hydrogen sulfide.

EXPERIMENTAL APPARATUS

Figure 1 shows the experimental apparatus used for determining the vapor-liquid equilibrium between acid gases and ethanolamine solutions. It is a constant volume type equilibrium cell patterned after the concept originally introduced by Woertz (1) and modified and improved by Akashah (2) and Adeyiga, et al (3). The volume of the equilibrium cell is determined precisely through calibration. For determinations with a single acid gas constituent such as those reported in this paper, the cell is evacuated and flushed with nitrogen three times. The evacuated cell is filled to the desired pressure with CO_2 or H_2S and the gas-filled cell is allowed to attain temperature equilibrium in the constant temperature bath. A known amount of the desired ethanolomaine solution is injected into the pressurized cell through the Ruska

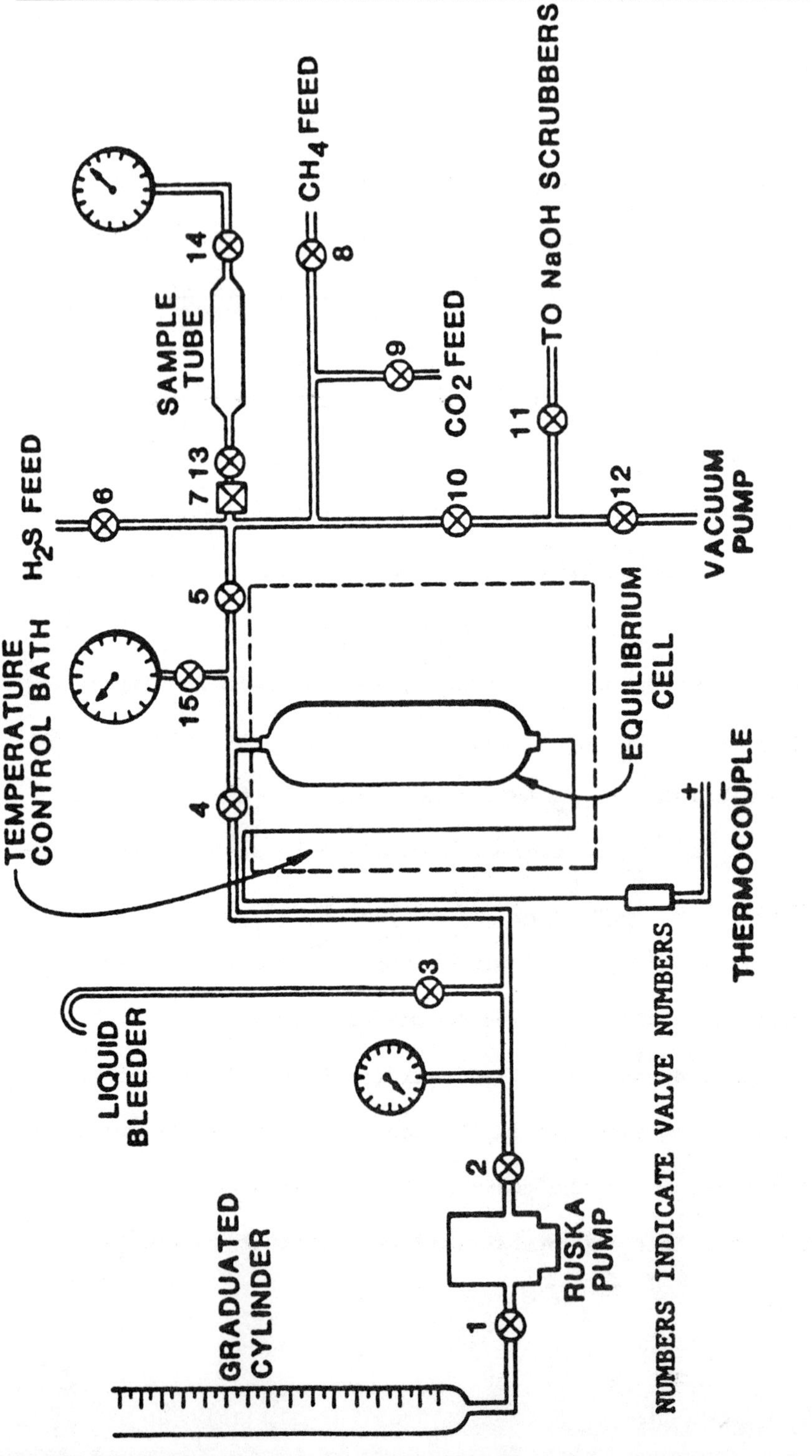

Figure 1. Schematic of experimental apparatus.

pump. The cell and contents are again allowed to reach temperature equilibrium. The pressure difference for the temperature equilibrated cell before and after ethanolamine injection, together with the known cell volume, are used to determine the quantity of CO_2 or H_2S which has gone into solution. The SRK equation of state is used to calculate the densities of the gas phase.

EXPERIMENTAL DATA

Figure 2 shows the results of equilibrium determinations for equilibrium between 60% by weight DGA solution and H_2S at 50 and 80 ^oC. Also shown are the data of Martin, et al (4) at 50^oC. The OSU and Martin data are in substantial agreement at low CO_2 pressures. At higher pressures, the OSU data show a higher loading.

Figure 3 shows experimental data for 40% by weight DGA and H_2S. Data were taken at 40 and 60 ^oC. The two sets of data appear to be consistent with similar shape lines fitting through the experimental points.

Figure 4 shows a comparison of the data collected in this work with that of Jou, et al (5) for 1 N MDEA and H_2S. The data were taken at 25^oC. Both sets of data appear to be consistent and in agreement.

Figure 5 shows data for 1 N MDEA and CO_2 at 25^oC.

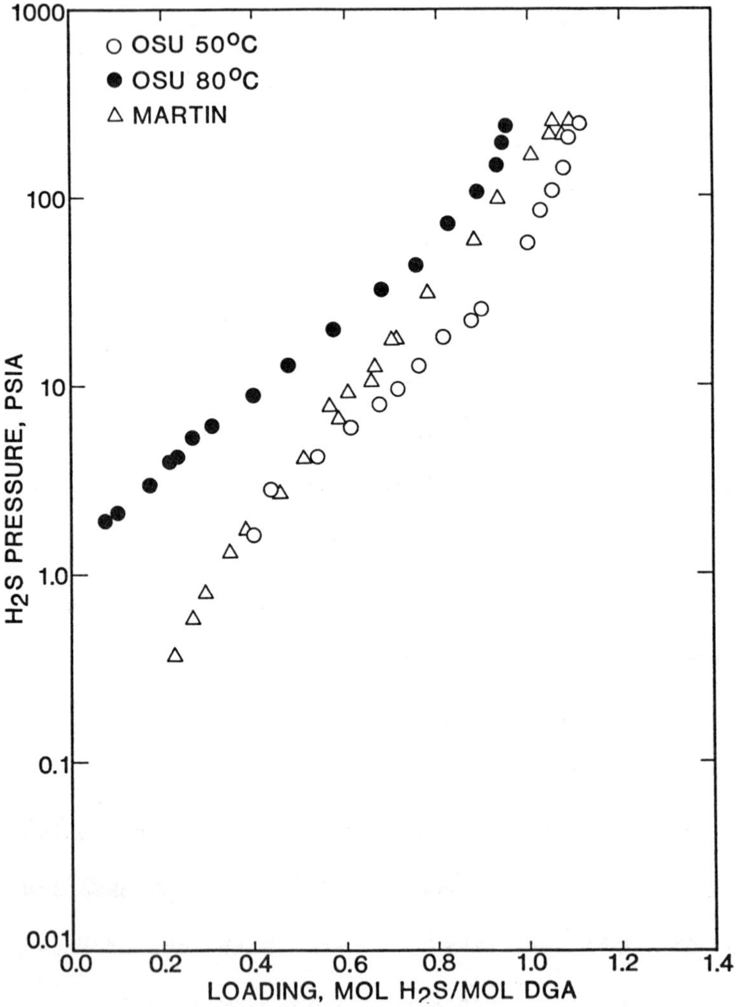

Figure 2. Equilibrium between 60% by wt. DGA and H_2S at 50 and 80°C.

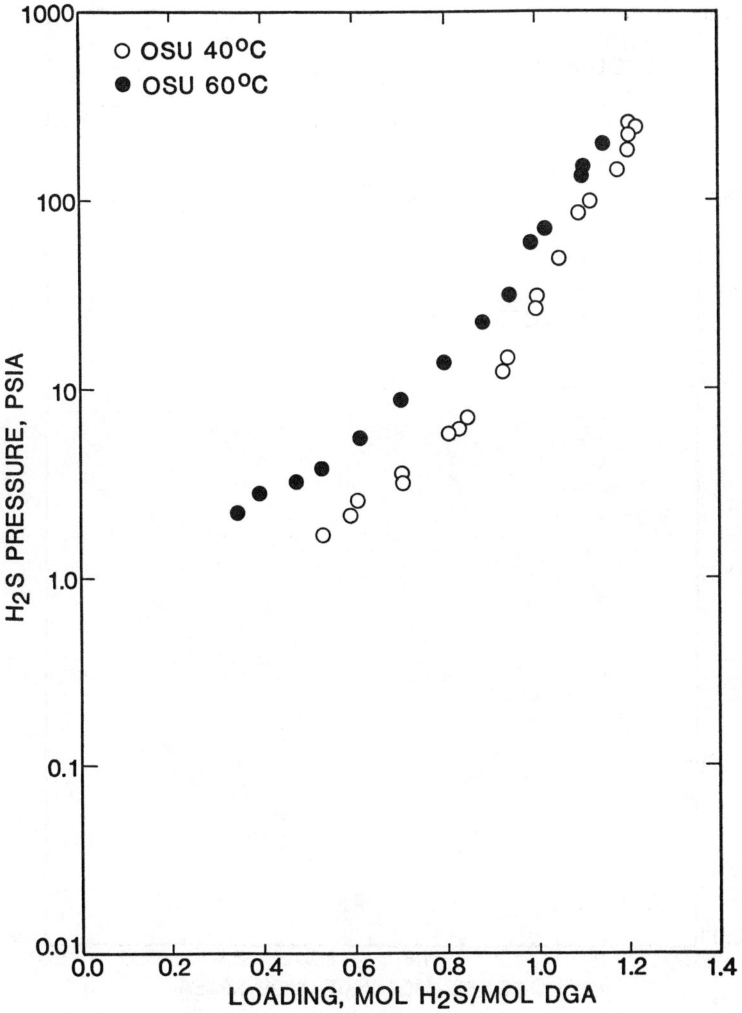

Figure 3. Equilibrium between 40% by wt. DGA and H_2S at 40 and 60°C.

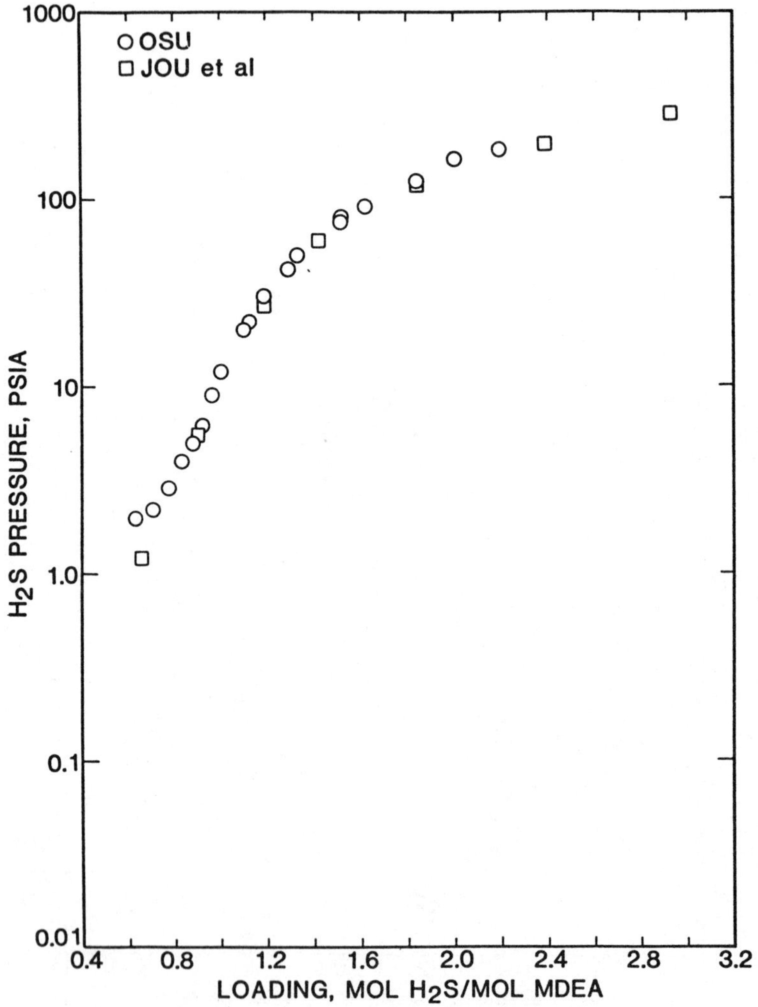

Figure 4. Equilibrium between 1N MDEA and H_2S at 25°C.

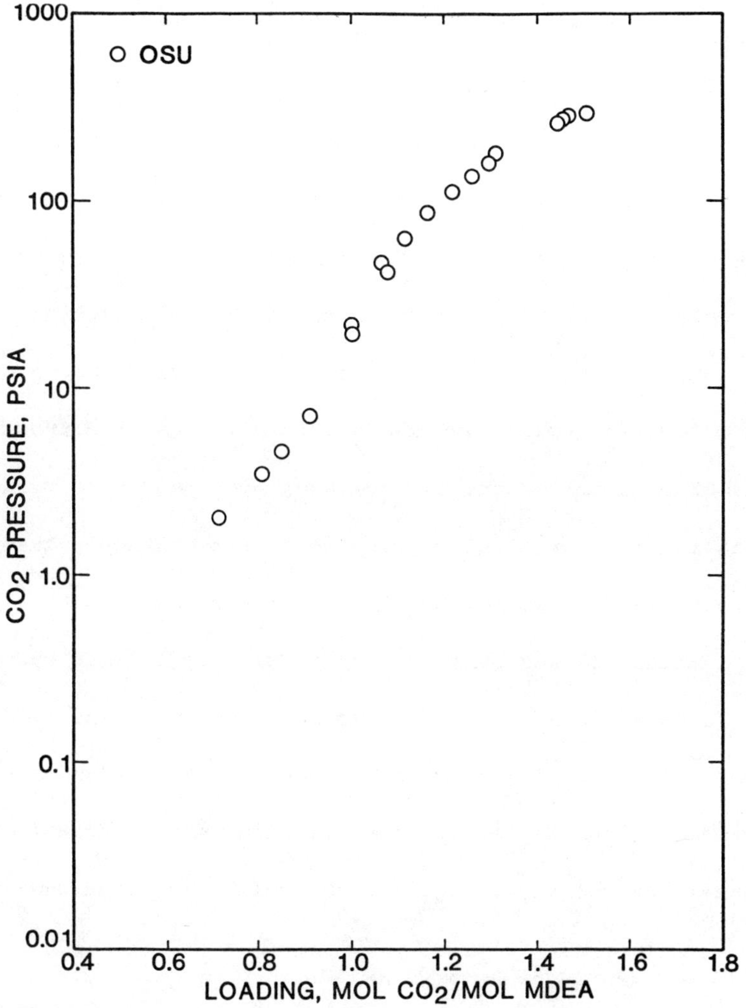

Figure 5. Equilibrium between CO_2 and 1N MDEA at 25°C.

Figure 6 shows a comparison of the data of Jou, et al with those taken in this work for 2 N MDEA and CO_2. The data at 25°C exhibit different characteristics. Also shown on Figure 6 is the data for 2 N MDEA at 50°C.

ACID GAS-ETHANOLAMINE REACTIONS

Over the years, there have been many efforts to correlate/predict the equilibrium behavior of acid gas-ethanolamine solutions. The various techniques tried have offered varying degrees of success and accuracy. The most successful technique appears to be the use of reaction equilibrium constants force-fitted through pure component H_2S or CO_2 data with the ethanolamines. Several different sets of chemical reactions have been proposed. The main qualification for a set of reactions is that they satisfy the necessary stoichiometry for the reactions.

Rahman (6) has shown that the reaction between hydrogen sulfide and a primary or secondary ethanolamine is a Lewis acid-base type reaction. Combining this with the physical solubility of hydrogen sulfide in the ethanolamine-water solution, suggests the following reactions:

$$RR'NH_2^+ \leftrightarrow H^+ + RR'NH \qquad (1)$$

$$H_2O \leftrightarrow H^+ + OH^- \qquad (2)$$

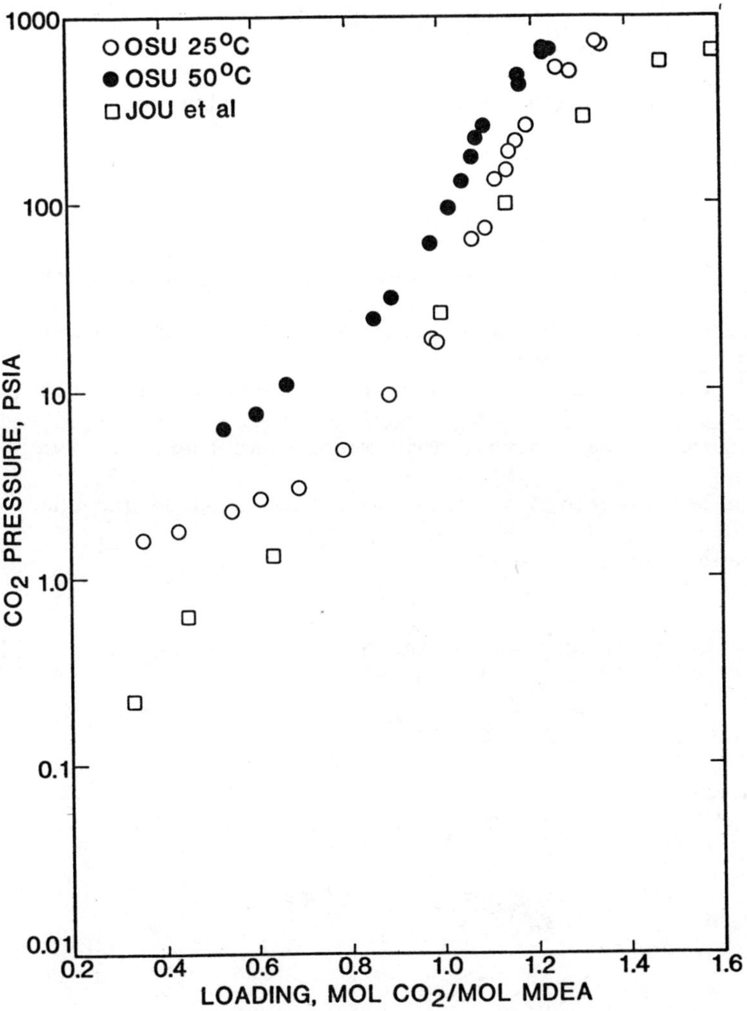

Figure 6. Equilibrium between CO_2 and 2N MDEA at 25 and 50°C.

$$H_2S \leftrightarrow H^+ + HS^- \tag{3}$$

$$HS^- \leftrightarrow H^+ + S^= \tag{4}$$

$$P_{(H_2S)} = H_{(H_2S)} [H_2S] \tag{5}$$

Carbon dioxide could undergo a similar set of reactions using the proton formed in the bicarbonate dissociation for protonation of the amine. However, Astarita et al, (7), Caplow (8), and Rahman (6) have all shown that, in addition to the Lewis acid-base adduct, CO_2 inserts into the structure of a primary or secondary ethanolamine to form the carbamate. This means that there are two possible reactions between the carbon dioxide and the ethanolamine:

$$RR'NH_2^+ \leftrightarrow H^+ + RR'NH \tag{1}$$

$$(CO_{2AQ}) + RR'NH \leftrightarrow RR'NCOO^- + H^+ \tag{6}$$

$$(CO_{2AQ}) + H_2O \leftrightarrow HCO_3^- + H^+ \tag{7}$$

$$HCO_3^- \leftrightarrow CO_3^= + H^+ \tag{8}$$

$$H_2O \leftrightarrow H^+ + OH^- \tag{2}$$

$$P_{(CO_2)} = H_{(CO_2)} [CO_2] \tag{9}$$

Kent and Eisenberg (9) used the reaction equilibrium approach to model the equilibrium between acid gases and ethanolamine solutions. The equations used by Kent and Eisenberg were not those shown above. However, their equations did satisfy the necessary stoichiometry for the chemical reactions outlined in equations 1-9.

Reactions 1-5 are used to model the equilibrium behavior for H_2S with ethanolamines and equations 6-9 (plus 1 and 2) are used to model the reaction of CO_2 with ethanolamines. The reader will note that the sets of reactions are linked together by reactions 1 and 2 and that the gas and liquid phases are linked together through equations 5 and 9. Making use of this linkage the reactions can be used for mixtures containing both H_2S and CO_2 as well as when only one of the acid gases is present.

USING THE CORRELATION

Kent and Eisenberg (9) fitted reaction equilibrium reaction constants for equations similar to 1 and 6 to pure CO_2 data, and equation 1 for pure H_2S data, and then used the resulting set of constants to predict equilibrium for mixtures of H_2S and CO_2 with ethanolamines. Figure 7 shows a comparison of the calculated and experimental determinations for 60% by weight DGA. The experimental

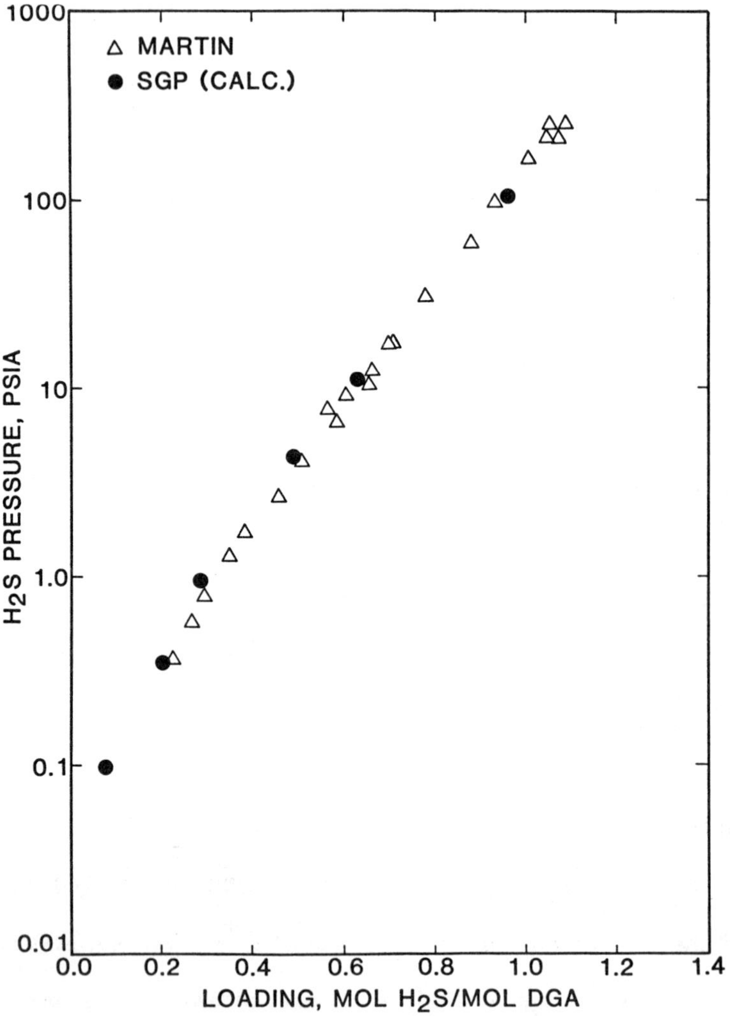

Figure 7. Comparison of experimental and calculated equilibrium data for 60% by weight DGA and H_2S at 50°C.

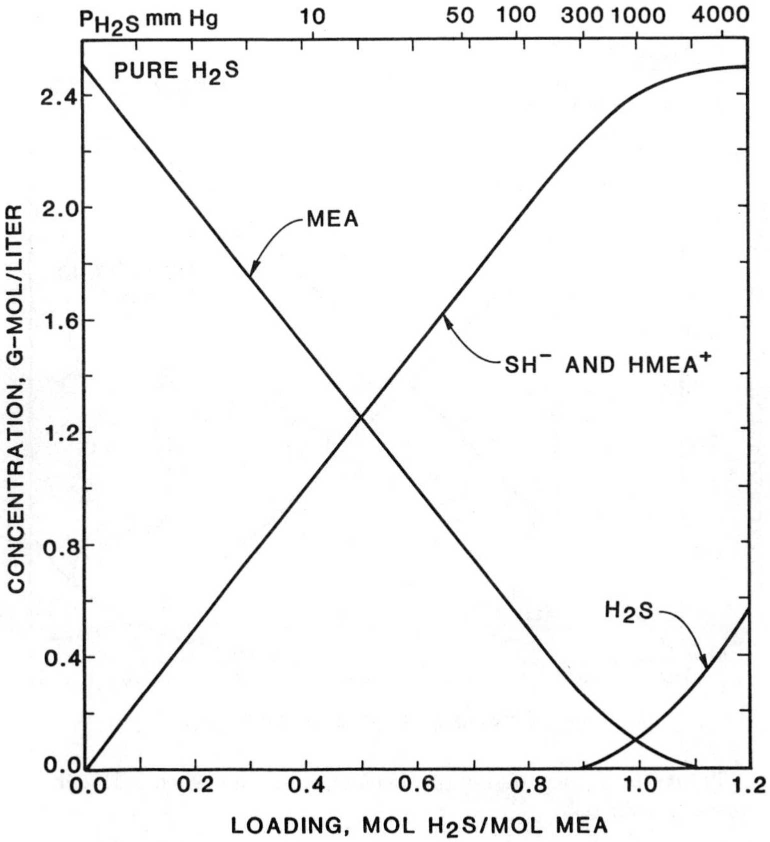

Figure 8. Concentration of major ionic species when 2.5N MEA reacts with pure H_2S.

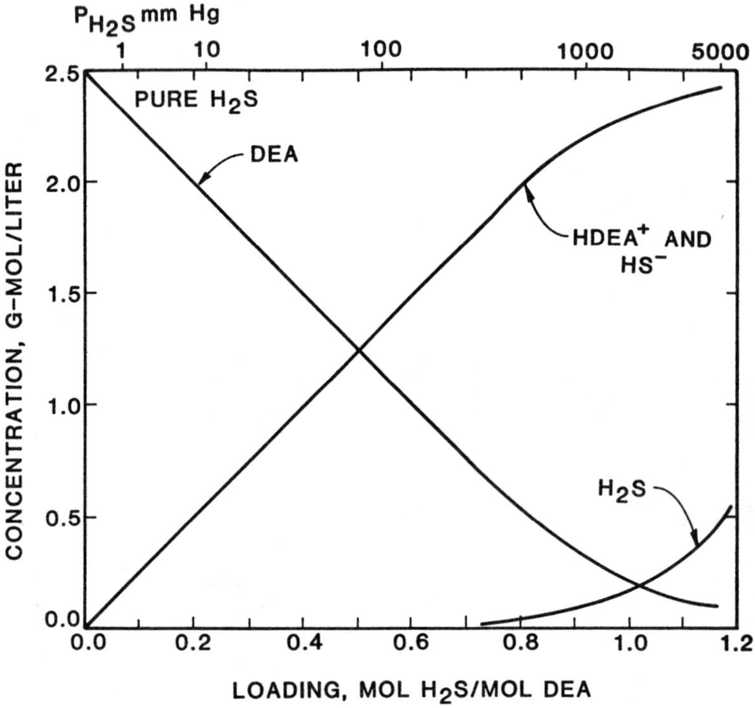

Figure 9. Concentration of major ionic species when 2.5N DEA reacts with H_2S.

data are those from Martin et al (4). The calculated data are from the SGP program (10).

Maddox and Mains (11) presented curves showing the concentrations of various ionic species in the ethanolamine solution as the reaction progressed for 2.5 N MEA at 100°F. Their graphs are reproduced in Figures 8, 10 and 12. The SGP program was used to calculate the same concentrations for 2.5 N DEA at 100°F. These results are plotted in Figures 9, 11 and 13.

Figures 8 and 9 show concentrations for 2.5 N MEA and 2.5 N DEA when only H_2S is present in the gas phase. The curves on the two graphs appear to be similar. However, closer observation will disclose some differences. Comparison of the pressure required for a given loading (reaction) of H_2S with ethanolamine shows that at an equilibrium partial pressure of 10 mmHg the MEA has reacted to a loading of slightly more than 0.4 mols H_2S/mol MEA, while the DEA solution has reacted to a loading of approximately 0.2 mols H_2S/mol DEA or approximately half that of the MEA solution. Though the ratios change, the same trend is discovered to hold throughout the range of loadings shown on Figures 8 and 9. The increased equilibrium loading of MEA at the same partial pressure is

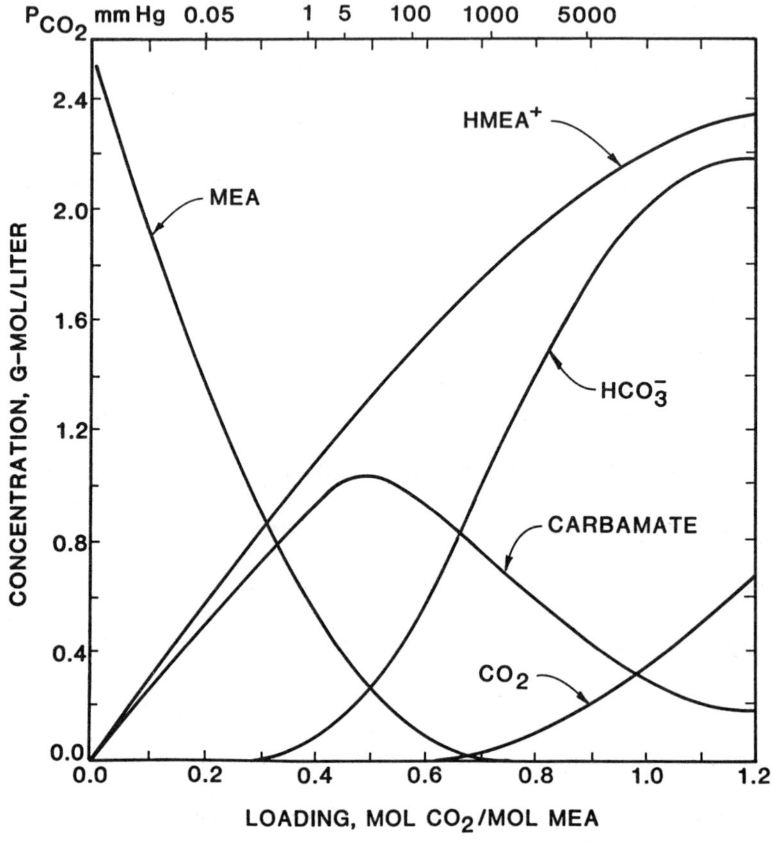

Figure 10. Concentration of major ionic species when 2.5N MEA reacts with pure CO_2.

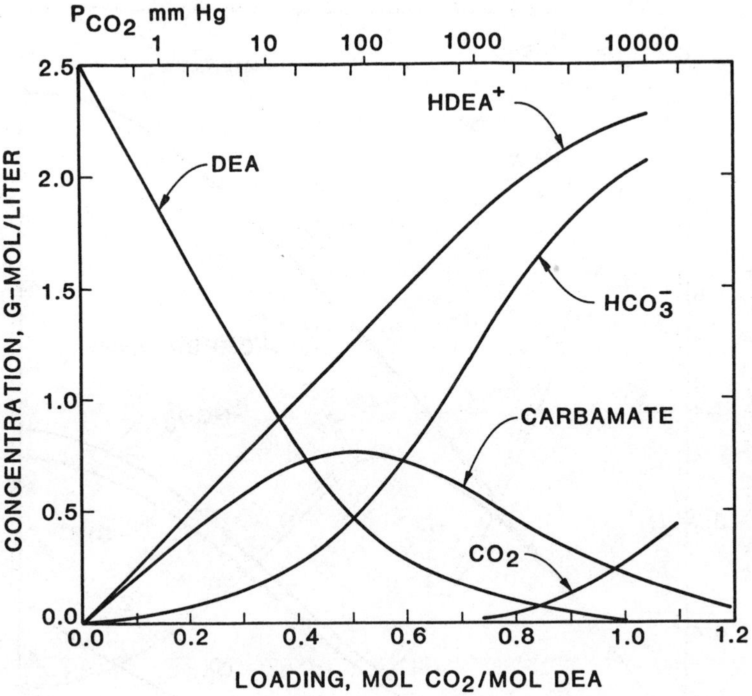

Figure 11. Concentration of major ionic species when 2.5N DEA reacts with CO_2.

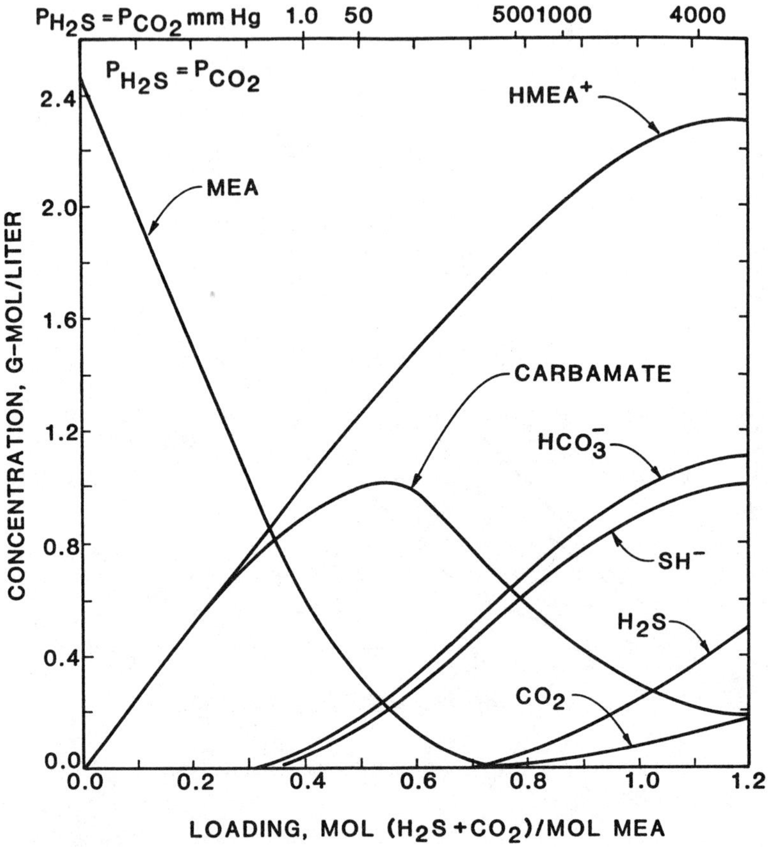

Figure 12. Concentration of major ionic species when 2.5N MEA reacts under equal partial pressures of H_2S and CO_2.

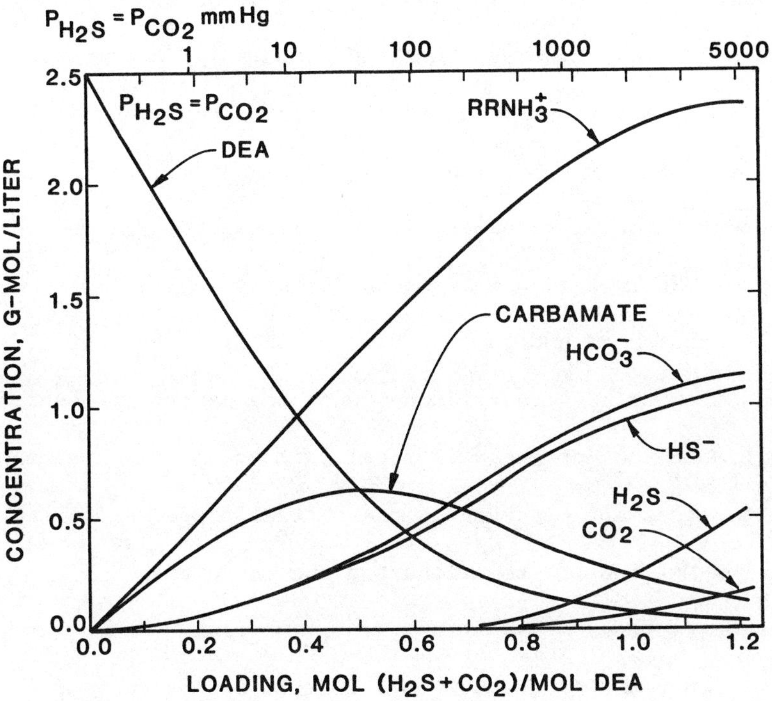

Figure 13. Concentration of major ionic species when 2.5N DEA reacts with H_2S and CO_2.

undoubtedly due to the fact that MEA, being a primary amine, is more basic than the secondary amine, DEA.

Figures 10 and 11 show the same graph when CO_2 only reacts with the ethanolamine. Again, the equilibrium loading in the MEA solution is higher than that for the DEA solution. However, there is another key difference. The maximum concentration reached by the carbamate in the MEA solution is approximately 1 g-mol/liter. The maximum concentration of carbamate in the DEA solution is only approximately 0.75 g-mol/liter, 25% less.

Figures 12 and 13 show the combined reaction of H_2S and CO_2 with the ethanolamines when the partial pressure of H_2S in the gas phase is equal to that of CO_2. The same differences are observed in the partial pressure-loading of the MEA and DEA solutions and the carbamate concentrations.

Figures 8-13 are only one example of the many uses that can be made of information gained through use of the reaction equilibrium approach to modeling acid gas-ethanolamine behavior. Figures 8-13 clearly show the complexity of the reactions taking place in the ethanolamine solution. The simplistic approach of "x mols pick - up per mol amine" used in so many instances for sizing ethanolamine sweetening systems obviously is not sufficient. Obviously, also, the complexities observed under contactor conditions can be expected to exist under regenerator conditions. The carbamate reaction undoubtedly causes part of the difficulty, at least, in regenerating CO_2 from spent ethanolamine solutions.

Other uses of the computer program include, but are not limited to: calculating equilibrium loadings for the amine at the bottom of the contactor; calculating the equilibrium partial pressures of the acid gases with the regenerated amine at the top of the contactor; estimating heats of reaction between the acid gases and ethanolamines; estimating temperatures of the rich amine leaving the bottom of the contactor. This flexibility makes the computer program based on the reaction equilibrium model a valuable tool for the process/design engineer engaged with sour gas.

ACKNOWLEDGEMENT

The authors gratefully acknowledge the partial support of the Gas Processor Association under project 841 "Experimental Determination of Vapor Liquid Equilibrium and Correlation of Behavior of Acid Gases in Solvents" in the collecting of the experimental data reported in this paper.

REFERENCES

1. Woertz, B.B., The Canadian Journal of Chemical Engineering, Vol. 50, (1972), p. 425.

2. Akashah, S.A., M. Sc. Thesis, Oklahoma State University, (1978).

3. Adeyiga, A.A., G.J. Mains, and R.N. Maddox, Proceedings of 30th Canadian Chemical Engineering Conference, Edmonton, Alberta, October 19-22, 1980.

4. Martin, J.L., F.D. Otto, and A.E. Mather, Journal of Chemical and Engineering Data, Vol. 23, (1978), p. 163.

5. Jou, F.Y., A.E. Mather, and F.D. Otto, Industrial and Engineering Chemistry Process Design and Development, Vol. 21, (1982), p. 539.

6. Rahman, M.A., "Study of Reactions of Carbon Dioxide and Sulfur Containing Compounds With Ethanolamines," PhD Thesis, Oklahoma State University, (1984).

7. Astarita, G., G. Marucci, and F. Gioia, Chem. Eng. Sci., Vol. 19, (1964), p. 95.

8. Caplow, M., J. Am. Chem. Soc., Vol. 90, (1968), p. 6795.

9. Kent, R.L., and B. Eisenberg, Hyd. Proc., (Feb. 1976), p. 87.

10. Maddox, R.N., Sour Gas Equilibrium Program, (1983).

11. Maddox, R.N., and G.J. Mains, Proceedings 63rd Annual Convention Gas Processors Association, Mar. 17-21, 1984.

9
Amine Guard Systems in Hydrogen Production-II

Daniel J. Kubek
Kenneth F. Butwell
Engineering Products & Processes
Union Carbide Corporation

ABSTRACT

This paper will explore the impact of Union Carbide's Amine Guard technology in aqueous MEA-CO_2 removal systems for the production of hydrogen via conventional steam reforming of hydrocarbons. The areas of application for this technology are in optimizations of existing MEA units, conversions of alternate CO_2 removal systems, and integration into grass roots hydrogen plant designs. The significance will be reviewed both with respect to the benefits afforded the CO_2 removal section and the global effect on the capacity/energy requirements of the hydrogen plant directly attributable to the purification train. Further enhancement of this latter concept will be demonstrated through the integration of other proprietary technologies in pressure swing adsorption and cryogenic processing when placed in sequential operation downstream of the Amine Guard unit. Finally, the important topics of

H_2 and CO_2 products purities and recoveries, reformer feedstocks and fuels, and environmental considerations will be examined.

INTRODUCTION

After several years of laboratory evaluation, the UCAR Amine Guard corrosion inhibition system was introduced to the ammonia (NH_3) and hydrogen (H_2) production industries (1) and has achieved a highly successful performance record over the past eight years (2, 3, 4, 5, 6). During this period, it became evident that corrosion as the dominant design and operating constraint of the monoethanolamine (MEA) - carbon dioxide (CO_2) purification system would be virtually eliminated through the judicious use of the Amine Guard inhibition technology. Consequently, major programs were initiated to optimize the MEA-CO_2 removal process in order to attain dramatic reductions in capital and operating costs both in existing units and grass roots design. These studies were also successfully implemented in the field, with the primary area of application to date being the ammonia industry (7, 8, 9, 10).

Recent investigations have revealed that these optimized Amine Guard-MEA systems have significant impact in the production of refinery and petrochemical H_2 via steam reforming. The enhanced efficiency of the CO_2 removal system with respect to its capital, energy, maintenance, chemical, and environmental costs is readily apparent from the prior work. More subtle and far-reaching are the economic implications afforded by the Amine Guard technology when integrated into the operation of the entire H_2 complex. The efficacy of the purification system favorably alters the throughput capacity, recovery and purity of products, and energy requirements of the entire H_2 processing train for both conversions of existing units and grass roots projects. This paper will deal with all these developments in H_2 manufacture/CO_2 purification as they reflect the changes in design and operating philosophy of such plants in realizing low cost and reliable H_2 production.

HYDROGEN PRODUCTION

H_2 manufacture based upon steam reforming of hydrocarbons has become the most widely accepted route during the last twenty years. Feedstocks may range from natural gas through saturated refinery gases to heavy naphthas. The conventional processing train for the production of H_2 comprises the following sequential steps, as depicted in the simplified flowsheet of Figure 1 (11, 12, 13):

- Desulfurization of hydrocarbon feedstock
- Steam Reforming ($CH_4 + H_2O \rightleftarrows CO + 3H_2$)
- High Temperature Shift Conversion
- Low Temperature Shift Conversion

  (both shift conversions: $CO + H_2O \rightleftarrows CO_2 + H_2$)
- CO_2 Removal
- Methanation $CO + 3H_2 \rightarrow CH_4 + H_2O$

  $CO_2 + 4H_2 \rightarrow CH_4 + 2H_2O$
- Compression

The first step in the operation is the residual desulfurization of the gaseous feedstock by activated carbon, molecular sieves, hot zinc oxide, etc. to protect against subsequent catalyst poisoning. The gas is then mixed with process steam and reformed at high temperatures (800^oC) over a nickel-based catalyst to yield an equilibrium mixture of H_2, carbon monoxide (CO), CO_2, methane (CH_4), and water (H_2O). This highly endothermic reforming reaction is propagated by the reformer furnace burners, with the critical parameters for CH_4 conversion and energy consumption being reaction temperature and steam/carbon ratio (defined as moles of steam per atom of carbon) (14). The reformer effluent is then cooled to approximately

375°C to undergo the exothermic high temperature shift reaction whereby the bulk of the CO is converted to supplemental H_2 and CO_2. The discharge gases are cooled to about 225°C and fed to the low temperature shift converter for residual CO conversion, ultimately attaining a CO concentration approaching 0.2 volume percent. The low temperature shift exit gases are cooled to about 70°C prior to introduction into the cyclic solvent-CO_2 removal system whereby CO_2 is selectively absorbed from an inlet level of approximately 20 volume percent to a crude H_2 product concentration of 0.01 volume percent. The residual carbon oxides (CO and CO_2) still present in the synthesis gas after shift conversion and CO_2 removal are methanated exothermically at 325°C to levels less than 5 ppm. The product H_2 is then cooled and sent to downstream uses, sometimes requiring compression if needed at higher pressure. The H_2 typically has a purity of 97 to 98 volume percent, with virtually the only impurity being CH_4 as a result of unconverted reformer feed and carbon oxides methanation.

The optimization of this cycle for the production of H_2 revolves around the proper utilization of the heat energy that is associated with the system. Large quantities of heat at extremely high temperatures must be supplied to sustain the reforming reaction. This makes available considerable waste heat in: 1) the furnace flue gases and 2) the process gas stream from the reforming step. The flue gas energy is converted to high pressure superheated steam in a waste heat boiler, with subsequent utilization via pressure letdown across the H_2 compressor turbine, followed by employment of this exhaust steam as process steam in the reformer. Similarly, the energy in the reformer process gas effluent is effectively used to generate additional steam and preheat reactants, as well as to drive the CO_2 removal system. As a result, the heat exchange integration, when applied to the simplified flowscheme in Figure 1, is extensive. The amount of steam provided to the process determines the completion of the reforming reaction, the completeness of the shift reactions, and the effectiveness of the CO_2 removal step and, thus determines the overall H_2 product purity. With an assessment then to the requirement of H_2 purity, it is possible to select an optimal combination of reforming temperature, steam/carbon ratio, and CO_2 purification system to produce the most efficient fuel balance for the

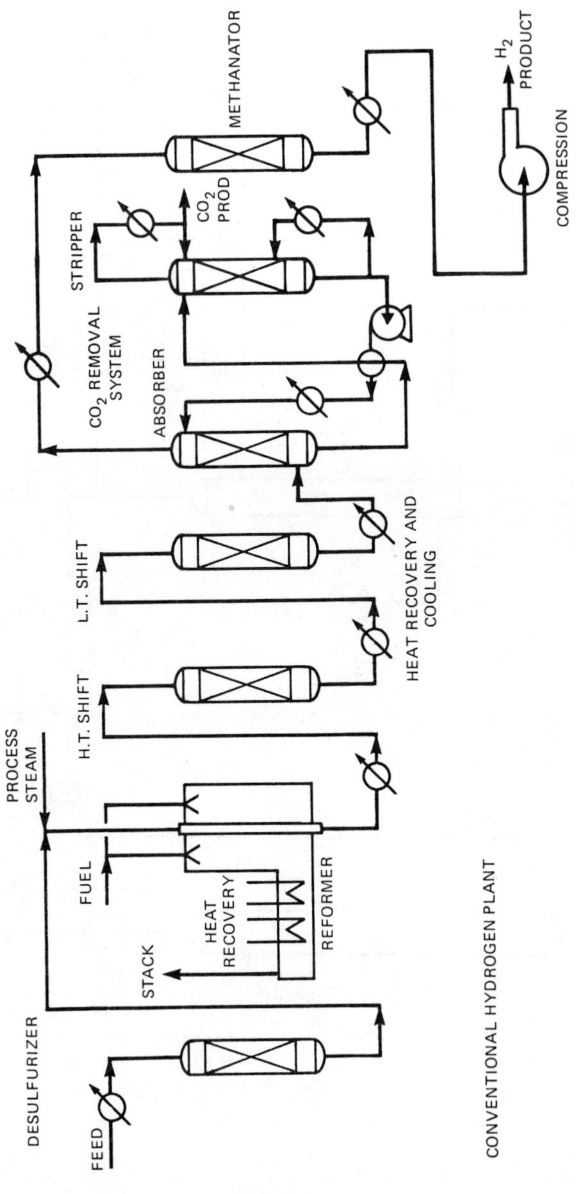

Figure 1. Simplified hydrogen production processing train.

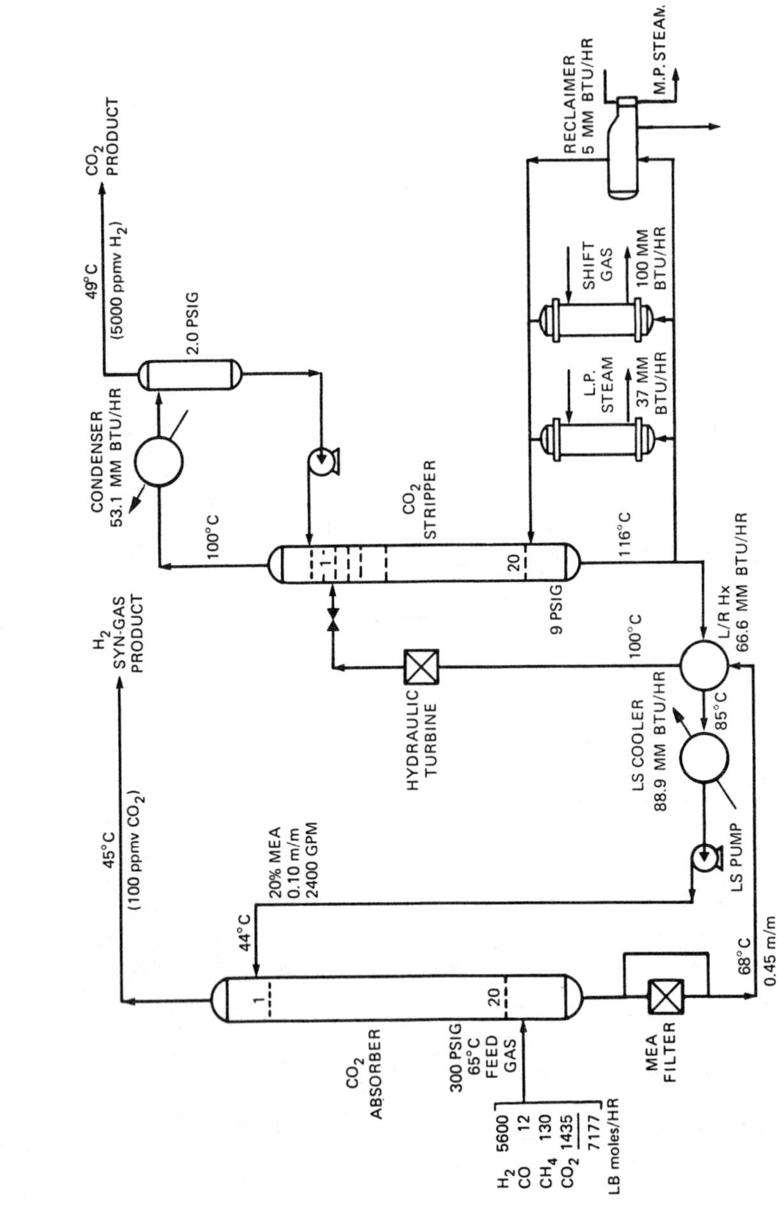

Figure 2. Conventional uninhibited MEA system typical CH_4-reformed 50MM scfd H_2 plant.

process. The goal is to have an overall system that is virtually self-sustaining in steam balance, i.e., one that requires neither additional fuel firing for extra steam production nor copious quantities of excess low pressure steam for export without sacrificing the recovery and purity of the H_2 and CO_2 products. Herein lies the key to the impact of the Amine Guard-CO_2 removal process as integrated into the H_2 complex.

<u>CONVENTIONAL MEA SYSTEM</u>

Figure II depicts the simplified flowsheet plus material and energy balance for a conventional uninhibited MEA system in a typical CH_4-reformed 50 MM standard-cubic-foot-per-day (SCFD) H_2 facility. Corrosion constraints dictate an MEA solution concentration of 20 weight percent maximum, with CO_2 solvent loadings increasing from a lean of 0.10 moles CO_2/mole MEA to a rich of 0.45 moles CO_2/mole MEA. The resultant solvent flowrate of 2,400 gpm removes 1,435 lb moles/hr of CO_2 from 5,600 lb moles/hr of H_2. These parameters ultimately yield a reboiler heat duty of 137 MM Btu/hr (95,500 Btu/lb mole CO_2), provided by both hot quenched synthesis gas and low pressure steam. In addition, a reclaimer is employed to continuously purify a slipstream of hot, lean solvent. Corrosion and degradation products are collected in the reclaimer bottoms liquid for periodic disposal, while pure MEA/H_2O overhead vapors are passed to the stripper where their subsequent condensation provides an incremental heat input. Despite the extensive use of stainless steel metallurgy throughout the MEA unit, as well as a functional reclaimer, corrosion still poses a serious operating problem.

AMINE GUARD IMPLEMENTATION IN EXISTING MEA UNITS

Corrosion Elimination

The incorporation of the Amine Guard inhibition technology into 150 operating units over the past 13 years has demonstrated the ease of its implementation and has verified that corrosion will be virtually eliminated for all carbon and stainless steel metallurgy. For the conversion of an existing MEA unit, this translates into the following apparent benefits:

- No unscheduled downtime
- No equipment replacement
- Reduced maintenance costs
- Enhanced worker safety
- Lower MEA losses via leaks

Reclaimer Elimination

The Amine Guard system is a closed one, not requiring continuous solvent reclaiming since there are essentially no corrosion or degradation products to be removed. As in all 150 plants, therefore, the reclaimer is idled, but is maintained in a standby mode in the unlikely event of gross system contamination. The resultant savings in reclaimer shutdown are as follows, with the values in parentheses representing estimates as applied to the 50 MM SCFD H_2 plant's MEA unit shown in Figure II:

- Elimination of medium pressure steam (5 MM Btu/hr of 150 psig steam)
- Elimination of MEA losses with bottoms liquid (50,000 lbs MEA/yr)
- Elimination of corrosion and maintenance costs costs associated with the reclaimer
- No environmental waste disposal problems with reclaimer liquids

Reboiler Duty Optimization

From a simplified perspective, the use of the patented Amine Guard corrosion inhibitor (15) in conjunction with higher MEA solution concentrations and increased lean/rich CO_2 solvent loadings affords a substantial reduction in lean solvent circulation rate and required reboiler duty. The resulting optimization of the typical 50 MM SCFD H_2 plant's MEA unit is illustrated in the flowsheet and material and energy balance of Figure III. The solvent flowrate and reboiler energy input have been dramatically lowered to 1,600 gpm and 81 MM Btu/hr (56,500 Btu/lb mole CO_2), respectively. In addition, the reclaimer has been idled and the cooling (air/water) requirements of the system have been proportionately reduced. Finally, no sacrifice has been made with respect to H_2 or CO_2 product purity or recovery.

The impact of the 40 percent reduction in reboiler heat duty is far-reaching. As a first step, it means the complete elimination of the low pressure steam reboiler and its associated steam consumption. Examination of the plant's steam balance will be necessary to explore whether this low pressure steam can be employed elsewhere or simply not be generated initially. The Amine Guard system is now driven by hot shift gas only. In fact, the reboiler load is so low that there is an "excess" of shift gas heat available. In this case, some synthesis gas heat may bypass the reboiler and be recovered via boiler-feed water (BFW) preheating.

More likely, however, is the realization that excessive steam/carbon ratios are being employed in the reformer section, not for incremental CH_4 conversion, but only to drive the energy-intensive conventional MEA-CO_2 removal system. The exaggerated steam/carbon ratios produce only a marginally purer H_2 product at the expense of greater reforming costs, all for the purpose of supplying the reboiler energy through placement of larger quantities of steam in the process synthesis gas exiting the low temperature shift. In some cases, the steam/carbon ratio is so high that no low pressure steam reboiler exists; the 95,500 Btu/lb mole CO_2 of reboiler heat is provided entirely by the hot shift gas. This was acceptable in the days of inexpensive energy, but it now represents a severe economic burden to the H_2 plant. The marginally-purer H_2 product produced in this manner is clearly not economically justified in relation to any benefits gained in downstream H_2-processing units (16). As a result, the large reduction in MEA reboiler heat duty afforded through Amine Guard optimization permits significant reductions in reformer steam/carbon ratio because of lower syn gas heat requirements. This translates into energy conservation and fuel cost savings for the entire H_2 facility due to increased thermodynamic efficiency, with virtually no sacrifice in product specifications. This effect of lower reformer steam/carbon ratios and its ramifications will undergo further examination later in the text.

As a final point, the extent of reboiler duty optimization is dependent upon the synthesis gas heat available, as well as its potential export value. The Amine Guard optimization depicted in Figure III resulted in an energy requirement of 56,500 Btu/lb mole CO_2 in the existing process and equipment configuration and matched the shift gas heat available after reformer steam/carbon reduction. Advanced Amine Guard process designs with equipment alterations are available to attain reboiler duties approaching 40,000 Btu/lb mole CO_2 if these low-energy systems are warranted in the context of the overall H_2 plant. This subject of reboiler duty will be further examined in subsequent sections.

Capacity Increases

The reduction in vapor (reboiler duty) and liquid

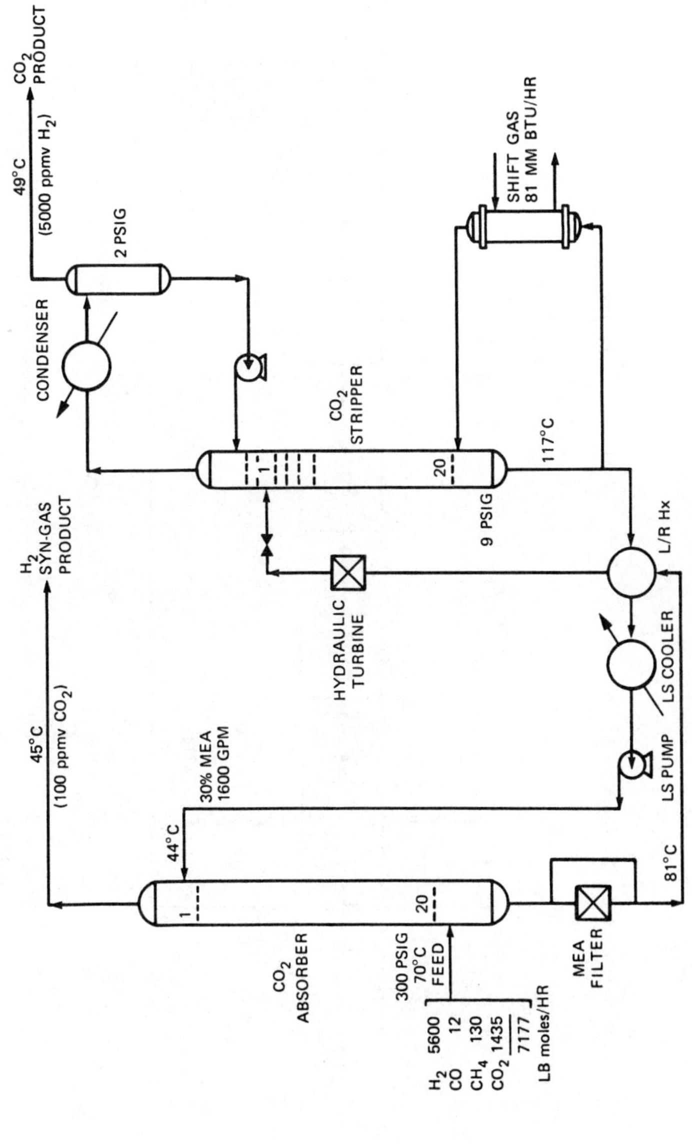

Figure 3. Amine guard optimization of existing MEA unit—typical CH_4-reformed 50 MM scfd H_2 plant.

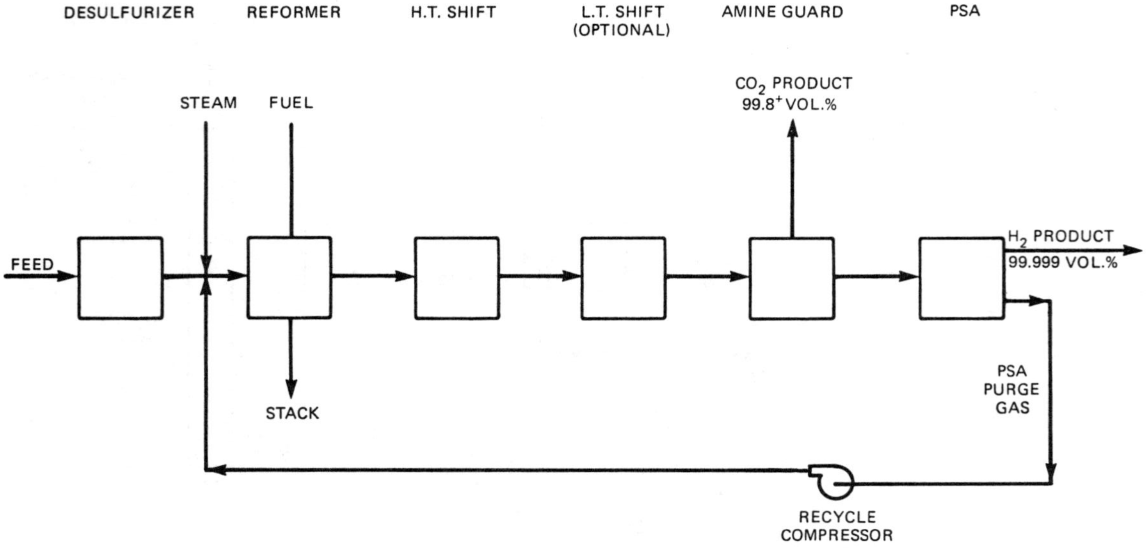

Figure 4. Simplified flow scheme of sequential operation of amine guard and HYSIV PSA.

(solvent circulation) flows throughout the MEA system via Amine Guard optimization hydraulically frees all towers, heat exchangers and coolers, vessels, and pumps, thereby permitting increased throughput. This debottlenecking of the MEA unit allows for increased CO_2 removal capacity, arising from either enhanced reformer feedstock flow or a transition in feedstock composition from CH_4 to heavier hydrocarbons. Proven capacity increases have been further complemented by equipment alterations, such as the retraying of towers with proprietary absorption and distillation trays (17) or the addition of a hydraulic turbine between lean and rich solvent streams to effect power recovery (18).

From a more global viewpoint, all processing units in the H_2 complex will be debottlenecked through the Amine Guard optimization of the MEA unit. In light of our prior discussion on lower shift gas reboiler duties and associated reformer steam/carbon reductions, the processing of less excess steam in the upstream reformer, CO shifters, and integrated heat exchanger network permits the greater utilization of all this equipment. The entire processing train will be debottlenecked by reductions in reformer steam/carbon ratio afforded by Amine Guard optimization of the shift gas reboiler duty in the MEA unit. The H_2 complex can now accept more feedgas to increase overall H_2 production.

Some of the major refiners are presently pursuing the installation of the Amine Guard technology in their H_2 facilities for this specific reason. Their examinations show that there is a major impact upon extra H_2 production, with the effect upon H_2 purity and downstream H_2-processing unit performance being minor.

Revamp Case History

The global impact that Amine Guard thermal optimization of the CO_2 removal unit can have on hydrogen plant production costs and capacity is best quantified in an actual case histroy (29). A large California refiner needed to significantly increase capacity of an existing steam-reformed H_2 plant to accommodate heavier and more sour crude oil feedstocks. The H_2 plant had a nameplate capacity of 51.5 MM SCFD H_2, but was pushed to produce nearly 60.0 MM SCFD H_2, admittedly with sacrifices in H_2 product purity and recovery. By working with a process engineering firm skilled in modern H_2 plant design, the following overall results were achieved after recommissioning the plant in the spring of 1983:

Parameter	Base Case	Revamp Case
H_2 Production, MM SCFD	51.5	75.0
Reformer S/C Ratio	5.9	4.5
Feed and Fuel Consumption, Btu/SCF-H_2	463	418
Reboiler Energy Consumption, Btu/lb mole CO_2	75,300	61,900

The existing Vetricoke/hot potassium carbonate-CO_2 system was converted to Amine Guard/MEA, as was a small auxiliary conventional MEA unit. Although the Vetricoke and conventional MEA units had been previously run in series, the Amine Guard retrofit placed them in parallel to handle the larger gas feed rates. In addition, the Vetricoke absorber intervals were modified for vapor capacity purposes. These combined changes permitted the CO_2 removal unit to comfortably process 45 percent more gas flow over nameplate design, while reducing specific heat consumption per mole CO_2 removed by 18 percent.

H_2 plant capacity was increased by about 45 percent over nameplate design, and specific feed and fuel consumption were reduced by approximately 10 percent. Steam/

carbon ratio was reduced from 5.9 to 4.5 with no appreciable loss in H_2 product purity, mainly by introducing a more active reformer catalyst and by reducing heat requirements in the CO_2 removal section. The more active catalyst results in reduced tube wall temperatures, even at expanded capacity, without retubing. The existing high and low temperature shift converters performed satisfactorily at the modified conditions with higher throughput and lower steam/dry gas ratio. Lastly, reformer efficiency was enhanced by installation of greater feedstock preheating as well as a new BFW preheat coil on the reformer stack to reduce exit flue gas temperature.

The success of this extensive H_2 plant revamp can be succinctly credited to three key items:

- Reformer catalyst and CO_2 removal process modifications.
- Integration of heat requirement for CO_2 removal process with heat availability from synthesis gas with lower reformer S/C ratio.
- Close communications that Union Carbide had with contractor and client.

PSA Effluent H_2 Treating

The potential installation of a pressure swing adsorption (PSA) unit to treat a fraction or all of the effluent H_2 from the Amine Guard-MEA unit for final purification allows even greater economics with respect to the energy efficiency and throughput capacity of the entire H_2 complex. The technology for a PSA treating unit has been described in considerable detail in the literature (19, 20); only a brief discussion is required here as it relates to steam-reformed H_2 upgrading. An important feature of PSA is that, in a single-step process, all residual constituents (CH_4, CO, CO_2, N_2,H_2O) except H_2 are simultaneously and completely removed from the Amine Guard effluent H_2 as it moves through the adsorbent bed. The dry H_2 product attains a total purity of 99.999 volume percent.

Regeneration of the PSA unit is accomplished by depressurization of the adsorbent bed followed by a purge of H_2 obtained from one of the other adsorbent beds undergoing depressurization. These purge gases (containing H_2, CH_4, CO, H_2O and trace CO_2) are compressed and recycled to the reformer feed inlet to insure essentially 100 percent recovery of H_2 product. (If some N_2 is present in reformer feedgas, a small slipstream of the purge gas is sent to the reformer fuel to prevent an N_2 buildup in the loop.) Of these recycled purge gas constituents, the CH_4 is reformed, the CO shifted, the CO_2 removed, and the H_2 recovered, thereby achieving maximum conversion and recovery of the reformer CH_4 to H_2 product.

Since PSA requires no heat input, has minimal pressure drop on the H_2 product, and is totally automated, it does not in any way burden the H_2 plant operation. In fact, since the H_2 product purity obtained from PSA is impervious to feed composition variations, it allows the reformer process conditions to be selected independently of the H_2 product purity. Specifically, further larger reductions in reformer steam/carbon ratios can be undertaken since the increasing unconverted CH_4 slippage is caught by the PSA unit and recycled without affecting final H_2 product purity. In light of our prior discussions on reducing reformer steam/carbon ration, this translates into major energy savings and potential capacity increases for the entire H_2 complex.

Figure IV is a simplified flowscheme of an H_2 plant wherein the synergistic effects of full-flow sequential operation of Amine Guard and PSA for the purification train are delineated. Separation of virtually all the CO_2 selectively first by Amine Guard allows for the subsequent removal of H_2 impurities by the PSA unit and recycle of the H_2 purge gas to the reformer inlet feed and generates the following concisely summarized benefits for the H_2 plant:

1. Large reductions in reformer stean/carbon ratio (to as low as 3/1) with dramatic global energy benefits and potential capacity increase;
2. Elimination of the methanator vessel, since residual CO and trace CO_2 are removed from the syn gas in the PSA unit for recycle. Without a methanator, H_2 consumption and CH_4 generation into the H_2 product via methanation are negated;
3. Generation of dry high-purity (99.999 volume percent) H_2 at 99.95 percent recovery. Without a methanator and with PSA purge gas recycle, virtually all incoming CH_4 reformer feed is ultimately converted to H_2 and recovered;
4. Generation of high-purity CO_2 product (99.8 volume percent), with essentially 100 percent recovery;
5. Elimination of the low temperature shift, if desired. Unshifted CO from the high temperature

shift effluent is recycled by the PSA unit to the reformer feed.

CONVERSIONS OF ALTERNATE CO_2 REMOVAL SYSTEMS

In addition to existing H_2 plants' conventional MEA units, conversions of competitive CO_2 removal systems to the Amine Guard-MEA technology are proceeding at various rates. A brief synopsis of this activity is as follows:

Sulfinol (*) Units

Eight Sulfinol conversions (NH_3 and H_2) have been successfully accomplished to date. One in particular allowed an H_2 plant to attain a 20 percent capacity increase (to 25 MM SCFD H_2) through conversion to Amine Guard-MEA and a stripper tower retrayed (17) with proprietary distillation trays.

Vetricoke (**) Units

The conversion of a refinery 60 MM SCFD H_2 plant's Vetricoke unit was accomplished in 1983.

Diglycolamine (***) Units

Twin 40 MM SCFD H_2 plants' Diglycolamine units in a refinery have been evaluated for conversion with resulting favorable economics.

Hot Potassium Carbonate Units

Although generally not amenable to Amine Guard-MEA

conversions, some older plants are compatible. A refinery 50 MM SCFD H_2 plant's Catacarb (****) system is presently being evaluated for retrofit to series treating with Amine Guard-MEA and Hysiv PSA for the reasons given in the preceding section.

In general, the incentives for conversion to Amine Guard may arise from any combination of the following:

- Corrosion elimination
- Reclaimer shutdown
- Reboiler duty optimization
- Capacity increase
- Reduced CO_2 slippage into the H_2 product
- Reduced H_2 content of CO_2 product
- Reduced solvent losses and makeup costs
- Environment safety

AMINE GUARD PROCESS FOR GRASS ROOTS HYDROGEN PLANTS

Amine Guard Process Options

The Amine Guard Processes for grass roots steam-reformed H_2 plants have the ability to selectively remove copious quantities of CO_2 from the feedgases to levels of approximately 100 ppmv for gases of virtually all pressures. The Amine Guard process configuration employed and the resultant energy requirement are dependent upon the feedgas pressure, feedgas CO_2 partial pressure (PP_{CO_2}), and CO_2 specification of the product. For a typical hydrogen plant's feedgas containing 15-25 volume percent CO_2 and necessitating removal down to approximately 100 ppmv, the reboiler duty would progress with changing Amine Guard process configurations as the feedgas pressure varies in the following fashion:

Pressure (psig)	PP_{CO_2} (psia)	Heat (Btu/lb mole CO_2)
50	8 -13	80,000
150	22.5-37.5	60,000
350	52.5-87.5	40,000

With respect to equipment, above-ambient process temperatures permit either air cooling or water cooling to be used extensively, with only final trim lean solvent cooling generally dictating cooling water. In addition, the relatively fast kinetics of the MEA-CO_2 reaction minimize required contact time which translates into smaller tower heights and internals. This effect is further enhanced by the optional employment of proprietary distillation trays (17) which utilize closer tray spacings and smaller tower diameters. Furthermore, the highly aqueous solvent system exhibits low viscosities and high thermal conductivities which promote high heat exchanger transfer coefficients. Finally, the low molecular weight of MEA (61) and its strong chemical affinity for CO_2 yield low solution circulation rates, thereby minimizing pump and piping sizes.

The experience generated from over 150 operating units combined with the performance record of the corrosion inhibition system provides a high level of reliability for all the Amine Guard Process options. The aforementioned strong affinity of MEA for CO_2 as integrated into the Amine Guard process design assures a wide range of flexibility capable of easily handling variations in feed composition and flowrate.

Amine Guard Process Integration into H_2 Plant Design

The CO_2-rich synthesis gas exiting the low temperature shift reactor at 225°C proceeds to the Amine Guard Process for CO_2 removal. It is first water quenched to 175°C prior to shift gas reboiler entrance to minimize skin temperatures. The Amine Guard process configuration employed is dependent on feedgas pressure but generally has a pattern similar to the simplified flowscheme seen earlier in Figure III. The condensation of steam and decrease in sensible heat of the gas across the reboiler provide the energy necessary to drive the Amine Guard system (40,000-80,000 Btu/lb mole CO_2). The synthesis gas exits the reboiler at 130+°C and is now water-saturated. Normally, there is sufficient or, in most cases, excess heat available from the synthesis gas between 175°C and 130°C to satisfy the requirements of the Amine Guard Process. This excess will depend on the initial steam/carbon ratio employed in the reformer. This difference in available and required heat is put to use in BFW preheating, initial methanator feed preheating, or heat conservation elsewhere, particularly at the high temperature end (175°C) of the syn gas.

The reboiler effluent is further cooled against BFW or methanator feed from 130°C to 70°C prior to absorber entrance. The CO_2 is selectively removed via countercurrent contact with the downflowing MEA solvent, and the synthesis gas discharges the absorber with 100 ppmv CO_2 and at approximately 45°C. The CO_2-laden rich solvent exits the absorber bottoms, is preheated by returning hot lean solvent, and is stripped of CO_2 in the regenerator tower. Further heat conservation is afforded by low temperature BFW preheating against the regenerator overhead condenser and lean solvent cooler.

The synthesis gas from the absorber at 45°C is pre-heated to about 300°C by heat exchange with methanator effluent and hot syn gas prior to methanator entrance. The exothermic conversion of CO (0.2-0.5 volume percent) and CO_2 (0.01 volume percent - 100 ppmv) to CH_4 increases the gas temperature by about 25°C. The methanation of a mole of CO_2 consumes four moles of H_2 and transforms 100 ppmv CO_2 into 100 ppmv CH_4 according to the stoichiometry:

$$CO_2 + 4H_2 \rightarrow CH_4 + 2H_2O$$

$$(\Delta H_{25°C} = -39.4 \text{ Kcal/mole})$$

The low CO_2 slippage into the synthesis gas afforded by the Amine Guard process is important because it minimizes the amount of H_2 consumption and CH_4 generation that occurs in the methanator. When compared to various competitive cyclic solvent CO_2 removal processes that often exhibit CO_2 slippages into the synthesis gas of 1000-5000 ppmv, the following conclusions can be drawn:

1. The methanator vessel for those processes will be incrementally larger due to the order of magnitude increase in CO_2 methanation;
2. The reformer feedgas quantity for these processes will be incrementally higher to produce the same absolute amount of H_2 due to increased H_2

consumption via CO_2 methanation;

3. The final overall H_2 purity for these processes will be incrementally lower as a result of greater CH_4 formation via CO_2 methanation. Conversely, if H_2 purity levels are held constant, marginally higher reformer steam/carbon ratios will be required for these processes in order that enhanced reformer CH_4 conversion counteracts greater CH_4 formation in the methanator.

After Amine Guard and methanation, the H_2 product is cooled via syn gas and BFW exchange and contains essentially CH_4 as the sole impurity. Of this CH_4, only 0.2-0.5 volume percent concentration comes from CO methanation. The remainder (1-3 volume percent) is unreformed CH_4 and the concentration from this source is dependent upon reformer steam/carbon ratio. The key point is that with Amine Guard for CO_2 removal and with an active low temperature shift catalyst, the H_2 purity is a direct function of reformer operation only. The Amine Guard systems permits the minimal required reformer steam/carbon ratio for proper CH_4 conversion, and the resulting synthesis gas heat available in the syn gas reboiler is more than sufficient to drive the Amine Guard-CO_2 removal section.

Hydrogen Product

With the Amine Guard-CO_2 removal system, the H_2 product recovery from the reformer feed is maximized. This is due to the minimal solubility of H_2 in the highly aqueous MEA solvent and the negligible methanation of CO_2 in the absorber exit synthesis gas. H_2 recovery is further enhanced, if desired, by the addition of a rich solvent flash tank for the removal recycle of the minute quantities of H_2 from the solvent proceeding to the regenerator. In addition, the H_2 product pressure is near that of the feed due to the minimal pressure drop across the Amine Guard absorber.

As previously mentioned, the H_2 product purity with Amine Guard is essentially a function of reformer steam/carbon ratio only. In most refinery applications with conventional hydrogen production (hydrocracking, hydrotreating), an H_2 purity of 97.0 volume percent is more than sufficient to economically satisfy downstream process performance requirements (14, 16). The reformer conditions (temperature, steam/carbon ratio) are, consequently, set to meet this H_2 purity only, thereby optimizing the thermodynamic efficiency for the H_2 production.

If greater H_2 purity is warranted in conjunction with

increased energy conservation, and/or CO_2 product, a smaller PSA unit can be added to partially or totally treat the Amine Guard synthesis gas effluent for final H_2 purification for the same reasons set forth earlier in our discussion of existing H_2 plant conversions. The H_2-rich purge gas from the PSA unit is again compressed and recycled to the reformer feed for H_2 product conservation. The advantages of a full flow series operation of Amine Guard and PSA (Figure IV) are:

- Large reduction possible in design reformer steam/carbon ratio;
- High purity dry H_2 product (99.999 volume percent) at high H_2 recovery (99.95 percent);
- High purity CO_2 product (99.8+ volume percent) at essentially 100 percent CO_2 recovery;
- Elimination of the methanator;
- Elimination, if desired, of the low temperature shift;
- Removal of N_2 impurity in H_2 product if present in the reformer feed.

Carbon Dioxide Product (21)

The Amine Guard Process integrated into the grass roots H_2 plant produces a secondary product, CO_2, at high purity and recovery. Considered in past years as a waste product to be vented, CO_2 is finding expanding markets and strong profit margins as a commodity chemical. Some of these markets are:

1. Urea: The importation of cheap foreign ammonia has adversely affected domestic ammonia production but spurred urea production, thereby requiring bulk pure CO_2;
2. H_2/CO Plants: In these types of plants, CO_2 is purposely recovered at considerable expense from reformer flue gases as reformer feed to enhance CO production;
3. Dry Ice: For cryogenic applications;
4. Food Processing: For beverage carbonation;
5. Tertiary Oil Recovery: For enhanced crude oil production via CO_2 injection.

If the operating costs of the H_2 plant are attributed to the chief product - H_2, the CO_2 may be viewed as a "free" raw material, the utilization of which will add to the overall profitability of the H_2 plant.

Reformer Feedstocks and Fuels

In many grass roots H_2 complexes under consideration, the materials to be utilized for the steam reformer reactor feedstocks and furnace fuels are different. The reformer feedstock in petrochemical plants is generally high-quality natural gas (22). In refineries, the feedstock may be high-quality, saturated hydrocarbon off-gas streams from the catalytic reformer or hydrocracking operation. Oftentimes, they are already rich in H_2 content (14, 23). In certain cases where gas availability is limited, desulfurized straight-run naphtha is chosen as the reformer feed (16).

In contrast, the fuel for the reformer furnace may well be a material of lower quality. In many refineries, the fuel gas is obtained from unsaturated off-gas streams (14), or it may be naptha in consideration of gas supply limitations (24). Another alternative receiving increased attention is fuel oil (No. 2), such as the VFO system developed by Allied Chemical Corporation and Heat Research Corporation (25, 26).

Commensurate with a difference in reformer feed and fuel quality and availability is a disparity in the value of these materials. Therefore, it is imperative in these situations that the feed and fuel sources be totally separated and relegated to their respective roles if the H_2 plant is to perform under optimal operating economics. This implies that the H_2 purification system must not expel any synthesis gas to the fuel system in the performance of its CO_2 removal function. To do so would sacrifice costly feedstock of limited supply at the expense of saving cheaper, more available fuel materials. In addition, the reformer itself would have to be increased in size to accommodate the increased H_2 generation required to counterbalance the lost H_2 expelled to the fuel system. Since the reformer cost dominates the capital investment of an H_2 plant, and since reformer failure is the leading cause of H_2 plant unscheduled downtime (27), this is clearly a disadvantage to be considered in any evaluation.

The ramifications of this discussion relate to some of the available H_2 purification systems, both of the cyclic solvent (absorption) and fixed bed (adsorption) types. With these processes, a reasonable quantity of a flash gas or purge gas is generated in performing the CO_2 removal task. This off-gas consists mainly of CO_2 and H_2 ($\simeq$ 50/50) at near-atmospheric pressure and of low caloric value; it is dumped to the furnace fuel system. In light of our discussion on feed/fuel values and increased reformer size, this off-gas is a disadvantage to be seriously considered in the overall economics of a new H_2 plant.

The Amine Guard Process makes a clearly-defined

separation between the H_2 and CO_2 products in the synthesis gas. As a result, no admixture of H_2 and CO_2 is generated for explusion to the fuel system and, therefore, the feed and fuel sources are always divided. Finally, H_2-rich reformer feedstocks yield less CO_2 in the synthesis gas, and this translates into substantially reduced capital investment and operating costs for the Amine Guard Process.

HYDROGEN/CARBON MONOXIDE PLANTS

In many chemical operations, the production of an H_2/CO synthesis gas is necessary as a raw material for other chemical processes (acetic acid, phosgene) (28). While H_2/CO-rich streams may be available from such sources as acetylene off-gases, the predominant route to their manufacture is steam reforming of hydrocarbons. Figure V represents the block processing steps required for H_2/CO synthesis gas. The feedstock is desulfurized and reformed, but not shifted since CO is now a highly-desirable product. After heat recovery, the synthesis gas is typically at 150-200 psig pressure and contains 10-12 volume percent CO_2 (PP_{CO_2} = 19 psia). Amine Guard processing for CO_2 removal reduces the CO_2 concentration to 50 ppmv. The CO_2 is compressed and recycled to the reformer feed for the enhancement of CO production by LeChatelier's Principle.

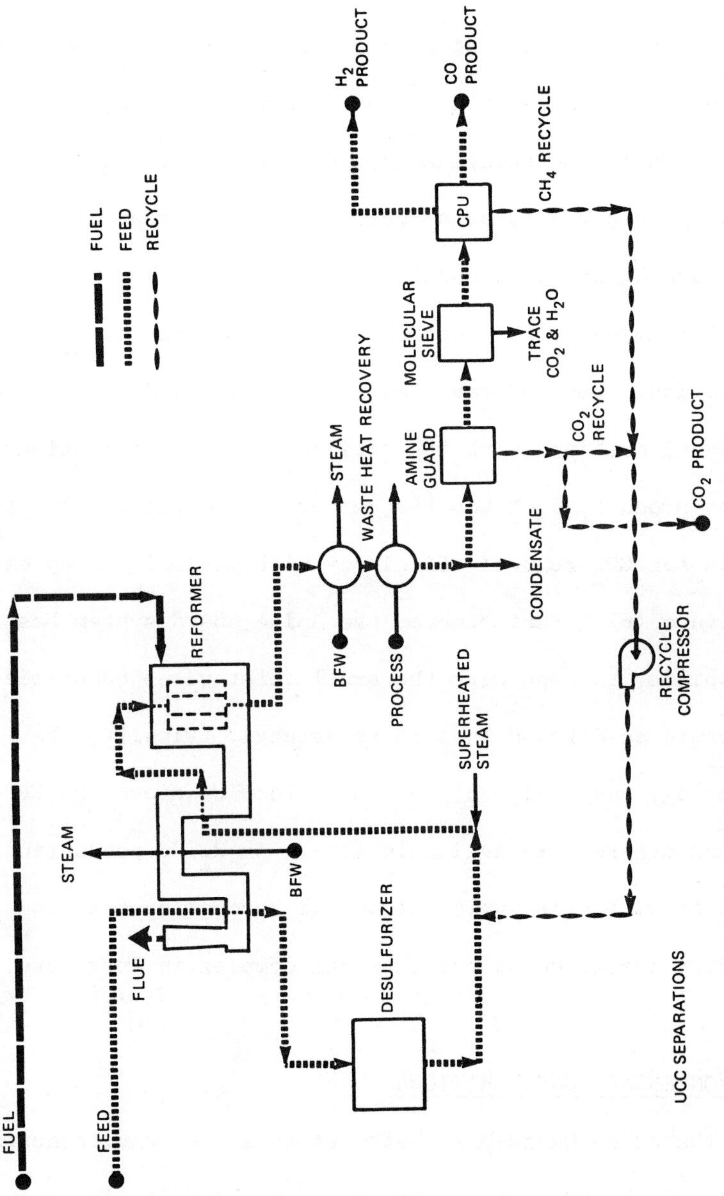

Figure 5. Block H_2/CO synthesis gas plant processing steps.

Final trace purification is accomplished by molecular sieve treatment which attains a maximum CO_2 + H_2O of 1 ppmv in the synthesis gas to prevent downstream freezeups. The H_2/CO stream then enters a cryogenic processing unit (CPU), commonly referred to as a "cold box", where separation into pure H_2 and CO occurs.

The Amine Guard Process is highly suited for H_2/CO production. The moderate feedgas pressure and CO_2 partial pressure, combined with the stringent CO_2 specifications of the product, dictates the use of an MEA-based solvent system for CO_2 removal efficiency with minimal energy expenditure (6). Furthermore, the Amine Guard system has the ability to cope with the small deleterious quantities of formic acid found in such feedstocks. Finally, the technology and equipment for the molecular sieve and CPU systems are readily available (19). An H_2/CO production facility with this purification and separation train was recently installed at our chemical complex in Institute, WV.

ENVIRONMENTAL CONSIDERATIONS

The Amine Guard-MEA system enhances the environmental safety of the CO_2 removal section for the following reasons:

1. With corrosion eliminated, fewer corrosion-

associated leaks and spills of solvent will occur which means minimal seepage to land and natural waters;

2. With the reclaimer eliminated from daily operation, disposal problems associated with the waste liquid bottoms are negated;
3. Major accidental contamination of the circulating solvent does not present a dramatic disposal problem for the large quantity of solvent inventory. Emergency operation of the standby reclaimer concentrates the purity into small volumes for disposal;
4. The Amine Guard corrosion inhibitors chemical levels are less than 500 ppm total in the circulating solution. This high level of dilution insures negligible environmental impact if solvent were to escape from the unit;
5. MEA is a relatively fast biodegrading organic chemical in contact with the environment. MEA spills that find their way to natural waters do not linger but are broken down to harmless elements in a relatively short time.

CONCLUSIONS

Thus, the impact of the Amine Guard technology is significant and far-reaching when applied to the production of refinery and petrochemical H_2 via steam reforming of hydrocarbons. The effects are obvious when viewed with respect to the CO_2 removal section and more subtle when examining the global implications on the throughput capacity, products specifications, and energy requirements of the overall H_2 comples. Series integration of Amine Guard with other separation technologies in PSA and CPU affords enhanced benefits for H_2 and CO production. Union Carbide is firmly committed to the continued development of their synthesis gas separation and purification technologies in order to attain systems with lower capital and operating costs without sacrifice in product specifications.

REFERENCES

(*) Shell Oil Development Company

(**) Giammarco-Vetricoke

(***) Texaco Chemical Company

(****) Eickmeyer and Associates

1. Hawkes, E. N. and Mago, B. F., "Stop MEA-CO_2 Unit Corrosion", Hydrocarbon Processing, August, 1971.

2. Wett, T., "Ammonia Producer Stymies Corrosion", Oil and Gas Journal, May 8, 1972.

3. Butwell, K. F., Hawkes, E. N. and Mago, B. F., "Corrosion Control in CO_2 Removal Systems", Chemical Engineering Progress, February, 1973.

4. Editor, "CO_2-Removal Cost Cut $550,000/Year", Oil and Gas Journal, March 17, 1975.

5. Editor, "Union Carbide's Amine Guard Systems Stop MEA Unit Corrosion", Nitrogen Journal, July/August, 1975.

6. Baker, G. H., "Corrosion-Free Gas Sweetening", Hydrogen Processing, August, 1975.

7. Editor, "Amine Guard Integrally Designed Into Eleven Ammonia and Six Hydrogen Plants", Nitrogen Journal, July/August, 1976.

8. Kelly, W. J., "How Amine Guard Saves Energy", Hydrocarbon Processing, July, 1977.

9. Butwell, K. F. and Kubek, D. J., "Optimize NH_3 With Amine Guard", Hydrocarbon Processing, October, 1977.

10. Butwell, K. F., Kubek, D. J. and Sigmund, P. W., "Amine Guard III - An Advanced CO_2 Removal System", Chemical Engineering Progress, February, 1979.

11. Cox, K. E. and Williamson, K. D., Hydrogen: Its Technology and Implications; Production Technology, Volume I, CRC Press, 1977.

12. Haddelard, G. E., Hydrogen, Stanford Research Institute, Process Economics Program, Report No. 32A, December, 1973.

13. Johnson, J. E. and Singman, T. L., "The Steam Reforming of Light Hydrocarbons to Product Hydrogen", Internal UCC/Tonawanda Report, Engineering Laboratory Note 675.

14. Hughes, D. W. J., "Steam-Hydrocarbon Reformer Variables Key to Cheaper H_2, Oil and Gas Journal, August 28, 1978.

15. U. S. Patent #3,808,140.

16. Uchida, H. and Kyogoku, "Treating Cost Sets Hydrogen Unit", Hydrocarbon Processing, April, 1977.

17. Union Carbide Distillation Technology Information Sheet, "Acid Gas Removal Systems", Bulletin No. F-3649 (Multiple Downcomer Trays).

18. Smith, R. C., "Hydraulic Turbines - Use in Synthesis Gas Purification Systems", Kellogg/Client Ammonia Club Meeting, Miami, FL, November 17, 1978.

19. "Advanced Hydrogen Purification and Upgrading", Union Carbide Bulletin No. F3427.

20. Stewart, H. A. and Heck, J. L., "Pressure Swing Adsorption", Chemical Engineering Progress, September, 1969.

21. Editor, "Key Chemicals - Carbon Dioxide", Chemical and Engineering News, June 27, 1978.

22. Buividas, L. J., Czuppon, T. A. and Cover, A. E., "Modern Ammonia Synthesis Use of Hydrogen", Kellogg/Client Ammonia Club Meeting, Miami, FL, November 17, 1978.

23. Internal Consultant Report, "Hydrogen Supply at U.S. Refineries", F. Kunreuther Associates, Inc., March, 1978.

24. Editor, "Chemical Plants Reform Energy Practices", Chemical Week, Jaunary 21, 1976.

25. Johnson, E. R., "Firing Oil On An Ammonia Reformer", Chemical Engineering Progress, March 1977.

26. Caldwell, D., Elfers, F. and Fankhanel, M., "The Vaporized Fuel Oil (VFO) System", Chemical Engineering Progress, Novenber, 1978.

27. Williams, G. P., "Causes of Ammonia Plant Shutdowns", Chemical Engineering Progress, September, 1978.

28. Internal Consultant Report, Synthesis Gas for Chemicals, Chem Systems, Inc., December 1977.

29. Berzins, V. and Udengaard, N. R., "Revamp Your Hydrogen Plant", Hydrocarbon Processing, May, 1983.

10

Solubility of H_2S and CO_2 in Triethanolamine Solutions

F. Y. Jou
F. D. Otto
A. E. Mather
Department of Chemical Engineering
University of Alberta

Triethanolamine (TEA) was one of the first amines used in aqueous solution for the removal of H_2S and CO_2 from gas streams (1). However, the fact that monoethanolamine (MEA) and diethanolamine (DEA) solutions had a greater capacity for the acid gases, H_2S and CO_2, and that the rate of reaction was greater for these amines caused TEA to be supplanted by MEA and DEA. Recently, interest in the use of TEA solutions for bulk CO_2 removal has led to reconsideration of the properties of TEA. The solution can be regenerated by the reduction of pressure and hence energy costs are minimized. This was the basis of the dual-solvent process used in ammonia plants (2). The use of CO_2 for enhanced oil recovery projects has also made TEA an attractive alternative. Goddin (3) showed that bulk removal of

CO_2 with TEA compared favorably with other possible processes. Selective removal of H_2S using tertiary amines is another reason for renewed interest in TEA. Since tertiary amines like TEA do not form a carbamate, the reaction rate with CO_2 is relatively slow, while the rate of reaction with H_2S is relatively fast. This leads to a kinetic selectivity, even though the overall rate of reaction with H_2S and TEA solutions is lower than that with MEA or DEA solutions (4).

To interpret kinetic data and to design separation processes, a knowledge of the equilibrium solubility of H_2S and CO_2 in the amine solution is necessary. Relatively few data for the solubility of H_2S and CO_2 in TEA solutions have been published. For CO_2 Mason and Dodge (5) presented data for five TEA solutions ranging in concentration from 0.5 M to 5 M at temperatures of 0°, 25°, 50° and 75°C. Partial pressures of CO_2 ranged from 0.2 to 101 kPa. Lyudkovskaya and Leibush (6) also measured the solubility of CO_2 in four TEA solutions ranging from 0.5 M to 5 M at temperatures of 25°, 50° and 75°C. Partial pressures of CO_2 ranged for 255 to 4124 kPa. Only a few scattered data points have been presented for the solubility of H_2S in TEA solutions (7). No data are available at temperatures higher than 75°C, about the lower limit for the regeneration of TEA solutions. To fill this gap in the experimental data for the solubility of the acid gases in TEA solutions, the present work was undertaken.

EXPERIMENTAL SECTION

The equipment used to measure the solubility data was that used previously in this laboratory (8). The apparatus consisted of a Jerguson liquid level gauge with a 250 cm^3 tubular gas reservoir mounted on the top. A magnetically-driven piston pump (9) was used to circulate the gas phase and bubble it through the liquid solution. The cell and pump were mounted in an air bath controlled to within $\pm$ 0.5°C. The pressure in the cell was measured with a calibrated Heise bourdon tube gauge.

Amine solutions were prepared from distilled water and commercially available triethanolamine having a purity of 99.4%. The TEA solution of the desired concentration was charged to the cell. Hydrogen sulfide or carbon dioxide were then added in an amount determined by observation of the pressure. If necessary, nitrogen was added so that the total pressure was always greater than atmospheric pressure. To ensure that equilibrium was reached, the vapor was bubbled through the liquid for at least 8 h prior to sampling of the vapor and liquid phases.

The vapor phase was analysed by gas chromatography. A known weight of loaded amine solution was drawn into a 1 N NaOH solution to form Na_2CO_3 or Na_2S. The concentration of CO_2 and H_2S in an aliquot of liquid sample was determined by titration using the procedure previously employed with MDEA solutions (10).

RESULTS

The solubility of CO_2 in 2.0, 3.5 and 5.0 M TEA solutions was determined at 25°, 50°, 75°, 100° and 125°C. Partial pressures ranged between 0.01 and 6360 kPa. The CO_2 data are plotted on Figure 1 for the 5.0 M solution for comparison with the published data. The new data are seen to be in good agreement with published values at conditions where they can be compared. The solubility of H_2S in 2.0, 3.5 and 5.0 M TEA solutions was measured at 25°, 50°, 75°, 100° and 125°C. Partial pressures of H_2S ranged from 0.01 to 6270 kPa. The data for the 3.5 M solution are plotted on Figure 2. Also shown on Figure 2 are the few data points of Atwood et al. (7) for similar TEA solutions. The agreement with the present work is not very good. The experimental data are to be published shortly.

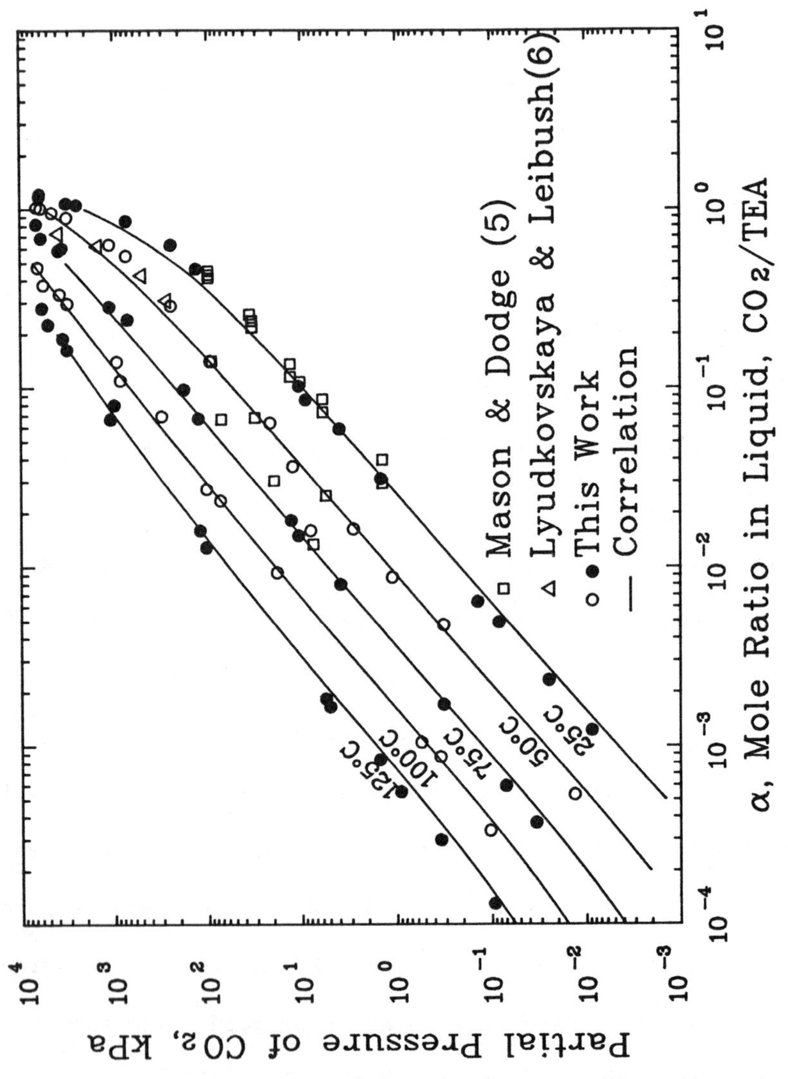

Figure 1. Solubility of CO_2 in 5.0 M TEA solution.

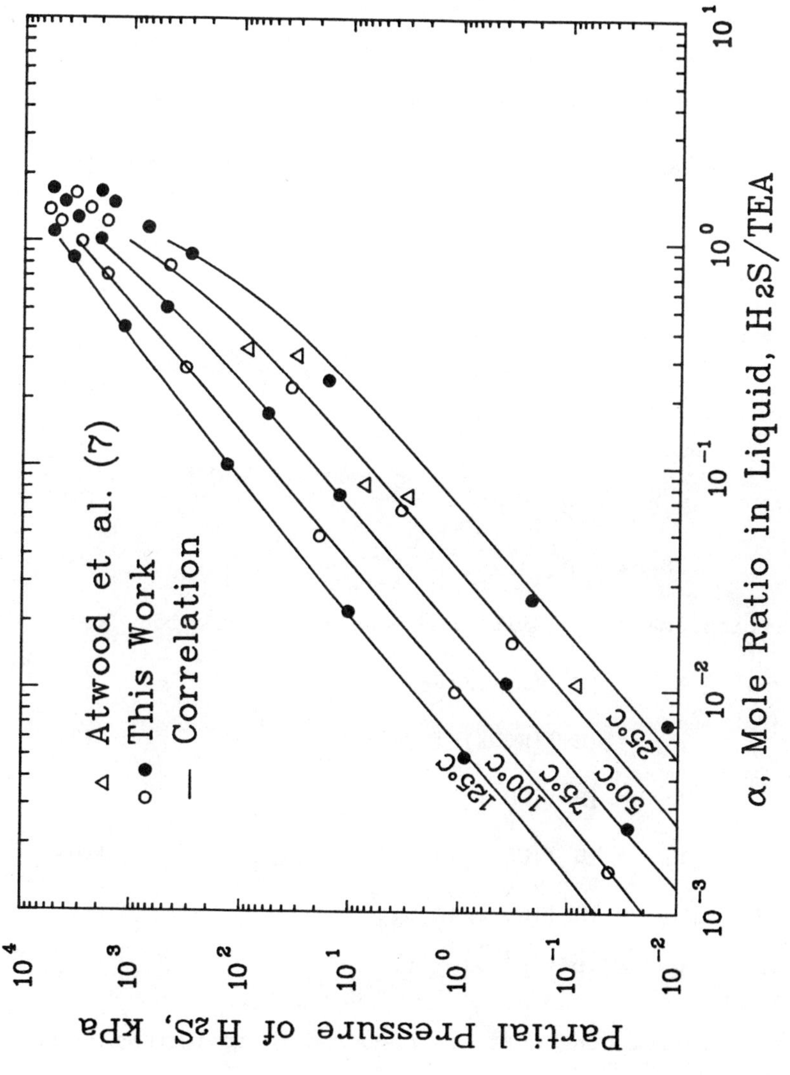

Figure 2. Solubility of H_2S in 3.5 M TEA solution.

CORRELATION OF THE SOLUBILITY DATA

The method of Kent and Eisenberg (11) was used to correlate the solubility data. In the case of TEA solutions, simplification is possible since no carbamate is formed. Hence this system is similar to that of MDEA which was studied by Jou *et al.* (10). The following reactions are postulated:

$$R_3NH^+ = H^+ + R_3N \tag{1}$$

$$H_2O + CO_2 = H^+ + HCO_3^- \tag{2}$$

$$H_2O = H^+ + OH^- \tag{3}$$

$$HCO_3^- = H^+ + CO_3^= \tag{4}$$

$$H_2S = H^+ + HS^- \tag{5}$$

$$HS^- = H^+ + S^= \tag{6}$$

The stoichiometric equilibrium constant expressions are:

$$K_1 = [H^+][R_3N]/[R_3NH^+] \tag{7}$$

$$K_3 = [H^+][HCO_3^-]/[CO_2] \tag{8}$$

$$K_4 = [H^+][OH^-] \tag{9}$$

$$K_5 = [H^+][CO_3^=]/[HCO_3^-] \tag{10}$$

$$K_6 = [H^+][HS^-]/[H_2S] \tag{11}$$

$$K_7 = [H^+][S^=]/[HS^-] \tag{12}$$

The equilibrium constants are numbered to conform with the system used by Kent and Eisenberg. The partial pressures of the acid gases are related to the molecular concentration in the liquid phase by Henry's law:

$$H_{CO_2} = P_{CO_2}/[CO_2] \tag{13}$$

$$H_{H_2S} = P_{H_2S}/[H_2S] \tag{14}$$

The mass and charge balances are:

$$m = [R_3N] + [R_3NH^+] \tag{15}$$

$$m\,\alpha_{CO_2} = [CO_2] + [HCO_3^-] + [CO_3^=] \tag{16}$$

$$m\,\alpha_{H_2S} = [H_2S] + [HS^-] + [S^=] \tag{17}$$

$$[R_3NH^+] + [H^+] = [HS^-] + 2\,[S^=] + [HCO_3^-] + 2\,[CO_3^=] + [OH^-] \tag{18}$$

where R represents C_2H_4OH, α is the mole ratio of acid gas to amine in the liquid phase, m is the amine concentration and the bracketed quantities are the molarities of the various species. As Jou *et al*. have shown, the partial pressures of the acid gases can be expressed in terms of the equilibrium constants:

$$P_{H_2S} = (H_{H_2S}[H^+]^2A/K_6K_7)(1/(1+[H^+]/K_7)) \tag{19}$$

$$P_{CO_2} = (H_{CO_2}[H^+]^2B/K_3K_5)(1/(1+[H^+]/K_5)) \tag{20}$$

$$[H^+] = A(1+K_7/(K_7 + [H^+]))/(1+m/K_1K') + B(1+K_5/(K_5 + [H^+]))/(1+m/K_1K') + K_4/([H^+](1+m/K_1K')) \tag{21}$$

where $A = \alpha_{H_2S}m-P_{H_2S}/H_{H_2S}$, $B=\alpha_{CO_2}m-P_{CO_2}/H_{CO_2}$,
and $K' = 1+[H^+]/K_1$.

The stoichiometric equilibrium constants are functions not only of temperature, but ionic strength and composition as well. Although values of pK_1 for TEA are available, these values apply at infinite dilution and if they are used in the model, the partial pressures calculated are not in good agreement with the experimental values. Following Kent and Eisenberg, the published values of K_3 - K_7 and the Henry's constants were accepted and the value of K_1 calculated from the experimentally measured pressures. As expected, the values calculated from CO_2 data are not exactly the same as those calculated from H_2S data, and the K_1 are functions of temperature, amine concentration and solution loading, α. The calculated values of K_1 were fit to the following empirical equation using a least squares criterion:

$$\ln K_1 = a_0 + a_1 m + a_2 m^2 + a_2/T + a_4\, T + a_5 \ln T + a_6 \ln \alpha + a_7 (\ln \alpha)^2. \qquad (22)$$

Values of the parameters, a_0 - a_7, for H_2S and CO_2 are presented in Table 1 and the correlation is compared with the experimental data in Figures 1 and 2. In determining the values of K_1 and the parameters in the correlating equation, both published data and the present experimental values were used. Hence the agreement shown on the Figures is a measure of suitability of Equation (22) to represent the values of K_1.

ACKNOWLEDGEMENT

The authors are grateful to Amoco Production Company, Tulsa, Oklahoma for financial support of this research. Thanks are due S.-S. Yee who carried out the correlation of the experimental data.

NOTATION

A	$= m\alpha_{H_2S} - P_{H_2S}/H_{H_2S}$, kmol m^{-3}
B	$= m\alpha_{CO_2} - P_{CO_2}/H_{CO_2}$, kmol m^{-3}
H	= Henry's coefficient, kPa/kmol m^{-3}
$K_1..K_7$	= stoichiometric equilibrium constants
K'	$= 1 + [H^+]/K_1$
m	= concentration of amine, kmol m^{-3}
P	= partial pressure of acid gas, kPa
T	= temperature, K
α	= mole ratio in the liquid phase, mol acid gas/ mol amine
[]	= molarity of a species, kmol m^{-3}

Table 1
Parameters In the Correlating Equation for K_1

	H_2S	CO_2
a_0	0.185 480 00 E+02	0.130 234 28 E+04
a_1	0.835 352 83 E-01	-0.354 168 52 E+00
a_2	0.170 699 45 E-01	0.103 409 81 E+00
a_3	-0.688 580 00 E+04	-0.447 969 25 E+05
a_4	-0.855 450 56 E-02	0.305 089 44 E+00
a_5	-0.229 339 96 E+01	-0.221 545 58 E+03
a_6	-0.329 340 14 E+00	-0.419 774 82 E+00
a_7	-0.106 611 26 E-01	-0.296 024 33 E-02

REFERENCES

1. Bottoms, R.R., "Organic Bases for Gas Purification", Ind. Eng. Chem., 23, 501-504 (1931).

2. Strelzoff, S., "Triethanolamine, Dual-Solvent Process", in Ammonia, Pt. II, ed. by A.V. Slack and G.R. James, pp. 201-203, Dekker, N.Y., 1974.

3. Goddin, C.S., "Pick Treatment for High CO_2 Removal", Hydrocarbon Processing, 61(5), 125-130 (1982).

4. Schingnitz, M., "Absorption von Schwefelwasserstoff und Kohlendioxid bei gleichzeitiger chemischer Reaction durch fallende Tropfen", Freiberger Forschungshefte, Reihe A, 413, 87-120 (1967).

5. Mason, J.W. and B.F. Dodge, "Equilibrium Absorption of Carbon Dioxide by Solutions of the Ethanolamine", Trans. A.I.Ch.E., 32, 27-48 (1936).

6. Lyudkovskaya, M.A. and A.G. Leibush, "Solubility of Carbon Dioxide in Solutions of Ethanolamines Under Pressure", Zhur. Priklad. Khim., 22, 558-567 (1949).

7. Atwood, K., M.R. Arnold and R.C. Kindrick, "Equilibria for the System, Ethanolamines-Hydrogen Sulfide-Water", Ind. Eng. Chem., 49, 1439-1444 (1957).

8. Lee, J.I., F.D. Otto and A.E. Mather, "Design Data for Diethanolamine Acid Gas Treating Systems", Gas Processing/Canada, 65(4) 26-34 (1973).

9. Ruska, W.E.A., L.J. Hurt, and R. Kobayashi, "Circulating Pump for High Pressure and -200 to +400°C Application", Rev. Sci. Inst., 41, 1444-1446 (1970).

10. Jou, F.-Y., A.E. Mather and F.D. Otto, "Solubility of H_2S and CO_2 in Aqueous Methyldiethanolamine Solutions", Ind. Eng. Chem. Process Des. Dev., 21, 539-544 (1982).

11. Kent, R.L. and B. Eisenberg, "Better Data for Amine Treating", Hydrocarbon Processing, 55(2) 87-90 (1976).

11

A Review of Current DIGLYCOLAMINE® Agent Gas Treating Applications

Tom F. Moore
Project Chemical Engineer
Texaco Chemical Company

John C. Dingman
Product Sales Manager, Gas Treating Chemicals
Texaco Chemical Company

Fred L. Johnson, Jr.
Senior Project Chemist
Texaco Chemical Company

DIGLYCOLAMINE® brand of 2-(2-aminoethoxy)ethanol is being used worldwide for the absorption of acid gases, such as hydrogen sulfide and carbon dioxide. The use of DGA™ gas treating agent can result in significant savings over other sweetening processes when the total acid gas content of the stream to be treated is greater than one mole percent. DGA sweetening units now in operation are treating gas streams with CO_2:H_2S ratios varying from over 500:1 to less than 0.1:1, respectively. These DGA units are processing sour gas feed streams having combined H_2S

and CO_2 contents ranging from 1 to greater than 50 mole %. Hydrogen sulfide recovered in these DGA units is being converted to bright sulfur via the Claus process.

DIGLYCOLAMINE is utilized in over 100 treating units worldwide. Individual plant locations have sour gas treating capacities ranging from less than 1 MM SCF/D to as much as 1500 MM SCF/D. The plants treating natural gas have operating pressures ranging from 60 to 1150 psig. Most of the plants processing associated or separator gas have operating pressures ranging from 15 to 300 psig. All of the plants processing natural gas, and most of those processing associated gas, are reducing the H_2S content to less than 0.25 grains/100 SCF.

DIGLYCOLAMINE agent is currently being used in six refineries with two additional start-ups slated for 1984. The DGA agent is being used to remove H_2S and CO_2 from a variety of both gas and liquid streams. H_2S concentration of low pressure fuel gas streams is being reduced to less than 150 ppmv.

Where carbonyl sulfide (COS) removal from liquid hydrocarbon streams is of primary importance, the Malaprop™ Process has been reported to reduce COS from levels as high as 1500 ppmv to less than 2 ppmv. The Malaprop Process is a proprietary process for the removal of COS from light liquid hydrocarbon streams which uses aqueous solutions of DGA.

DGA is also an effective treating agent for the removal of carbonyl sulfide from gas streams during normal gas sweetening operations. The level of COS removal increases with increasing temperatures in the DGA absorbers and with greater contact time. Depending on temperatures, pressures, and L/V ratios in the DGA absorbers, COS levels of the gas streams can typically be reduced by 75% or more with little, if any, change in process conditions used for sweetening.

NATURAL GAS APPLICATIONS

DGA concentrations ranging from 40 to 70 wt.% are typical of most natural gas sweetening applications. All of the natural gas sweetening plants using DGA are reducing the hydrogen sulfide concentration to less than 0.25 grains of H_2S/100 SCF. CO_2 concentrations of less than 100 ppmv are usually obtained in most medium and high pressure applications. Lower levels of CO_2 can be achieved when needed.

Conditions found in natural gas plants are shown in Table 1.

Table 1
Natural Gas Sweetening with DGA

Range of Feed Gas Compositions:
0.5-25 Mole % CO_2
0-33 Mole % H_2S

Range of Treating Pressures:
60-1150 PSIG

Range of Feed Gas Temperatures:
65-120°F

Typical Treated Gas Quality:

	Low Pressure	Medium Pressure	High Pressure
H_2S, ppmv	<4	<4	<4
CO_2, ppmv	>100	<100	<100

Acid Gas Loading

The acid gas loadings of the rich DGA solution in natural gas sweetening operations will vary significantly in terms of standard cubic feet of acid gas per gallon of DIGLYCOLAMINE solution. The major factors influencing the differences in acid gas loadings are DGA concentration, CO_2:H_2S ratio, and materials of construction. Most of the plants using DGA to process natural gas with high CO_2:H_2S ratios operate with rich amine loadings of 0.35 to 0.40 moles of total acid gas/mole DGA. Residual acid gas loadings of the lean DGA in these plants will normally be about 0.1 mole of acid gas/mole of DGA.

Plants processing natural gas with low CO_2:H_2S ratios generally have lower residual acid gas contents in the lean DGA solutions ranging from 0.03-0.07 moles of total acid gas/mole DGA. The lower acid gas loadings of the lean DGA solution is the result of higher reflux ratios typically used for low CO_2:H_2S gas streams.

Natural gases containing low CO_2:H_2S ratios are considered less corrosive to process in amine sweetening systems. DGA plants, treating this type gas, typically use higher acid gas loadings in the rich DGA which, in conjunction with lower lean solution loadings, results in a significant increase in the net acid gas pickup per mole of DGA.

Regeneration Heat Requirements

As is true with acid gas loadings, the regeneration heat requirements for DGA solutions are affected by the CO_2:H_2S ratio of the acid gas contained in the inlet gas stream. Natural gas plants containing high CO_2:H_2S ratios require 1.0-1.5 moles of water vapor per mole of acid gas in the regenerator overhead. Plants operating with low CO_2:H_2S ratios in the inlet gas require more heat for solution regeneration. Natural gas plants operating with CO_2:H_2S ratios of 0.2:1 may require 2.5-3.5 moles of water vapor per mole of total acid gas in the regenerator overhead to obtain adequate solution regeneration. A reflux ratio of this magnitude is not unusual for low pressure treating plants which are required to produce natural gas with less than 0.25 grains of H_2S/100 SCF.

Reclaimer Operation

The major chemical degradation product of DGA in gas treating solutions is N,N'bis(hydroxyethoxyethyl)urea, (BHEEU). It is formed by the reaction of two moles of DGA with one mole of either CO_2 or COS. A second degradation product can be formed by the reaction of one mole of carbon disulfide or possibly carbonyl sulfide, with two moles of DGA to form N,N'-bis(hydroxyethoxy-ethyl)thiourea (BHEETU). Experience indicates the dominant reaction of DGA with COS forms BHEEU. The reactions between CO_2, COS, and CS_2 with DGA are reversible in the reclaiming process at temperatures ranging from 360-380°F. These reactions are shown in Table 2.

As reclaiming temperatures are increased, the rate of conversion of BHEEU to DGA increases. This reaction is slow at temperatures below 340°F and increases to acceptable rates at temperatures of 360-380°F. Experience has shown that thermal decomposition of the DGA is not significant unless reclaiming temperatures in excess of 380°F are employed.

A supplemental water feed to the reclaimer is to be used only when necessary to prevent reclaiming temperatures from exceeding 380°F. The reclaimer temperature can be limited to a maximum temperatuer of 380°F automatically, without continuous operator attention, by using a temperature controller in the reclaimer kettle to feed

Table 2
Degradation Reactions of DIGLYCOLAMINE® Agent with CO_2, COS, and CS_2

```
                                   H O H
2R-NH2+CO2      ________         R-N-C-N-R + H2O or H2S
       or       ________
       COS
```

N,N'bis(hydroxyethoxyethyl)urea + H_2O or H_2S
(BHEEU)

```
                                   H S H
         COS
2R-NH2 + or     ________         R-N-C-N-R + H2O or H2S
         CS2    ________
```

DGA + COS or CS_2 N,N'bis(hydroxyethoxyethyl) thiourea + H_2O or H_2S
(BHEETU)

where R = $HO-CH_2-CH_2-O-CH_2-CH_2-$

supplemental water when required. The temperature controller regulates a supplemental water feed, usually reflux, to prevent the reclaiming temperature from exceeding 380°F.

Heat duty to the reclaimer is set on flow control and is determined by the desired rate of reclaiming. Lean DGA feed to the reclaimer is flow controlled by a liquid level controller on the reclaimer kettle.

Reclaimer feed rates of 0.5 to 1.0% of the circulating solution are usually adequate to keep BHEEU levels in the lean DGA solution below 1-2%. If significant levels of COS are present in the sour feed gas, higher reclaiming rates may be required.

The length of reclaiming cycles vary widely from plant to plant. It is difficult to predict the length of a typical reclaimer cycle. If a plant is heavily loaded, reclaiming cycles may be as short as two months. Plants operating lightly loaded and using low DGA concentrations may experience reclaiming cycles of six months or more. Reclaimer operations need to be evaluated for the actual operating conditions used. Refusal to periodically dump the reclaimer in an effort to minimize DGA consumption can be counter-productive.

Process control for a typical DGA reclaimer is shown in Figure 1.

Materials of Construction

Most of the DGA sweetening facilities utilize carbon steel as the material of construction. When natural gas streams with high CO_2:H_2S ratios are treated with DIGLYCOLAMINE in concentrations of 65-70 wt. % DGA, and rich DGA solution acid gas loadings are 0.4 moles of acid gas per mole of DGA, the following equipment is frequently made of stainless steel:

1. Lean to rich DGA heat exchanger tube bundle.
2. Absorber and regenerator tray decks and valves or caps.
3. Regenerator reboiler tube bundle.

DGA Make-up Requirements

DGA losses vary significantly from plant to plant. DGA usage of less than three pounds per MM SCFD of sour gas processed can be obtained. Abnormal losses of DGA are usually the result of poor housekeeping. Poor filter operation, excessive reclaimer dumps, tight plant design, entrainment, and foaming problems represent the most common causes of excessive losses. To minimize losses, leaking

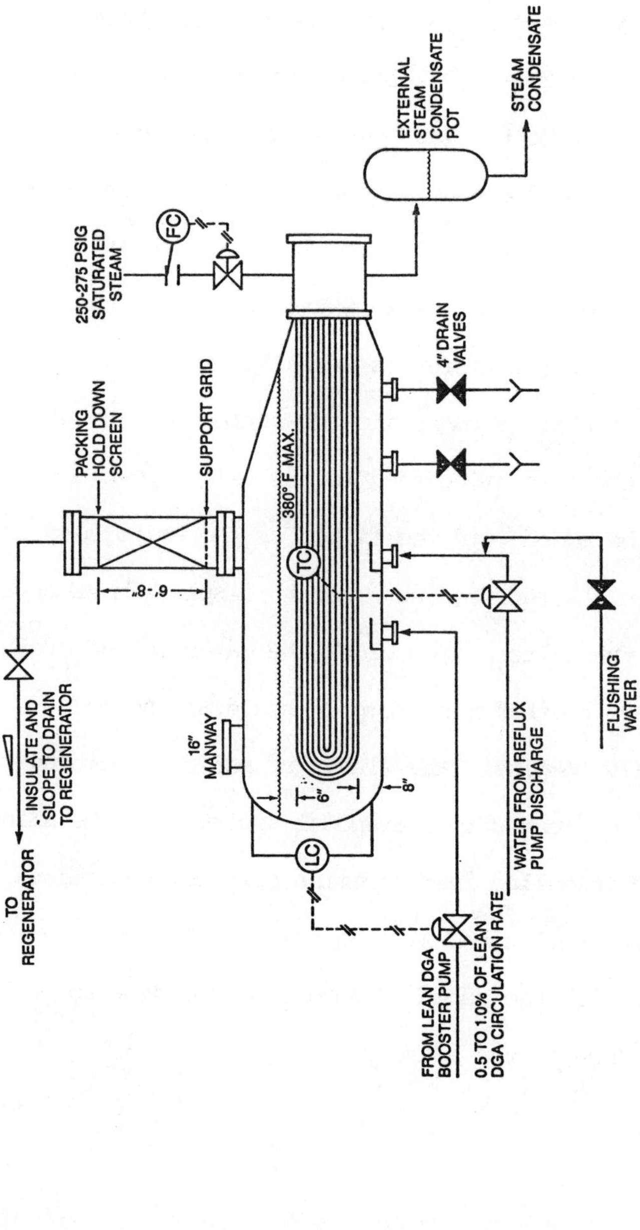

Figure 1. Typical process flow for DIGLYCOLAMINE® agent reclaimer.

pump seals, open drains, foaming, and level control dump valves must be given constant attention. When processing high temperature gases, a water wash of the treated gas is effective in reducing DGA losses due to equilibrium vaporization into the gas leaving the absorbers.

Incomplete drainage of amine solution from filter cases at the end of a filter cycle is a loss of amine that can be avoided with a little extra attention from personnel. If the filter is installed on the lean solution, install a connection downstream of the block valve on the lean amine feed to the filter to allow make-up water to be added into the filter as the amine feed is blocked in. After feeding water to displace most of the amine in the filter case, block the water in and drain the filter case to the amine sump. When draining is essentially complete, any sludge or solids remaining in the case can be flushed to a disposal system. If the amine filter is installed in rich amine service, the amine regenerator reflux can also be used for this displacement operation with the added advantage of not disturbing the water balance in the system.

Dumping the DGA reclaimer before necessary is another cause of high amine losses. When the decision has been made to terminate the reclaiming cycle, discontinue lean DGA feed to the reclaimer. Instead of lean DGA, feed regenerator reflux while maintaining normal levels of heat duty to the reclaimer. Continue operation in this manner until either the DGA content of the solution in the reclaimer kettle is reduced to approximately 10 wt.% or the DGA content of the reclaimer overhead is reduced to approximately 1 wt.%. At this point, discontinue heat to the reclaimer and discard the remaining solution to the reclaimer sump for disposal. This will minimize unnecessary DGA losses associated with reclaimer cycles.

Corrosion

Severe or abnormal levels of corrosion have not been a problem in properly designed and operated gas treating facilities using DGA agent. Tube failures in the regenerator reboiler and reclaimer have generally been traced to poor operations or chloride contamination. Using a heating medium above the maximum recommended temperature will, in most cases, cause regenerator reboiler tube failure.

Maximum spot heat flux should generally be limited to approximately 9000 BTU/Hr/Ft^2 for carbon steel and 12000 BTU/Hr/Ft^2 for stainless steel reboiler and reclaimer tube bundles. Insufficient heat to the regenerator has also caused reboiler tube failures. Inadequate heat input to the regenerator reboiler can result in the stripping of acid gas on reboiler tubes. The flashing of acid gas from solution on reboiler tube surfaces will result in a pitting type corrosion and subsequent tube failure. Sufficient regeneration heat should be provided to the regenerator reboiler to maintain a reflux ratio of at least 1.0-1.5 moles of water per mole of acid gas in the regenerator overhead to minimize the quanity of acid gas released on regenerator reboiler tubes.

DGA agent reclaimer tube failure has generally been due to either excessive heat flux rates, poor reclaiming procedures, or buildup of chlorides. One major cause of reclaimer tube failures has been the result of a poor "boil-out" procedure where, prior to dumping, the kettle fluid was distilled overhead without the continuous addition of water. Without the continuous addition of water, the liquid level in the reclaimer decreases, progressively uncovering the tube bundle while the solids and salts are being concentrated in the remaining liquid.

Regenerator reboiler vapor return nozzles and lines have experienced high corrosion rates when velocities of the vapor were excessive. This is an erosion-type of corrosion and can be minimized by the installation of multiple vapor return outlets and larger diameter vapor return lines.

ASSOCIATED OR SEPARATOR GAS APPLICATIONS

DIGLYCLOLAMINE agent is a primary amine which makes it an ideal treating agent for low pressure, high temperature, associated gas. The largest associated gas plants which utilize DGA® are located in Saudi Arabia. At the present time nine plants are in operation with a total capacity in excess of 4.5 billion SCF/Day. These DIGLYCOLAMINE plants are some of the largest in the world. The majority of the associated gas being treated in Saudi Arabia has a molecular weight as high as 28 with total acid gas concentrations of 10-15%. The associated gas is available at low pressures, ranging from 110-180 psig. The inlet gas temperatures range from 110-125°F. Using DGA, all of the associated gas plants operating at pressures above 80 psig are producing a residue gas containing less than 0.25 grains of hydrogen sulfide/100 SCF. No problems are being encountered in producing bright yellow sulfur from the acid gas via the Claus process.

DIGLYCOLAMINE agent is used in Saudi Arabia for the removal of CO_2 and H_2S from a very rich associated gas. It is interesting to note that hydrocarbon contents of the acid gas are typically less than 0.5 mole %. This confirms that hydrocarbon solubility is basically no more a problem with DGA than for MEA or DEA. Where hydrocarbon contents in the acid gas are high, it is usually the result of entraining liquid and gaseous hydrocarbons in the rich amine and the absence of a rich DGA flash tank. The installation of a filter coalescer upstream of the contactor is a necessity to prevent liquids from entering the DGA facility. The lean DGA temperature must be maintained 5-10°F above the hydrocarbon dew point of the residue gas. Always install the necessary nozzles and connections on the rich amine flash tank and reflux accumulator to permit the removal of any hydrocarbon liquids which may accumulate over a period of time.

Typical associated gas plant conditions are shown in Table 3.

Acid Gas Loadings

Most of the DGA units which are removing hydrogen sulfide and carbon dioxide from associated gas are

Table 3
Associated Gas Sweetening with DGA

Range of Feed Gas Compositions:	
	1-50 Mole % H_2S
	1-14 Mole % CO_2
Range of Treating Pressures:	
	15-400 PSIG
Range of Feed Gas Temperatures:	
	80-130°F

operating with a net pick up of from 0.2-0.35 moles of total acid gas/mole DGA across the contactor. Residual acid gas loadings of the lean DGA solutions vary from 0.02 to 0.10 moles of total acid gas/mole DGA. The acid gas loading of the lean solution varies inversely with the reflux ratio and pressure used in the DGA regenerator. Most associated gas plants are operating at moderate to low CO_2:H_2S ratios.

Intercoolers on DGA absorbers have been installed in some plants to reject part of the heat of reaction. The partial removal of the heat of reaction between the acid gas and DGA reduces the equilibrium partial pressures of H_2S and CO_2 over the rich DGA solution. This results in both extending the loading range and improving the equilibrium conditions in the absorber. One DGA plant in Saudi Arabia, which uses intercoolers, has reduced the H_2S in the residue gas to less than 0.25 grains of H_2S/100 SCF and less than 100 ppmv of CO_2 with rich DGA acid gas loadings as high as 8.5 SCF/gallon. Using a 50 wt.% solution, this is equivalent to a gross acid gas loading of 0.55 moles of total acid gas per mole of DGA. The use of intercoolers can be particularly beneficial when insufficient acid gas partial pressure is available to permit normal solution loadings.

Regeneration Heat Requirements

Because of the low treating pressures and unfavorable $CO_2:H_2S$ ratios in associated gas streams, it may be necessary to operate the DGA regenerators at greater reflux ratios than for a typical high pressure natural gas plant. Most of the DGA plants processing associated gas are operating with reflux ratios in the range of 1.2-3.0:1, in the regenerator. The reflux ratio needed is determined by the $CO_2:H_2S$ ratio, the total acid gas content, the temperature and pressure of the sour feed gas to the absorber, and the treated gas specifications.

Reclaimer Operation

Reclaiming operations in associated gas plants are basically the same as those in natural gas plants. Reclaimers are operating with flow rates ranging from 0.5 to 2% of the DGA circulation rate. Problems encountered in associated gas DGA plants have been similar to those experienced in natural gas facilities.

Materials of Construction and Corrosion

Most major equipment and piping are constructed of carbon steel. Stainless steel is usually used for services such as centrifugal pump shafts and impellers, pressure letdown valves and control valve internals. The general use of stainless steel for absorber and regenerator tray decks and valves has become quite common in recent years for many of the aqueous amine sweetening systems.

Carbon steel is usually satisfactory for lean-rich DGA heat exchanger tube bundles if the temperature of the rich DGA leaving the exchanger is limited to 190°F. Stainless steel should be used if rich amine preheat exceeds 190°F. Even stainless steel becomes ineffective in controlling corrosion due to acid gas flashing from solution if the rich DGA from the exchanger is over 220°F. Regenerator reboiler tube bundles designed for heat flux rates exceeding recommended limits for carbon steel will usually be constructed of stainless steel. Carbon steel is satisfactory for reclaimer tube bundles if heat flux rates are limited. If a build up of chlorides in the reclaimer kettle solution is anticipated, do not use the austenitic stainless steels, such as 304 and 316, which are subject to attack by chlorides. If stainless steel must be used to minimize problems due to excessive heating medium temperatures, use a chloride resistant stainless steel.

DIGLYCOLAMINE Agent Losses

The causes of DGA losses from associated gas plants are the same as those experienced in natural gas plants. It is imperative that housekeeping and plant maintenance be emphasized at all DIGLYCOLAMINE agent sweetening facilities. Some of the largest associated gas plants in operation have reduced DGA losses to levels of 3-5 lbs. DGA/MM SCFD.

REFINERY APPLICATIONS

Because of the low treating pressures involved in refinery applications, DIGLYCOLAMINE agent is ideally suited for removing H_2S and CO_2 from refinery streams. At the present time, five domestic refineries are using DGA, one of which is Texaco's refinery located at Convert, LA. Two more are scheduled for startup during 1984. DIGLYCOLAMINE is currently being used in concentrations ranging from 25 to 50 wt.%. Concentrations of DGA above 50 wt% have been avoided to reduce the solubility of hydrocarbons in the treating solution. All of the refineries are operating satisfactorily and producing bright-yellow sulfur. Table 4 lists typical refinery sweetening conditions.

Table 4
Refinery Sweetening with DGA

Range of Feed Gas Compositions:	
	0.5-10 Mole % H_2S
	0-4 Mole % CO_2
Range of Treating Pressures:	
	50-1400 PSIG
Range of Feed Gas Temperatures:	
	90-125°F
Typical Treated Gas Quality	
	H_2S, ppmv <130
	CO_2, ppmv >100

Acid Gas Loadings

Most refineries are operating with gross acid gas loadings varying from 0.31 to 0.40 moles total acid gas/mole DGA. All vapor-liquid applications are reducing the H_2S concentration to less than 130 ppm.

Regeneration Heat Requirements

Most refinery applications are using reflux ratios of 1.5-3.0 moles water/mole acid gas in the regenerator overhead. In refinery applications, the CO_2:H_2S ratio, as well as the treating pressure, is usually low. These conditions necessitate the use of larger quantities of heat for regeneration as compared to high pressure natural gas treating facilities.

Reclaimer Operations

Reclaiming of DGA solutions in refinery applications is an important aspect of maintaining good operations. Refinery gas streams can contain such compounds as hydrogen cyanide, thiosulfuric acid, naphthenic acids, formic acids, carbonyl sulfide, carbon disulfide, etc. These compounds will form either heat stable salts or degradation products which can contribute to abnormal DGA losses and corrosion problems. Therefore, it is necessary to continuously remove these heat stable salts and degradation products from the circulating solution. The addition of low chloride soda ash to the reclaimer to react with the heat stable salts will liberate DGA which will be distilled into the circulating DGA solution. Reclaiming cycles vary from one month to as long as six months depending upon the concentration of undesirable compounds in the various gas streams.

Materials of Construction

Most DGA systems in refinery applications use carbon steel as the material of construction. Only in special cases is 304 stainless steel recommended.

DIGLYCOLAMINE Agent Losses

In most refinery applications, three to eight contactors are in service with one regenerator. Because of the wide variation in feed composition, feed rates, pressures, and physical location, the losses of treating agent have been much larger than in natural gas or associated gas plants. In fact, one of the most difficult problems encountered in refinery applications is reducing losses. Losses in refinery applications have ranged from 7 to 20 gallons of DGA per MM SCF acid gas removed per day. These losses are not unique to DGA treating; they occur regardless of the amine being used.

Corrosion

Corrosion has not been a problem in refinery applications. The one exception was abnormal corrosion in a regenerator. And, in this case, the mechanical design of the regenerator was responsible for the abnormal corrosion.

LIQUID-LIQUID TREATING

DGA is used in concentrations ranging from 25-50 wt% for the removal of H_2S and CO_2 from liquid hydrocarbon streams. Most plants are using a water wash downstream of the contactor to minimize DGA entrainment and solubility losses. Liquid-liquid treating systems, using the Malaprop™ process, have been installed to reduce the carbonyl sulfide content of light liquid hydrocarbon streams. One liquid-liquid treating system is reducing the carbonyl sulfide content of propane from 300-1200 ppm to <2 ppm. The design of liquid-liquid treating systems for the removal of carbonyl sulfide must include sufficient capacity in the DGA agent reclaimer to convert the additional degradation product back to DGA. The liquid-liquid treating conditions for the removal of COS, CO_2, and H_2S are listed in Table 5.

Table 5
Liquid-liquid Sweetening with DGA

	Plant A	Plant B	Plant C
Inlet Composition			
COS, ppmw	300-1200	Nil	
H_2S, ppmw	Nil	Nil	25,000
CO_2, ppmw	Nil	8300	9,000
Type Hydrocarbon	Propane	Ethane-Propane	Mixed Olefin and Aliphatic
Outlet Composition			
COS, ppmw	<2	Nil	
H_2S, ppmw	Nil	Nil	<5*
CO_2, ppmw	Nil	<100*	?
Pressure, PSIG	350	325	200
Temperature, °F	90-100°F	90-100°F	100°F

*Lower limit of Detection

Acid Gas Loadings

Gross acid gas loadings of rich DGA solutions are typically limited to about 0.35 moles of total acid gas/mole DGA. Higher loadings in the range of 0.6 moles total acid gas per mole of DGA have been seen, but are not recommended because of possible high corrosion rates. Since most liquid-liquid systems are operated in conjunction with vapor-liquid systems, the residual acid gas content of lean DGA solutions has varied from 0.02 to 0.10 moles total acid gas/mole DGA.

Regeneration Heat Requirements

As previously mentioned, most liquid-liquid treating systems are operated in conjunction with vapor treating systems. Because of the wide variation in the $CO_2:H_2S$ ratio, regeneration heat requirements vary widely.

Reclaiming Operation

Reclaiming of DGA solutions used in liquid-liquid treating is important to keep the solution free of contaminants. Reclaimers in liquid-liquid treating systems are operating at feed rates of 0.5 to 1.0% of the circulating solution. Systems which contain significant quantities of carbonyl sulfide require higher reclaiming rates based on the quantity of carbonyl sulfide entering the system.

Materials of Construction

The liquid-liquid treating systems now in operation are using carbon steel as the material of construction.

Corrosion

There have been no significant corrosion problems reported in the liquid/liquid treating systems.

ACKNOWLEDGEMENT

The authors wish to acknowledge the management of Texaco Inc. for permission to publish this paper.

LITERATURE REFERENCES

1. Anon.: "GPA Explores H_2S Removal Methods", Oil Gas J. 76, No. 29, 66, 71-73 (1978).

2. Anon.: "Berri NGL Plant Nearly Complete", Oil Gas J. 75, No. 20, 82-84 (1977).

3. Bucklin, Robert W.: "DGA - A Workhorse for Gas Sweetening", Oil Gas J. 80, No. 45, 193-197 (1982).

4. Bucklin, Robert W.: "Removal of Hydrogen Sulfide from Natural Gas by DIGLYCOLAMINE", Proc. Annu. Conv. - Gas Processors Assoc. 57, 145-146 (1978).

5. Dingman, J.C. and Moore, T.F.: "Compare DGA and MEA Sweetening Methods", Hydrocarbon Process. 47, No. 7, 138-140 (1968).

6. Dingman, J.C.: "Gas Sweetening with DIGLYCOLAMINE Agent" Presented at Third Iranian Congress of Chemical Engineering, Pahlavi University, Shiraz, Iran. Nov. 6-10, 1977.

7. Dingman, J.C., Jackson, J.L., Moore, T.F. and Branson, J.A.: "Equilibrium Data for the H_2S-CO_2-DIGLYCOLAMINE Agent-Water System" Proc. 62nd Annual Convention of Gas Processors Assoc., March 14-16, 1983, pages 256-268.

8. Freireich, E. and Tennyson, R.N.: "Process Improves Acid Gas Removal, Trims Costs, and Reduces Effluents", Oil Gas J. 74, No. 34, 130-132 (1976).

9. Goar, B. Gene: "Today's Gas-Treating Processes-1", Oil Gas J. 69, No. 28, 77-79 (1971).

10. Griffith, T.E.: "Gas Sweetening in Field Operations", Gas 46, No. 5, 77-70 (1970).

11. Harbison, J.L. and Dingman, J.C.: "Mercaptan Removal Experiences in DGA Sweetening of Low Pressure Gas", Proc. Gas Conditioning Conf. 1972: C.A. 83 82356w.

12. Hikita, H., Asai, S., Ishikawa, H, and Honda, M.: "The Kinetics of Reactions of Carbon Dioxide with Monoisopropanolamine, DIGLYCOLAMINE, and Ethylenediamine by a Rapid Mixing Method", Chem. Eng. J. (Lausanne-Neth.) 14, 27-30 (1977).

13. Holder, Howard, L.: "DIGLYCOLAMINE - A Promising New Acid Gas Remover", Oil Gas J. 64, No. 18, 83-86 (1966).
13. Holder, Howard, L.: "DIGLYCOLAMINE - A Promising New Acid Gas Remover", Oil Gas J. 64, No. 18, 83-86 (1966).
14. Huval, Malcolm and Van de Venne, Harry: "DGA Proves Out as a Low-Pressure Gas Sweetener in Saudi Arabia", Oil Gas J. 79, No. 3, 91-103 (1981).

15. Kohl, Arthur L. and Riesenfeld, Fred C.: Gas Purification, 2nd Ed. Gulf Publishing Co., Houston 1974.

16. Martin, Joel L., Otto, Frederick D., and Mather, Alan E.: "Solubility of Hydrogen Sulfide and Carbon Dioxide in a DIGLYCOLAMINE Solution", J. of Chem. & Eng. Data, 23, No. 2, 163-164 (1978).

17. Mason, J.R. and Griffith, T.E.: "MEA-DGA Switch Saves Steam", Oil Gas J. 67, No. 23, 67-69 (1969).

18. McClure, George and Morrow, David C.: "Amine Process Removes COS from Propane Economically", Oil Gas J. 77, No. 27, 106-108 (1979).

19. Tennyson, R.N. and Schaaf, R.P.: "Guidelines Can Help Choose Proper Process for Gas Treating Plants", Oil Gas J. 75, No. 2, 78-80, 85-86 (1977).

20. Weber, Steve and McClure, George: "New Amine Process for FCC Desulfurized Light Liquid Streams", Oil Gas J. 79, No. 23, 160-163 (1981).

21. Williams, W.W.: "Treatment of Gas Plant Liquids with DIGLYCOLAMINE", Proc. Gas Conditioning Conf. 1973; C.A. 81 123945s.

12

FLEXSORB Gas Treating: New Amine Technologies

A. M. Goldstein
Manager, Hydrotreating and Gas Processes Division
Exxon Research and Engineering Company

A. M. Edelman
Manager, Gas Treating Licensing
Exxon Research and Engineering Company

P. A. Ruziska
Chief Engineer, Agricultural Chemicals, Technology Division
Exxon Chemical Company

A new family of gas treating agents is now available which offers significant energy and investment credits relative to conventional processes. Three technologies have been developed by Exxon Research and Engineering Company scientists and engineers and are now available for licensing. These technologies are FLEXSORB SE, FLEXSORB PS, and FLEXSORB HP. The technologies are based on a new class of amines developed by ER&E, called hindered amines (1). Each of the technologies uses a separate hindered amine designed to economically handle a specific gas treating application commonly found in the petroleum, petrochemical, and production industries.

FLEXSORB SE agent selectivity removes H_2S and is a significant improvement over other selective H_2S removal technologies, such as aqueous MDEA. FLEXSORB PS achieves the removal of CO_2 plus H_2S in an economically preferred manner. It is an improvement over other amine-solvent systems now available. Lastly, FLEXSORB HP for the removal of CO_2 offers advantages over the common hot potassium carbonate gas treating systems. Each of these systems can be run in conventional absorber-regenerator equipment and is readily retrofitted to obtain energy savings or debottlenecking credits. This paper describes each of these technologies.

FLEXSORB SE AGENT

FLEXSORB SE is the new, proprietary gas treating agent for the selective removal of H_2S (2). Technology for selective removal of H_2S is economically preferred when there is a need to remove essentially all of the H_2S and there is no process requirement to remove the other acidic gaseous components such as CO_2. This type of service is encountered in both refinery and producing applications, including Claus sulfur plant tail gas cleanup units, FLEXICOKING low BTU gas desulfurization, enhanced oil recovery operations, and the enrichment of acid gases from natural gas plants.

The FLEXSORB SE molecule features good H_2S selectivity. It is capable of high acid gas loadings and good cleanup. The molecule has high solubility in water and low volatility. Excellent stability is obtained because there is negligible direct CO_2-amine reaction. Therefore, no amine degradation reactions are observed. These generally proceed through a CO_2-amine intermediate.

Development of FLEXSORB SE Agent

FLEXSORB SE was invented by Exxon Corporate Research scientists. As shown in Figure 1, its development then passed into a series of laboratory, pilot plant, and engineering studies. This work established a process data base used to define economic incentives and potential applications for FLEXSORB SE. These studies indicated that FLEXSORB SE was a potentially attractive replacement for selective H_2S removal processes now in commercial use including both

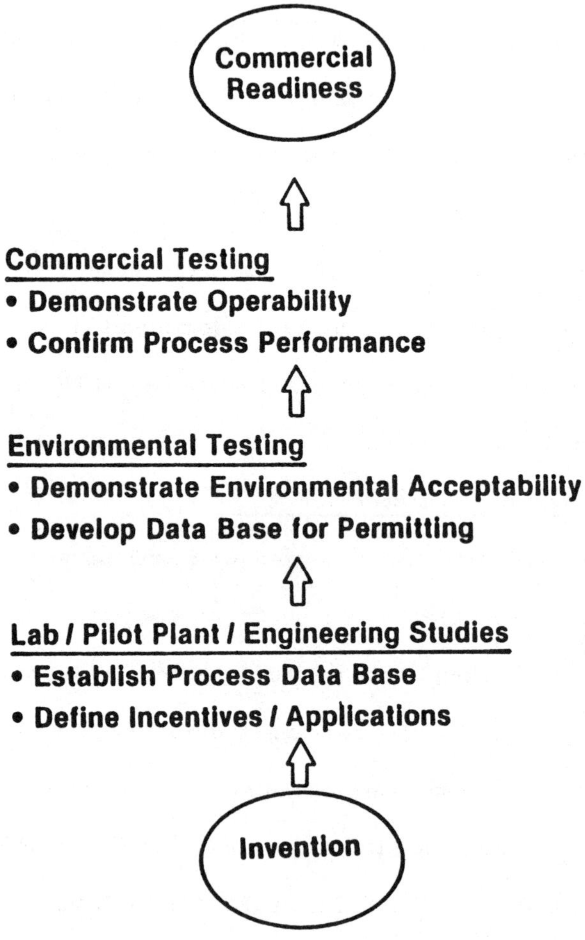

Figure 1. FLEXSORB SE commercial readiness achieved.

methyldiethanol amine (MDEA) and direct conversion processes. The pilot plant work, which was carried out at the Exxon Research and Development Laboratories in Baton Rouge, Louisiana, generated a substantial data base in large pilot plant equipment. The process data base covered a wide range of conditions including amine circulation rate, regeneration steam to amine ratio, gas pressure, and composition.

A series of environmental tests were then undertaken to demonstrate the environmental acceptability of the new amine and to prepare a data base for eventual environmental permitting. This work demonstrated that FLEXSORB SE is comparable to other amines commonly used in industrial gas treating practice as to to its environmental impact. FLEXSORB SE is biodegradable and can be handled with waste treatment facilities and processes commonly found in a refinery environment.

The preceding work had continued to demonstrate the attractiveness of FLEXSORB SE in selective H_2S service. Hence, the decision was made to pursue a commercial test in a refinery setting. The objective was to demonstrate the operability of FLEXSORB SE in a commercial operating environment and to confirm the scale-up of the process from the pilot plant stage. The following sections discuss applications for selective H_2S removal and present the commercial test data and incentives for FLEXSORB SE.

Applications for selective H_2S removal

Figure 2 shows that FLEXSORB SE can be used in both refining and producing applications. One refinery application is the desulfurization of sour, low BTU gas produced in Exxon's FLEXICOKING petroleum residuum conversion process. The process objective is to sweeten the low BTU gas prior to its combustion within the refinery, thereby reducing sulfur emissions.

Claus sulfur plants are used to convert sour gases into elemental sulfur. Conversion of about 95% of the incoming sulfur species to elemental sulfur is common. Where required by environmental constraints, it is general practice to utilize a tail gas cleanup unit to achieve an overall sulfur capture of about 99+%. Tailgas cleanup using selective amine treating is one of the more commonly used processes. This process

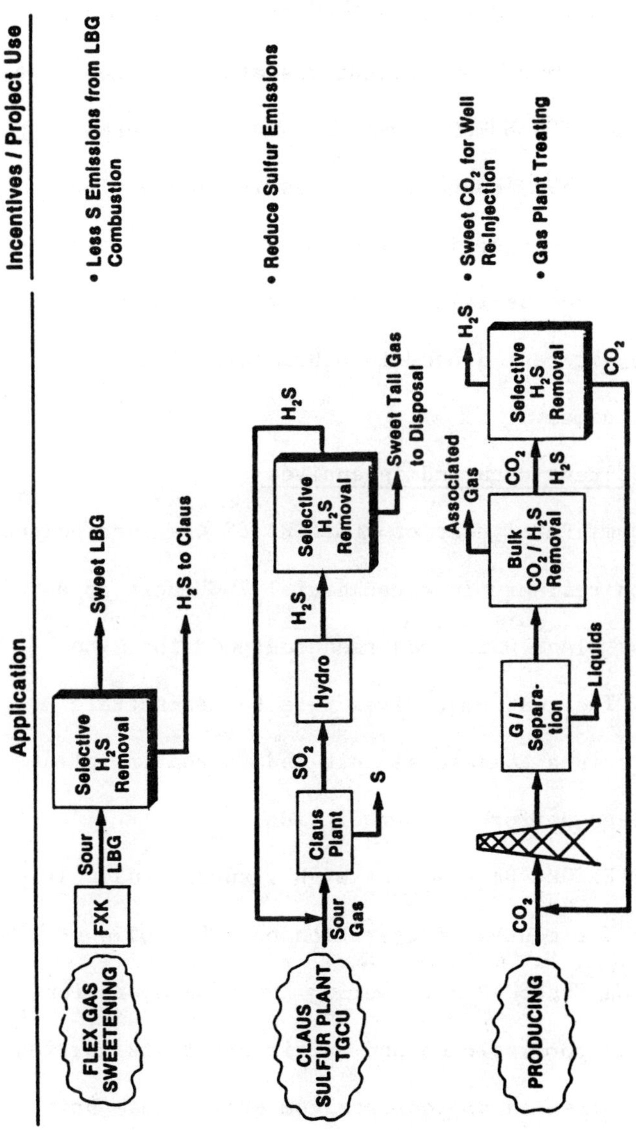

Figure 2. FLEXSORB SE for selective H_2S removal.

involves hydrogenation of sulfur species in the Claus tail gas to H_2S and its absorption via a selective H_2S removal step and recycle back to the Claus plant.

Several producing applications are feasible, also. For example, FLEXSORB SE can be used in an enhanced oil recovery (EOR) operation to desulfurize the CO_2 used in well re-injection, thereby mitigating corrosion and safety concerns related to the recycle of the H_2S. Also, it can be used in other natural gas treating schemes.

FLEXSORB SE test data and incentives

The commercial test of FLEXSORB SE was carried out at an Exxon refinery in a commercial TGCU unit on a large Claus plant which had been designed for MDEA operation. The test objectives were to demonstrate operability in a commercial unit and to confirm pilot plant process performance predictions.

The FLEXSORB SE test was highly successful. It involved three months of operation on MDEA followed by operation on FLEXSORB SE. During the MDEA operation, the unit was commissioned and sufficient tests carried out to establish an appropriate commercial base point for ultimate comparison with FLEXSORB SE. All test objectives for FLEXSORB SE were achieved. No corrosion problems were indicated and no foaming problems encountered. Operators reported that the amine unit was stable and required no special attention. Hence, operability was proven. In addition, pilot plant performance and the process model predictions derived from the pilot plant were confirmed. The unit has continued in operation with FLEXSORB SE and is in its second year with the new solvent.

Figure 3 shows the test data. FLEXSORB SE shows a significant capacity advantage versus MDEA. FLEXSORB SE requires a 40% lower circulation rate to achieve the same degree of cleanup as does MDEA. The FLEXSORB SE solution is significantly more dilute than the MDEA solution. The data shown are at the same regeneration steam to solution ratio. Hence, the 40% capacity advantage translates to about a 40% energy savings.

As mentioned earlier, we obtained extensive pilot plant data prior to the commercial test. These data were modeled using a fundamental mass transfer model and predictions made as to the anticipated commercial

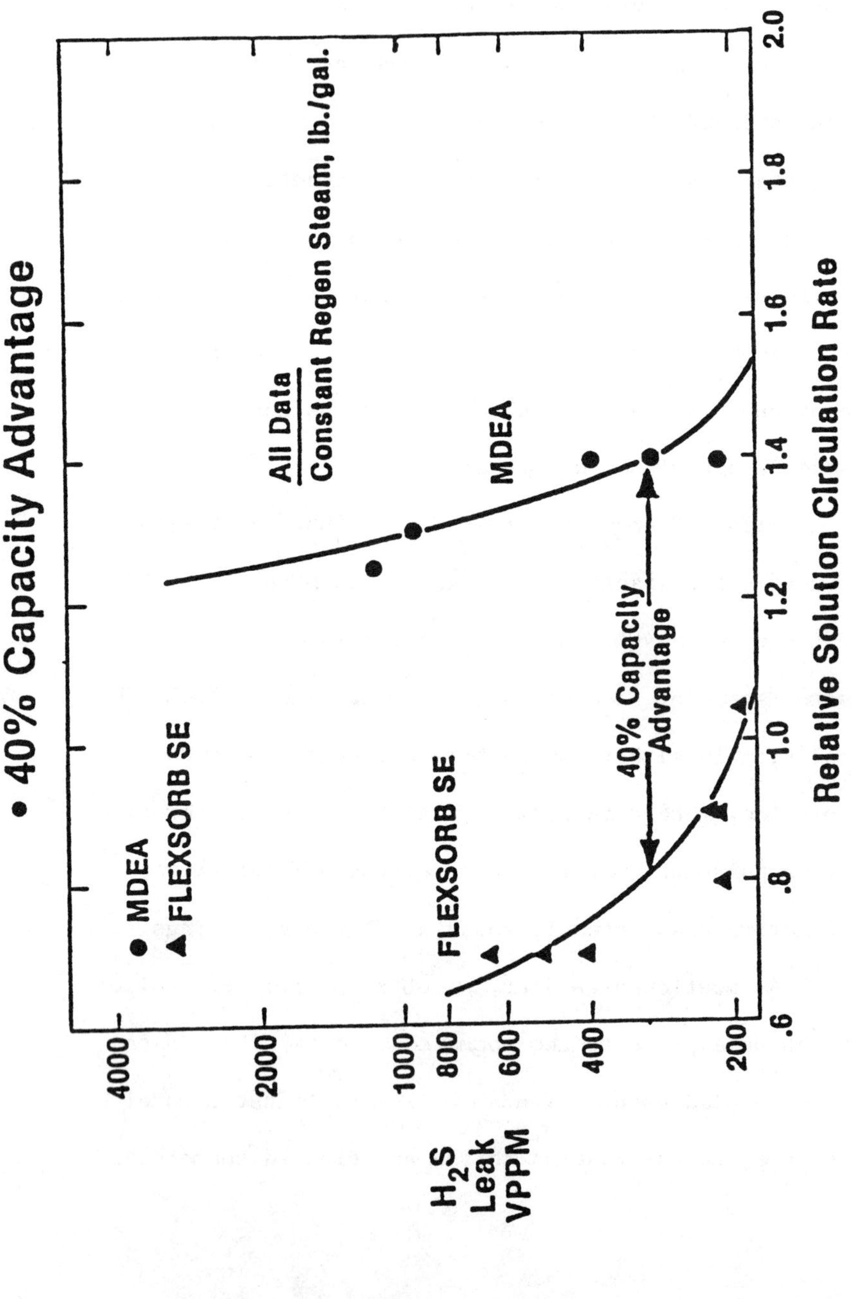

Figure 3. FLEXSORB SE commercial data shows expected advantage.

performance. This parity plot in Figure 4 compares the pre-test predicted H_2S leak (abscissa) versus the commercial measurements. Excellent agreement was achieved over a wide range of process conditions.

Engineering studies have shown significant incentives for FLEXSORB SE. Use of FLEXSORB SE for desulfurization of FLEXICOKING low BTU gas and Claus sulfur plant tail gas can result in major reduction in investment and operating costs for these gas treating facilities. Figure 5 shows an example of such an application in which 123 T/D of sulfur are being removed.

The figure compares FLEXSORB SE with MDEA. FLEXSORB SE requires 41% of the MDEA circulation rate and 51% of its energy for regeneration. This lower flow rate contributes to the regenerator being both smaller in diameter and shorter in height. The net effect is more than a 25% reduction in investment and more than a 40% reduction in operating costs. Because of advantages such as these, the technology has been selected for use in several new projects and retrofits.

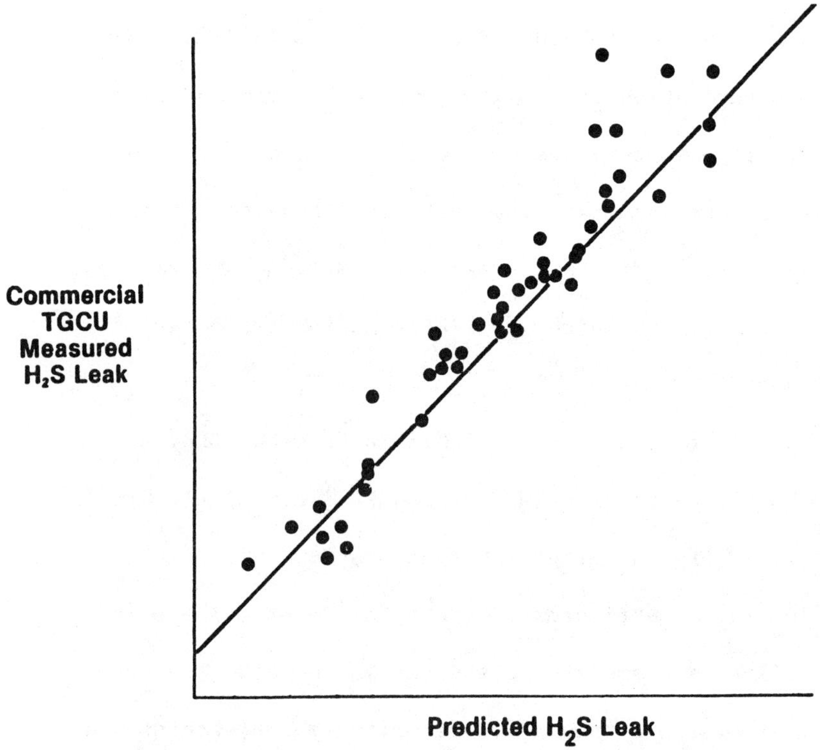

Figure 4. Pilot plant-based model accurately predicts FLEXSORB SE commercial process performance.

BASIS:

- Desulfurization of 280 MSCF/D low joule gas
- Desulfurization of 35 MSCF/D Claus tail gas
- Total sulfur removed, 123 T/D

	FLEXSORB SE, % of MDEA
• Amine circ rate, gpm	41%
• Steam rate, KLB/HR	51%
• Regenerator	
—Diameter, ft.	69%
—Height, ft.	87%
• Investment, M$	74%
• Operating Cost, M$/yr	57%

Figure 5. FLEXSORB SE is economically attractive.

FLEXSORB PS AGENT

FLEXSORB PS is superior technology for bulk removal of CO_2 and H_2S. It is an improvement over conventional amine-solvent processes in several important areas. First, it exhibits higher capacity for acid gas removal which results in smaller equpment for grass roots designs. Second, FLEXSORB PS agent has significantly greater stability than conventional amine-solvent processes. This enhanced stability results in significantly lower amine degradation and lower amine makeup costs. Third, improved acid gas removal capability is achievable with the FLEXSORB PS agent. Lastly, it can be used for retrofits; easy debottlenecking or energy savings can be achieved on existing units. Most, or all, of these credits also are achievable in units where conventional processes are used for removal of CO_2 in the absence of H_2S.

Figure 6 shows the significant capacity advantage enjoyed by FLEXSORB PS versus conventional amine-solvent technology. As shown here, use of FLEXSORB PS results in a 30% reduction in amine circulation rate. Alternatively, the improved

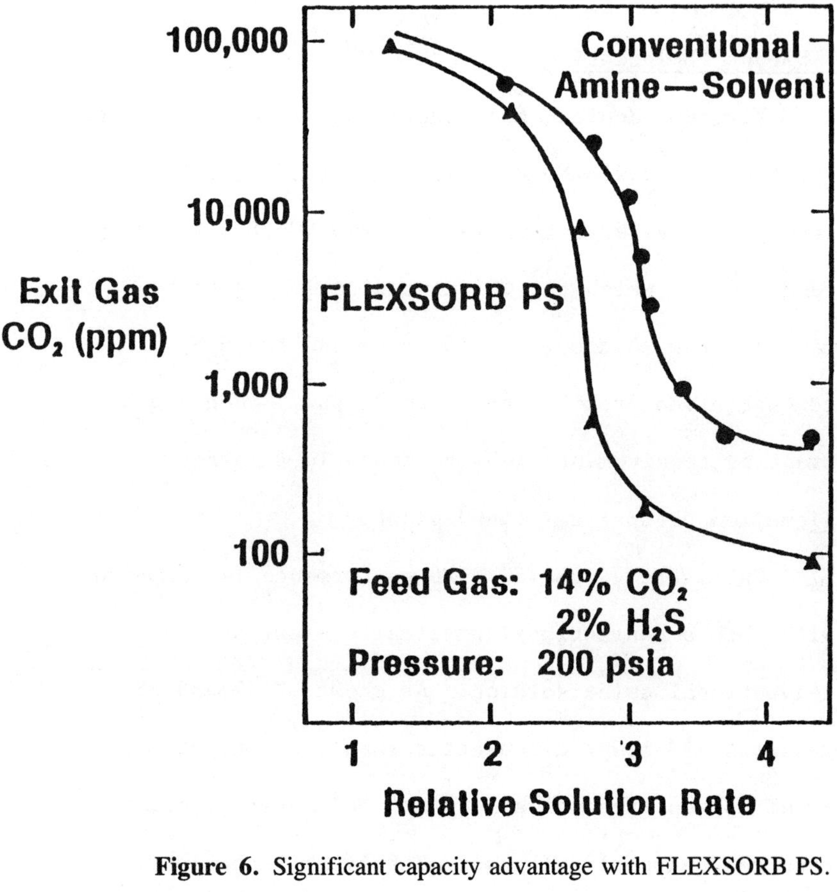

Figure 6. Significant capacity advantage with FLEXSORB PS.

performance of FLEXSORB PS can be used to achieve significantly lower exit gas CO_2 (and/or H_2S) concentrations, when there are economic incentives to do so.

FLEXSORB PS Incentives

Figure 7 depicts the significant economic credits for the use of FLEXSORB PS in a natural gas plant. The case is for a large gas plant (1.7 G SCF/SD, 800 psig feed gas) in which the CO_2 concentration is to be reduced from 18.5 mol & to 50 vppm and the H_2S concentration from 0.1 mol % to 4 vppm. Such a gas treating requirement might be found in a large liquefied natural gas (LNG) plant.

On Figure 7, FLEXSORB PS performance is compared with both a conventional physical solvent and a conventional amine-solvent. As shown, FLEXSORB PS requires 24% lower circulation than the conventional amine-solvent technology and and 98% lower circulation than the conventional physical solvent technology.

In a grass roots design, the lower circulation rate would result in significantly smaller equipment. Equipment savings of 34 and 58 M$, respectively, were

- **Large Plant**

BASIS: 1700 MSCF/SD, 850 psig Feed Gas (Grass Roots)

	Feed Mole %	Treated Gas VPPM
CO_2	18.5	50
H_2S	0.1	4

	Conventional Amine – Solvent Technology	Conventional Physical Solvent Technology	FLEXSORB PS
PROCESS			
• Solution GPM	124% Base	198% Base	Base
• Steam	104% Base	15% Base	Base
• Power	120% Base	290% Base	Base
ECONOMICS			
• Investment, M$	+34	+58	Base
• Operating Costs, M$/Yr	+6.3	+27.8	Base

CONCLUSION: FLEXSORB PS provides energy savings and investment credits in grass roots applications

Figure 7. FLEXSORB PS has significant economic credits in natural gas plants.

calculated for FLEXSORB PS relative to these two technologies. Also, energy usage is significantly reduced. Annual operating costs for FLEXSORB PS are 6.3 M$ and 27.8 M$ per year lower than the amine-solvent technology, and physical solvent technology, respectively, typically found in the industry today. These lower operating costs are a direct result of the reduced circulation rate which is possible with FLEXSORB PS.

FLEXSORB PS is much more stable than conventional amine-solvent technology, thereby affording additional savings from both reduced inventory replacements and reclaimer operation. FLEXSORB PS is in commercial use in an Exxon plant for the treatment of a chemical synthesis gas.

FLEXSORB HP AGENT

FLEXSORB HP agent is a significant improvement over conventional promoted hot potassium carbonate technology for the removal of CO_2 (3). Promoted hot carbonate processes commonly are used downstream of hydrogen plant steam reformers and for CO_2 removal in ammonia plants.

FLEXSORB HP has a 30% higher cyclic capacity than does conventional technology as shown in Figure 8. It can be retrofitted to provide debottlenecking or energy savings. In grass roots applications, FLEXSORB HP agent offers significant operating cost and investment reduction credits.

Figure 9 provides an economic comparison of FLEXSORB HP agent vs. conventional promoted hot potassium carbonate technology for CO_2 removal in a steam reforming hydrogen plant. The hydrogen plant produces 20 MSCF/SD of hydrogen. CO_2 is reduced from 19.9 mol% to 0.20 mol%. Lower CO_2 concentration in the treated gas can be achieved if desired. FLEXSORB HP, in this example, requires about 19% lower circulation and 11% lower reboiler duty than conventional promoted hot potassium carbonate technology. These process advantages result in an investment savings of about 1 M$ for a grass roots application and an annual operating cost reduction of 0.4 M$ per year.

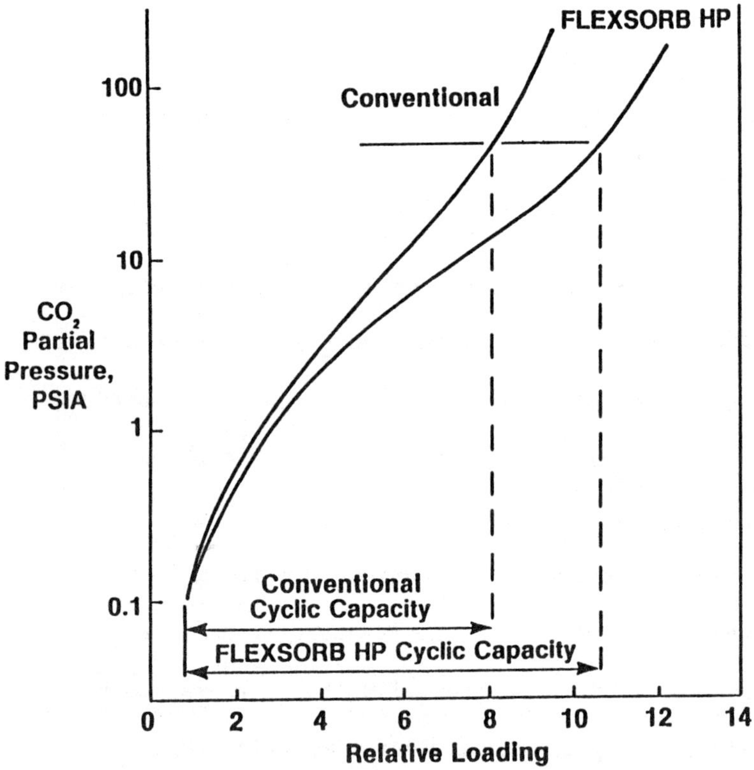

Figure 8. FLEXSORB HP has 30% higher capacity than conventional technology.

BASIS: 20 MSCF/SD H_2 Plant (Grass Roots)
250 psia Absorber

	Mole %	
	Feed	Treated Gas
CO_2	19.9	0.2

	Conventional Hot K_2CO_3 Technology	FLEXSORB HP
PROCESS		
• Solution GPM	119% Base	Base
• Reboiler Duty	111% Base	Base
ECONOMICS		
• Investment, M$	+1	Base
• Operating Cost, M$/Yr	+0.4	Base

CONCLUSION: FLEXSORB HP provides energy savings and investment credits in grass roots applications

Figure 9. FLEXSORB HP has significant economic credits in hydrogen plants.

In retrofit applications, using equipment sized for conventional processes, energy credits would likely be greater than the above. This technology is currently in the pilot plant operation stage and, following successful development, will be followed by a large scale commercial demonstration.

SUMMARY

Exxon has made significant progress in developing and commercializing new gas treating agents. FLEXSORB SE is a major improvement in the selective removel of H_2S and is being widely applied. FLEXSORB PS offers significant advantages over conventional amine-solvent technology. Its first commercial application is proceeding well and other applications are under study. Lastly, the first application of FLEXSORB HP for the bulk removal of CO_2 is planned for 1984. These technologies are available for license to the industry.

References

1. Weinberg, H. N., et al, "New Gas Treating Alternatives for Saving Energy in Refining and Natural Gas Production," presented at the 11th World Petroleum Congress, London, August 1983.

2. Goldstein, A. M., et al, "Commercialization of a New Gas Treating Agent: FLEXSORB SE," presented at the 1984 Spring National AIChE Meeting, Anaheim, California, May 1984.

3. Say, G. R., et al, "Development of a New Hindered Amine-Promoted Hot Carbonate Scrubbing Solution for Simultaneous CO_2 and H_2S Removal from Gases," presented at AIChE meeting, Atlanta, March 1984.

technically sophisticated unit to minimize operating costs.

The Benfield Process is based on the use of activated, inhibited hot potassium carbonate solutions to scrub CO_2, H_2S, and other acid gas components. There have been many innovations incorporated in the Benfield Process since its initial development over 20 years ago, and improvements continue to be developed today which will keep it at the forefront of acid gas removal technology. One such new development is an inorganic activator system, which is not sensitive to oxygen concentrations of ten percent and higher. This has resulted in highly efficient performance in treating blast furnace off-gas, petrochemical off-gases, and other oxygen-bearing streams.

Another recent innovation is the Benfield LoHeat Process in several configurations. Using ejectors, the net heat consumed in a Benfield unit can be reduced by over 30 percent in a CO_2 removal unit in a typical hydrogen or ammonia plant. Using mechanical vapor compression, LoHeat allows reducing the regeneration heat required in an ammonia plant CO_2 removal unit to as low as 15,000 Btu/lb-mole of CO_2 scrubbed, a reduction of about 70 percent. Process innovations such as this are attractive not only in Benfield grass roots units and revamps but as low-cost conversions from other carbonate systems such as "Catacarb" and "Vetrocoke" units.

The following sections of this paper provide an overview of Benfield Process technology, including background, chemistry, flowsheets, and recent process improvements. In addition, generalized cost curves are presented which allow preliminary economic assessment of the various Benfield options.

BACKGROUND OF THE PROCESS

The original hot potassium carbonate process was developed at the U. S. Bureau of Mines by Benson, Field, and co-workers in the early 1950's (1, 2, 3, 4). Their work was prompted by the need to economically remove acid gases from coal-derived synthesis gas. Upon leaving the Bureau of Mines, Benson and Field formed the Benfield Corporation which was later acquired by Union Carbide Corporation.

Research and development work by the Benfield Corporation and Union Carbide has significantly improved the economics and extended the applicability of the original process. The addition of proprietary activators to enhance the reaction kinetics and to alter the vapor-liquid equilibrium relationship has led to reduced tower sizes, a greater degree of purification, and improved thermal efficiency. The development of effective corrosion inhibitors and control methods has practically eliminated corrosion problems so that mostly carbon steel construction can be used.

Additional innovations have been on the process equipment to reduce tower sizes and enhance the thermal efficiency of the process (5, 6). An extensive development program was undertaken to evaluate over 30 commercially available tower packings for mass transfer efficiencies in Benfield service, using research equipment larger in size than some commercially operating units.

Various flowschemes were developed such as the Benfield HiPure Process, which allows very high purity product gases (7). Other recent developments have led to selectively removing sulfur from gases containing both CO_2 and H_2S (8), and to an economical method of purifying low pressure gases.

The Benfield LoHeat Process has allowed reducing the regeneration heat requirements by an additional 25 to 50 percent over already reduced consumptions (9). The latest flowsheets under development can reduce operating expenditures to near the range of efficiency presently enjoyed only by physical solvents, particularly when considering process gas losses frequently encountered in physical solvent systems, such as "Selexol" process.

Benfield-designed acid gas removal systems are employed in over 520 units either operating or in the design/construction phase. Industrial activities which use one variation or another of the Benfield Process include:

1. production of ammonia;
2. purification of hydrogen for refinery use;
3. purifying natural gas to pipeline quality or prior to cryogenic liquification;
4. production of ethylene oxide;
5. production of vinyl acetate;
6. purifying coal-derived synthesis gases, etc.

To indicate the application and flexibility of the process: gas stream pressures in operating Benfield units range from under 690 kPa (100 psia) to near 14,000 kPa (2000 psia); acid gas contents in feed streams from a few percent to 50 percent; and feed gas volumes from less than 28,000 sM^3/d (1 MMSCFD) in some semicommercial units to 935,000 sM^3/d (330 MMSCFD) per train in some large multiple train units purifying natural gas.

PROCESS CHEMISTRY

The Benfield Process is based on chemical reactions of potassium carbonate and acid gases, which can be represented as follows:

$$K_2CO_3 + CO_2 + H_2O \rightleftharpoons 2\ KHCO_3 + \text{heat} \qquad (1)$$

$$K_2CO_3 + H_2S \rightleftharpoons KHCO_3 + KHS + \text{heat} \qquad (2)$$

Because these are reversible reactions with the potassium carbonate, we must consider not only the kinetics of the reactions but also the degree of conversion from carbonate to bicarbonate and the heat of reaction. The following are a few of the key Benfield Process features.

1. The heats of reaction for Equations 1 and 2 above are less than half those of comparable amines used for acid gas removal. This means less heat is generally required for regeneration of HPC.

2. These chemical reactions are specific for CO_2 and H_2S, thus the process inherently gives a clean separation of the acid gases from other feed components in most industrial gas streams. The entire volume of scrubbed acid gases is available at nearly 100 percent concentration for sulfur processing or for recovery of CO_2 at high purity.
3. As the absorber pressure and acid gas partial pressure increases, the rate of reaction increases and solution absorption capacity increases until at very high pressure near-equilibrium conditions are achieved. Further increases in pressure will not increase the solution capacity, which is a maximum of 4 to 8 SCF of acid gas per gallon of solution. This is due to the chemical nature of the absorption, as opposed to physical solvents.
4. Valuable process gases including hydrogen, nitrogen, and hydrocarbons are essentially insoluble in the Benfield solution. Hence, there is no loss of these components as compared to physical solvent systems such as "Selexol" where

losses can be severe.

Other gases may also be present in trace quantities and need to be removed. These include COS, CS_2, HCN, NH_3, and RSH (mercaptains), all of which are scrubbed by Benfield solution with no loss of efficiency in removal of the bulk acid gases CO_2 and H_2S. COS and CS_2 are hydrolyzed by the HPC solution and are regenerated as mixtures of CO_2 and H_2S. See Reference 12 for further information.

Process Flowsheet Descriptions

Benfield hot potassium carbonate processes employ conventional packed or trayed towers for countercurrent contact of gas and liquid. Figure 1 shows a process flow diagram of the simplest configuration, a "single-stage" design. Note that there is no transfer of heat between lean and rich solutions; the absorber operates at nearly the same temperature as the reboiler, thus the designation hot potassium carbonate (HPC) process.

Benfield units with this flowscheme are usually designed for a purified gas with 1 to 5 percent CO_2 plus H_2S. Typical applications are purification of natural gas, substitute natural gases, or catalytic reactor

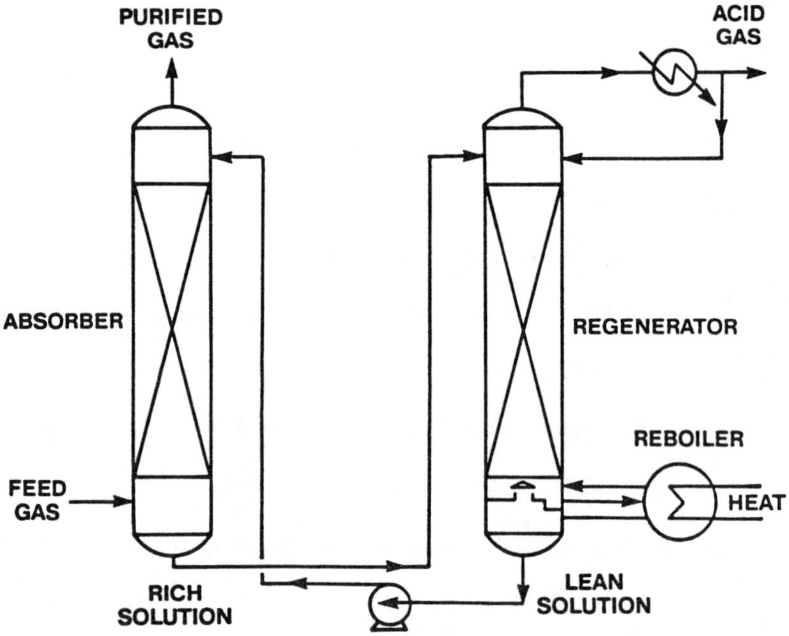

TYPICAL APPLICATIONS

Natural Gas

SNG

Ethylene Oxide

Figure 1. Benfield process single-stage absorber, typical purity to 1% CO_2.

recycle gas streams such as those used in ethylene oxide manufacture.

For a higher degree of purification, the flowscheme shown in Figure 2, a "split-flow absorber" design is used. The bulk of the hot, regenerated lean solution is introduced to the absorber at an intermediate point, while a small trim solution is cooled and sent to the top of the tower. The lower-temperature trim stream has a reduced equilibrium pressure which permits higher product gas purity to be achieved. Because the bulk of the absorber still operates hot, the rich solution stream is hot and regeneration efficiency is not impaired. Typical purities and process applications are noted.

A "two-stage" design can be used when still higher purity is required: see Figure 3. Here, the lean trim solution is more thoroughly regenerated in a second regeneration stage where it is contacted with all the regeneration steam. After cooling, the reduced equilibrium of CO_2 and H_2S permits higher product gas purities. This system is slightly more complex to operate and may require marginally more capital investment than the split-flow absorber design.

The Benfield HiPure Process is illustrated in Figure

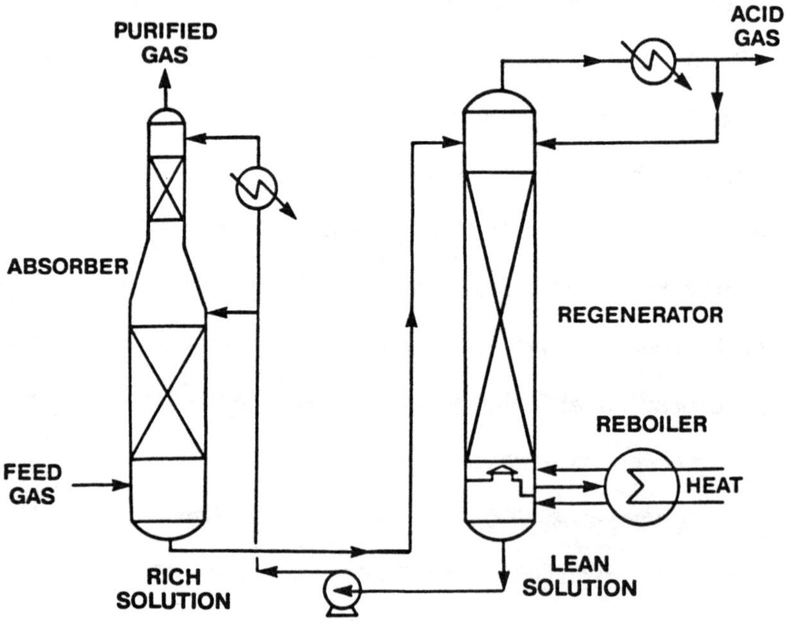

TYPICAL APPLICATIONS

Ammonia Synthesis Gas

Hydrogen

Natural Gas

Figure 2. Benfield process split-flow absorber, typical purity to 0.1% CO_2.

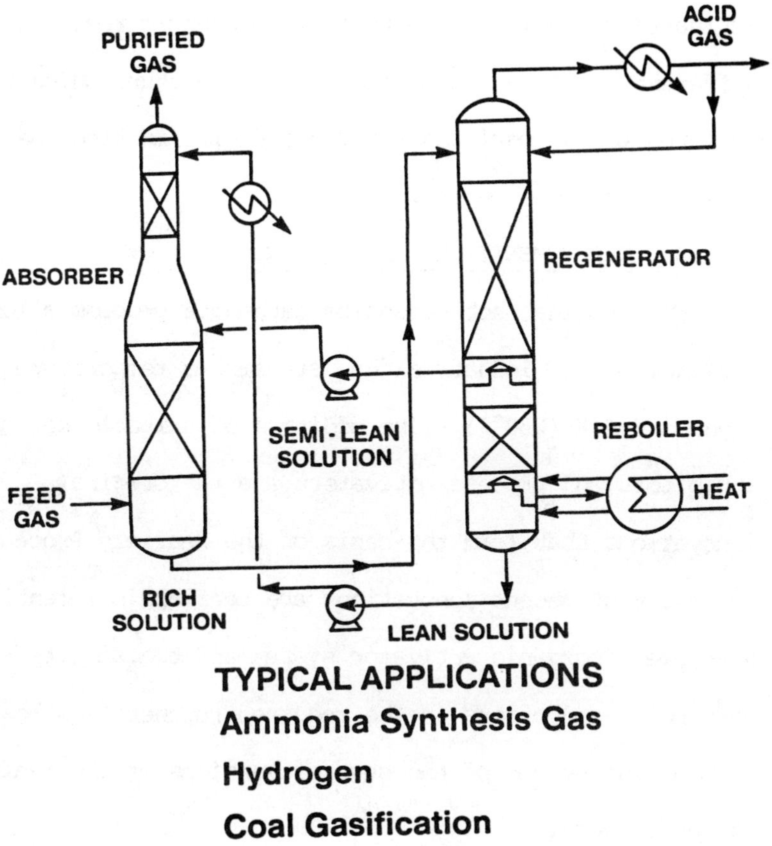

Figure 3. Benfield process Two-Stage Regenerator, typical purity to 0.05% CO_2.

4. This employs two independent solution circuits, arranged to have multiple use of the regeneration steam from the secondary circuit. The primary circuit employs activated HPC solution, with the secondary circuit using a different solution composition and temperature which are tailored to the purification required for meeting the final product gas purity.

PROCESS IMPROVEMENTS AND INNOVATIONS

The original hot potassium carbonate process allowed purification of industrial gas streams at temperatures as high as 394°K (250°F). The addition of suitable corrosion inhibitors and process activators are the original innovations that form the basis of the Benfield Process. The two most recent innovations are Benfield's recently developed inorganic activator system and the ability to make major reductions in the required regeneration heat consumption by use of the several versions of the Benfield LoHeat process.

Certain industrial gas streams needing CO_2 removal present special difficulties in that they contain oxygen in more than trace quantities. This is a highly oxidizing atmosphere which prohibits the use of most organic solvents or organic activators in HPC solutions, as the

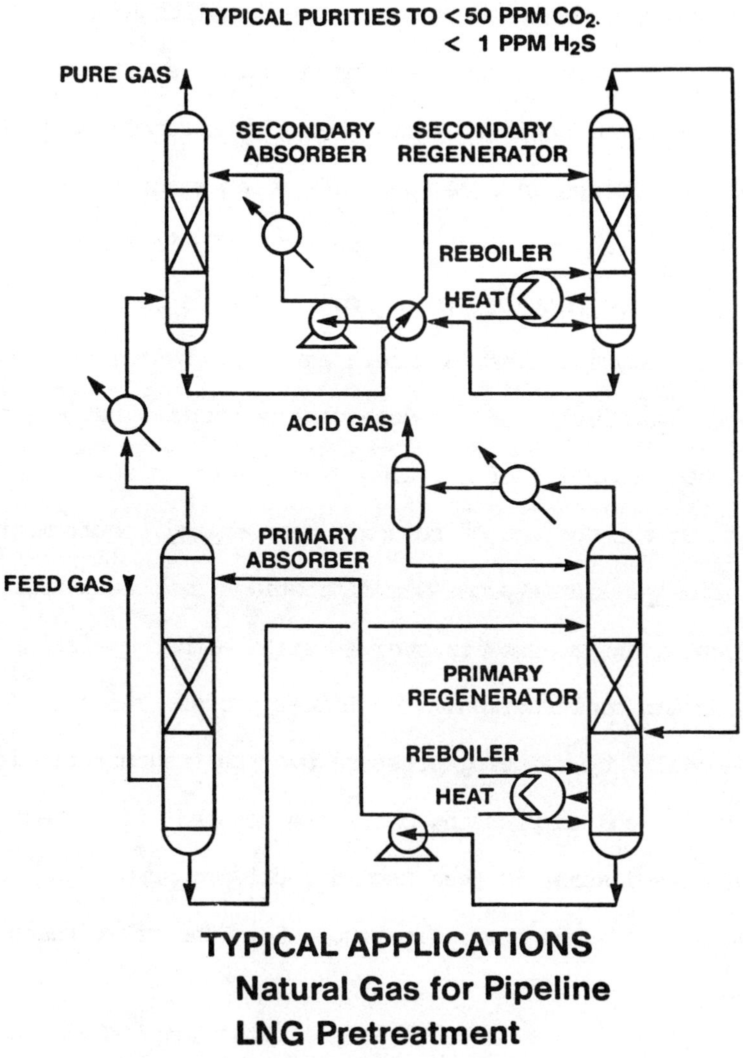

Figure 4. Benfield HiPure Process Unit.

organics would be rapidly oxidized and destroyed. Benfield has developed an inorganic activator for hot potassium carbonate which gives the benefits of activation (shorter towers, faster rate of reaction, etc.) and is not subject to oxidation. This patented activator system (10) has been proven in several plants which manufacture ethylene oxide or vinyl acetate where CO_2 is removed from reactor recycle streams containing as much as 5 to 10 percent oxygen. The inorganic activator system can be most effectively used in designing new units, but can also increase capacity of existing unactivated units.

On the subject of reducing heat energy requirements, Benfield's LoHeat process modifications have been well accepted and are now in over 40 units either operating or in design/construction. The LoHeat process recovers heat internally by reducing pressure over the regenerated lean solution and compressing the steam flashed off. The compressed steam is then used directly to supply part of the regeneration heat (See Figure 5). The net effects are:

1. Less outside heat required for the same regeneration and the same product gas purity.

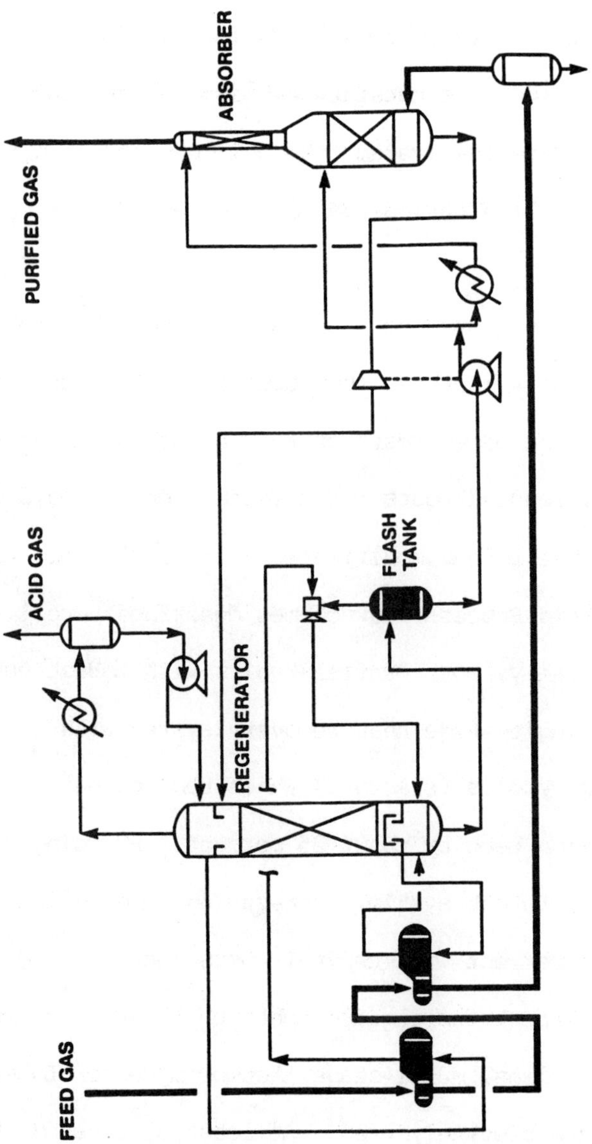

Figure 5. Benfield LoHeat Process for CO_2 removal, utilizing steam ejectors.

2. Required reboiler surface area is decreased, typically to two-thirds former size.
3. Since less outside heat is put into the unit to drive the system, the acid gas and lean solution cooling duties are also decreased.

For most grass roots units, incorporating LoHeat does not measurably increase the capital costs of acid gas removal. The added costs of the LoHeat equipment are offset by reduced costs for the reboiler and acid gas cooler. The LoHeat modifications are applicable to any of the Benfield Process flowschemes described here.

Initially, most Benfield units with LoHeat used a simple, single-stage heat recovery system with thermocompressors (ejectors) which reduced net regeneration heat by about 25 percent. By going to a multistage LoHeat system, heat savings can be increased to around 35 percent with capital costs increased by only about 1 percent for a grass roots unit. For the most current designs, multi-stage thermocompressor LoHeat systems are routinely used. At least seven such units are now operating with 8 to 10 more in design or under construction.

As energy costs have continued to increase, there has been greater acceptance of mechanical compressors in the Benfield LoHeat system. Compared to a non-LoHeat design, energy savings can be considerable: 40 to 70 percent of the regeneration heat can be recovered. Capital investment costs are increased by less than 10 percent, including the cost of the compressor and reduced sizes of reboiler and acid gas condenser. The Northern Illinois Gas Company SNG plant at Morris, Illinois operates the first two such units using Benfield LoHeat and reported the payout for the compressors as about 16 months (11). On a pure energy basis, electricity to operate the compressor is equivalent to some 7 percent of the Btu's in the recovered heat for the Illinois plant.

Following are a few examples of the significance of heat savings in Benfield acid gas removal units:

Example 1

In an ammonia plant, the steam-to-carbon ratio in the feed to the primary reformer can be reduced, from 4.0/1 to about 3.0/1, by using thermocompressors in a multi-stage Benfield LoHeat Process. Using a mechanical compressor, S/C can be 2.5/1 or even lower. Of the LoHeat units operating or under construction, two-thirds are in ammonia plants. One U. S. ammonia plant, built about 20 years ago, recently completed an extensive revamp for energy savings which included a LoHeat retrofit. The final design was for a 3/1 steam-to-carbon ratio with a 26 percent increase in CO_2 removal capacity. The unit is now operating with the increased capacity, and at a reduced CO_2 slip and much lower energy consumption than it was designed for 20 years ago.

Example 2

In natural gas purification units, product gas is consumed in generating steam for regeneration of the HPC solution. Depending on the feed gas pressure and the quantity of CO_2 and H_2S to be removed, regeneration heat requirements can be 35,000 to 45,000 Btu/lb-mole of acid gas removed (81-104 Mega-Joules/Kilogram-mole). With thermocompressors and Benfield LoHeat schemes, energy requirements can be reduced to 21,000 to 27,000 Btu/lb-mole (49-63 M-J/Kg-mole). With mechanical compression, our studies show regeneration efficiencies as low as 12,000 to 15,000 Btu/lb-mole (28-35 M-J/Kg-mole) are achievable with commercially available equipment.

Example 3

In a study for conversion from another proprietary

HPC process to the Benfield Process with LoHeat for purifying a refinery hydrogen stream, capacity could be increased by 25 percent without increasing the heat input or cooling duty. Thus, there was no need to replace the heat exchangers; and with only minor mechanical changes, the unit would operate at 125 percent of original CO_2 removal capacity.

BUDGETARY ESTIMATING

Feasibility and preliminary budget assessments of acid gas removal via the Benfield Process described here can be made using the following graphs. Note that this estimating procedure is based on the assumption that CO_2 is the major acid gas component to be removed.

Figure 6 indicates the approximate applicability ranges for the flowschemes previously described. For high sulfur loadings, the next more complex unit should be used. As with all acid gas removal processes, the key parameters are partial pressures of the acid gas components at inlet and outlet of the absorber.

Figure 7 indicates typical costs of Benfield units for acid gas removal. The costs indicated are for total battery limits capital costs of a Benfield unit installed at a U. S. Gulf Coast location. The costing basis is Oil

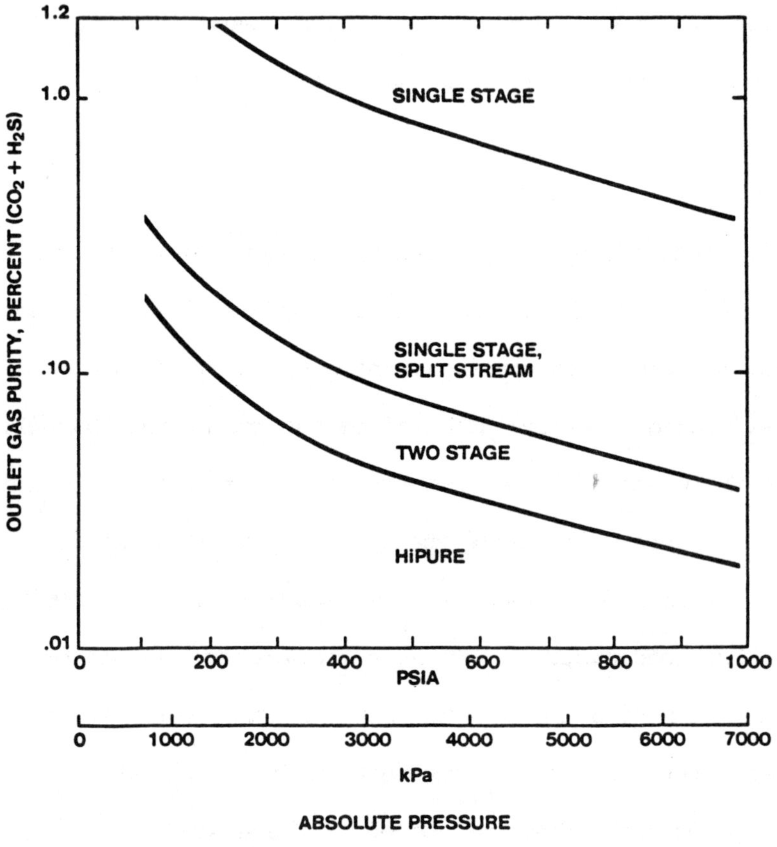

Figure 6. Benfield process, typical outlet purities using various Benfield flow schemes.

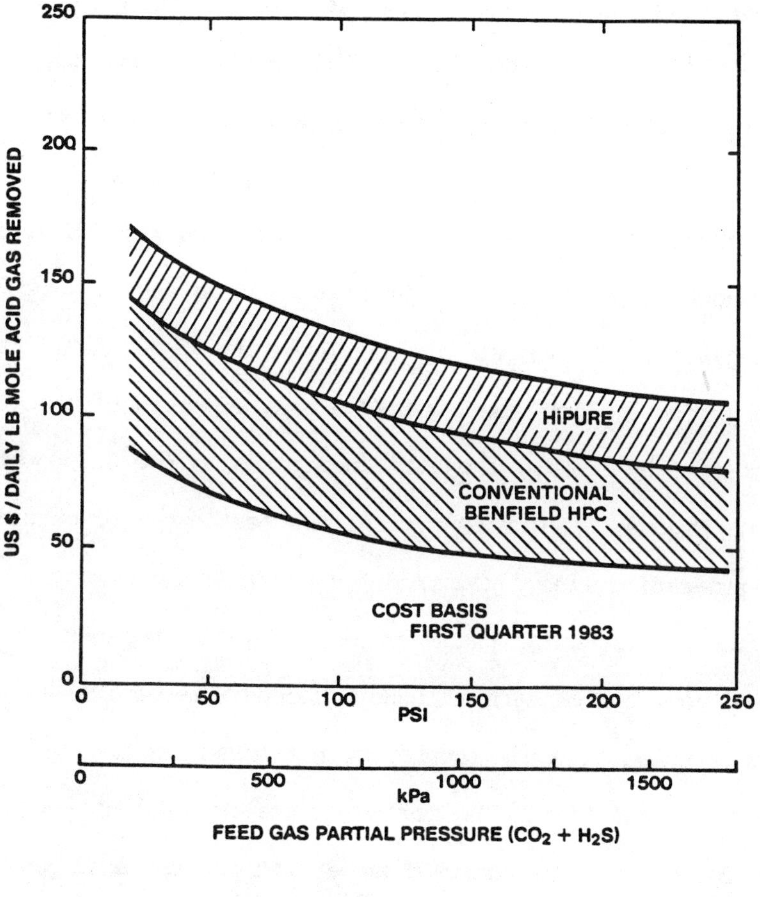

Figure 7. Benfield process, typical capital costs for installed unit.

and Gas Journal's Nelson Indices, updated for first quarter 1983. On this figure, the area indicated "Conventional Benfield HPC" includes single-stage, split-flow, and two-stage absorber units, with or without Benfield LoHeat. The bottom of the range is used for units with high partial pressure in feed and purified gas; the top of the range is used for low acid gas partial pressures or high sulfur contents to be removed. Similar comments apply to the HiPure range. These curves were prepared for units costing approximately $6,000,000. For larger or smaller sized units, a good approximation is obtained by scaling the estimated costs to the 0.7 power.

Figure 8 indicates typical net heat duty, or outside regeneration energy required for Benfield units as a function of partial pressure of acid gas components in the feed gas, and of type of LoHeat chosen: mechanical compression, steam ejectors, or non-LoHeat as a conventional unit. The bottom of the appropriate range is used for a single-stream absorber with a high acid gas slippage, and the upper portion of the range for a more complex unit or higher purity requirements or high sulfur contents. Note the requirements for 20 percent more heat for a HiPure unit.

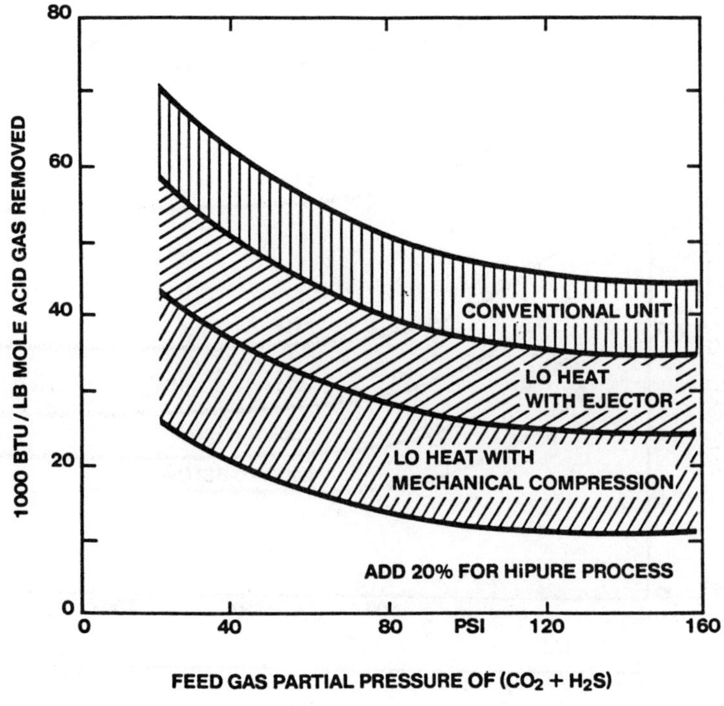

Figure 8. Benfield process, typical regeneration energy requirements.

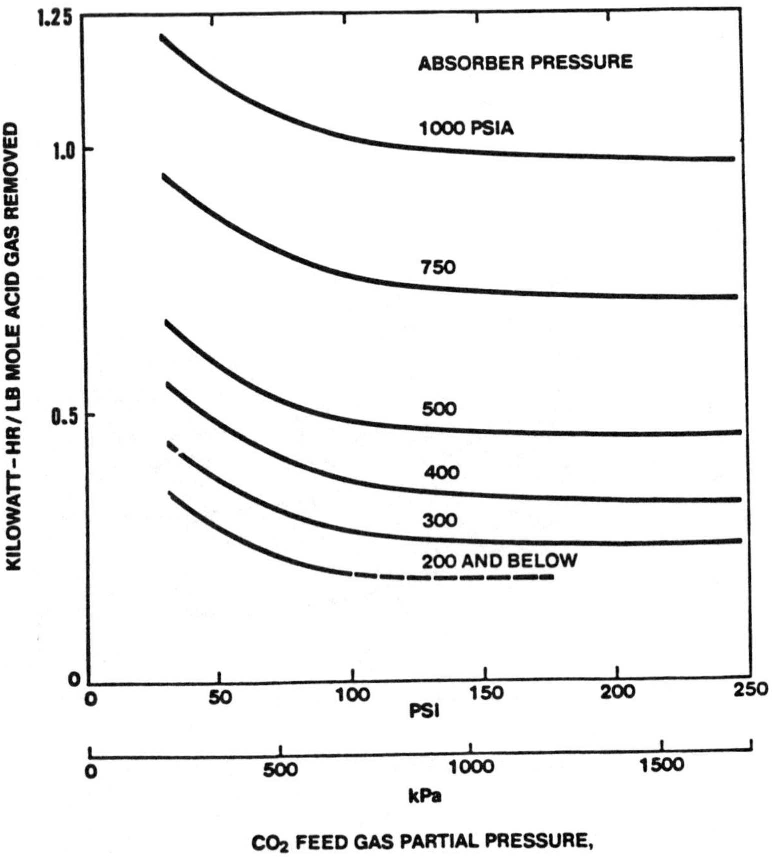

Figure 9. Benfield process, typical process pumping energy requirements.

Figure 9 indicates typical solution pumping energy requirements for Benfield Process units. This does not include any credit for power recovery from rich solution pressure let-down.

Estimates made based on these graphs cannot, of course, replace estimates prepared from a fully optimized design. They are, however, sufficiently accurate for initial feasibility and process screening of applicable hot potassium carbonate processes.

CONCLUSION

The Benfield Process for acid gas removal is a versatile process having found acceptance in many different industries and over a broad range of application. Commercial experience is demonstrated by over 520 units licensed to operate. With the improved thermal efficiency of the LoHeat modification, heat economics as low as 12,000 to 15,000 Btu/lb-mole (28 to 35 Mega-Joules/Kilogram-mole) of acid gas can be achieved.

REFERENCES

1. Benson, H. E., and Field, J. H., "Method of Separating CO_2 and H_2S from Gas Mixtures", U. S. Patent 2,886,405, May 1959.

2. Tosh, J. S., Field, J. H., Benson, H. E., and Haynes, W. P., "Equilibrium Study of the System Potassium Carbonate, Potassium Bicarbonate, Carbon Dioxide, and Water", Bureau of Mines Report of Investigations 5484, 1959, 23 pp.

3. Tosh, J. S., Field, J. H., Benson, H. E., and Anderson, R. B., "Equilibrium Pressures of Hydrogen Sulfide and Carbon Dioxide Over Solutions of Potassium Carbonate", Bureau of Mines Report of Investigations 5622, 1960, 25 pp.

4. Field J. H., Benson, H. E., Johnson, G. E., Tosh, J. S., and Forney, A. J., "Pilot Plant Studies of the Hot Carbonate Process for Removing Carbon Dioxide and Hydrogen Sulfide", Bureau of Mines Bulletin 597, 1962, 44 pp.

5. Benson, H. E., "Separation of CO_2 and H_2S from Gas Mixtures", U. S. Patent 3,563,695, February 1971.

6. Benson, H. E., "Separation of CO_2 and H_2S from Gas Mixtures", U. S. Patent 3,823,222, July 1974.

7. Benson, H. E., and Parrish, R. W., "HiPure Process Removes CO_2/H_2S", Hydrocarbon Processing, Vol. 53, No. 4, pp. 81-2.

8. Rib, D. M., Kimure, S. G., and Smith, D. P., "Coal-Gas Clean-up Facility", Environment Progress, Vol. 2, No. 2 (May 1983), pp. 86-90.

9. Baker, R. L., and McCrea, D. H., "The Benfield LoHeat Process: An Improved HPC Absorption Process", Paper Number 22b, AIChE 1981 Spring National Meeting, Houston, Texas, April 5-9, 1981.

10. Benson, H. E., and Field, J. H., "Separation of CO_2 From Gas Mixtures", Patent 3,907,969, September 1975.

11. Galstaun, L. S., McCrea, D. H., and White, R. J., "Recovery of Usable Energy from Low Level Heat Sources", Fuel Conversion Symposium, Ottawa, Canada, November 1977.

12. Parrish, R. W., and Neilson, H. B., "Synthesis Gas Purification Including Removal of Trace Contaminants", from 167th National Meeting of American Chemical Society, Los Angeles, April 1974.

14

The CATACARB Process for Acid Gas Removal

H. A. Gangriwala
Sr. Process Engineer
Eickmeyer & Associates

I-Meen Chao
Process Engineer
Eickmeyer & Associates

EXPERIENCE

Eickmeyer and Associates, the licensors of the CATACARB Process, have had experience in acid gas removal for over 35 years. They trouble-shooted most of the early Hot Potassium carbonate (HPC) installations, which were based on the process developed by the U.S. Bureau of Mines in the 1950s. During this period they became familiar with the advantages and also the drawbacks of the HPC process. The big advantage was the lower heat consumption than the amine process (1) and the drawback was that the absorption/desorption rates were slow. The equipment was, therefore, large. Moreover, the solution was somewhat corrosive. To overcome the drawbacks of the HPC process, Eickmeyer and Associates researched to find a catalyst and corrosion inhibitor. After 3 years of research and pilot plant trials the first commercial installation was put into operation in 1961 by Escambia Chemical Company. This plant is still operating successfully.

Soon thereafter, in 1963, Esso (now Exxon) took a worldwide license for the use of the CATACARB Process. Esso also tested the CATACARB Process in a sour natural gas treating plant and after an evaluation of the results applied it successfully to Esso's sour natural gas plants in Libya. The latest CATACARB application for Exxon has been at their Canadian ammonia plant. At 1600 MT/D ammonia capacity, this is probably the world's largest single train ammonia plant. Here Exxon has used the LOW-HEAT CATACARB technology to get a very low steam requirement of about 15000 BTU/lb-mol of CO_2 removed. Worldwide Exxon and its affiliates have 23 CATACARB plants.

In addition to Exxon, CATACARB has gained worldwide acceptance from the largest companies. Over 100 plants have been built in 27 countries for 51 companies by 18 major contractors to remove over 1.5 billion SCFD of acid gas to purify 10 billion SCFD of gas. Gas streams containing CO_2, H_2S and COS can be purified to contain less than 4 PPM H_2S and 0.005% to 2% CO_2.

Typical applications for the CATACARB Process include:

1. Natural gas purification
2. Substitute natural gas
3. Ethylene oxide recycle gas
4. Liquefied natural gas
5. Ammonia synthesis gas
6. Hydrogen production
7. Mole sieve regeneration gas
8. CO_2 from stack gas
9. CO_2 flood recycle
10. Amine conversions
11. Other HPC Process conversions

THE CATACARB PROCESS

The CATACARB Process employs a modified potassium salt solution containing a very active, stable and nontoxic catalyst and corrosion inhibitor as covered by U.S. Patents and foreign equivalents. The chemistry is very simple:

$$K_2CO_3 + CO_2 + H_2O \rightleftharpoons 2\ KHCO_3 + \text{heat}$$

$$K_2CO_3 + H_2S \rightleftharpoons KHCO_3 + KHS + \text{heat}$$

The Process is also capable of removing trace amounts of other acid gases such as COS, CS_2 and RSH.

Several catalysts and corrosion inhibitors are used in the CATACARB Process. The choice depends on the gas compositions to be treated by the Process.

Catalysts

CO_2 Removal. This has been the most common application of the Process and our standard catalyst is used. The effect of this catalyst on the pure K_2CO_3 solution is dramatic as indicated in Figure 1 (2).

CO_2 + H_2S Removal. In this application the proportion of the components used in the previous catalyst is altered to increase the H_2S removal capacity of the solution. In the extreme case the Process becomes capable of removing H_2S alone, even in the absence of CO_2. In this case the Process is not truly a potassium carbonate based process but instead uses potassium borate.

Oxygen bearing gases/ethylene oxide plants. For this application a special CATACARB catalyst has been developed and patented (3). It has been in successful commercial application for over 3 years. This, being an inorganic catalyst, is not oxidized/degraded by oxygen in the gas. Also since organics are absent the catalyst has essentially no

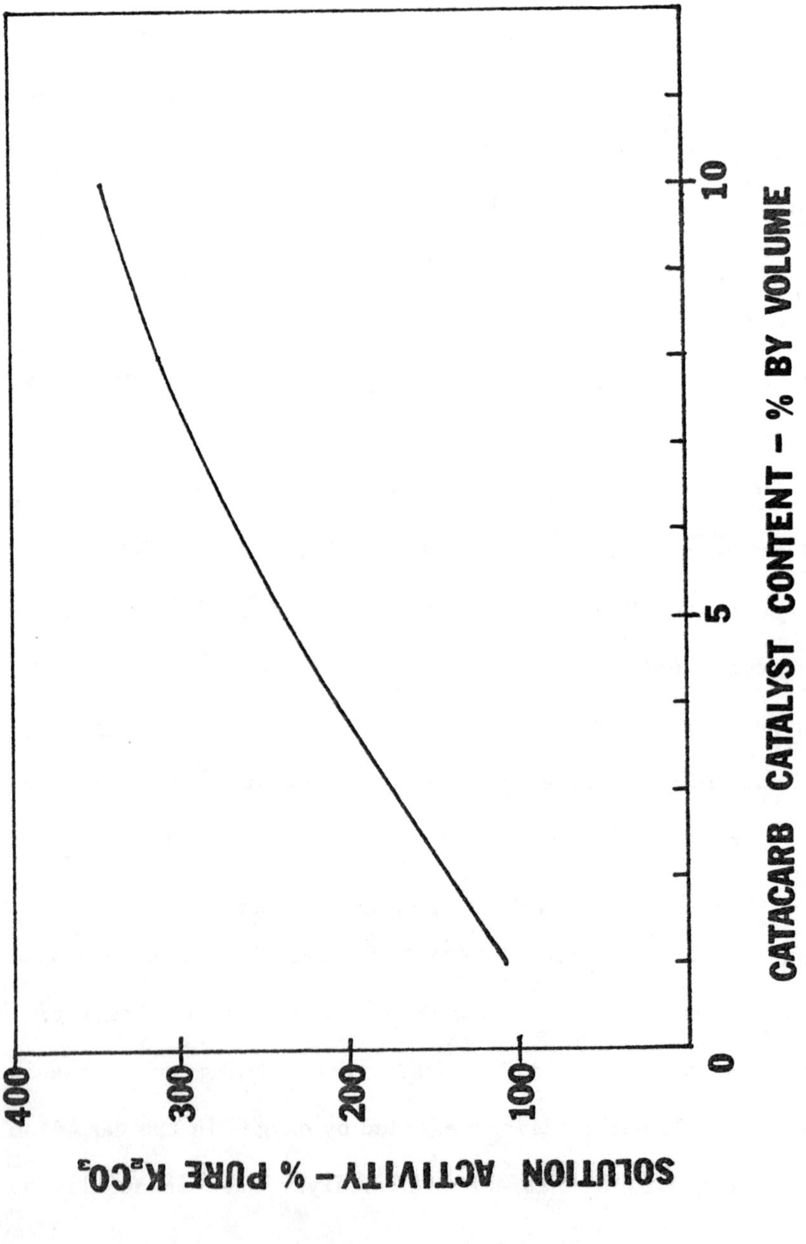

Figure 1. Activity of catalyzed carbonate solutions (2).

vapor pressure, which is of particular importance in ethylene oxide plants where the downstream E.O. catalyst could be poisoned by contamination with even trace amounts. This catalyst was specially developed for treating EO recycle gas and can be added directly to the existing HPC solution during operation to reduce steam consumption by 15-50% and/or increase CO_2 removal capacity.

Corrosion Inhibitor

CO_2 Removal. This application uses our standard CATACARB corrosion inhibitor which is vanadium based. For this inhibitor to be effective it requires to be oxidized by a proprietary oxidation scheme (4). It then forms a protective film on metal surfaces which prevents corrosion due to CO_2 attack permitting carbon steel to be used in most of the plant, except in areas of high erosion.

CO_2 + H_2S Removal. If H_2S is present in small amounts it requires a special inhibitor to prevent corrosion. Such an inhibitor has been developed for use in the CATACARB Process and has been successfully used for several years in CATACARB applications. Of course when H_2S is present in a large amount no corrosion inhibitor is required since the H_2S itself acts as a corrosion inhibitor by forming an iron sulfide film over the carbon steel surfaces, which in turn prevents further corrosion.

PROCESS CONFIGURATIONS

The CATACARB Process is very versatile and finds application over a broad range of acid gas partial pressures. The purity of the treated gas particularly affects the configuration to be used in the application of the Process.

Single Stage Design

A simplified flowsheet of this configuration is shown in Figure 2. This is the simplest CATACARB flowsheet in which the gas is scrubbed countercurrently by the lean solution in the absorber. The rich solution from the absorber is then regenerated countercurrent to the steam from the reboiler in the regenerator. Both the absorber and the regenerator can be either packed towers or trayed towers.

This flowsheet can be used when a very high purity is not required in the treated gas. In general, it can be used when the treated gas has a CO_2 partial pressure of about 2 PSI or more.

Split Cooled Lean Design.

A typical flowsheet is shown in Figure 3. In this configuration a small portion of the lean solution is cooled

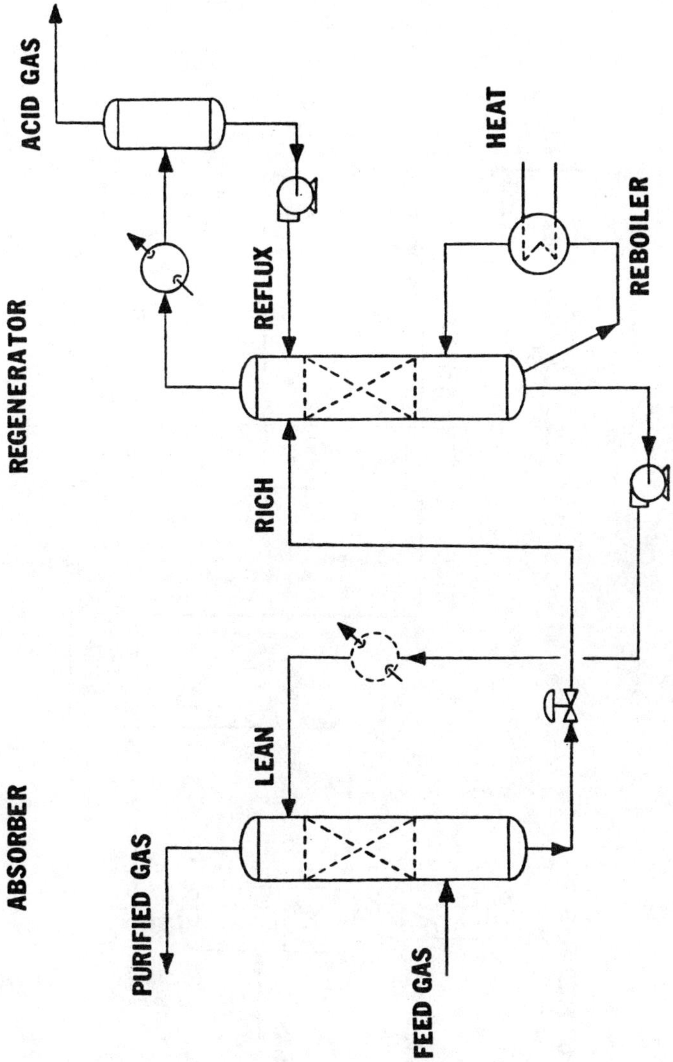

Figure 2. CATACARB single-stage design.

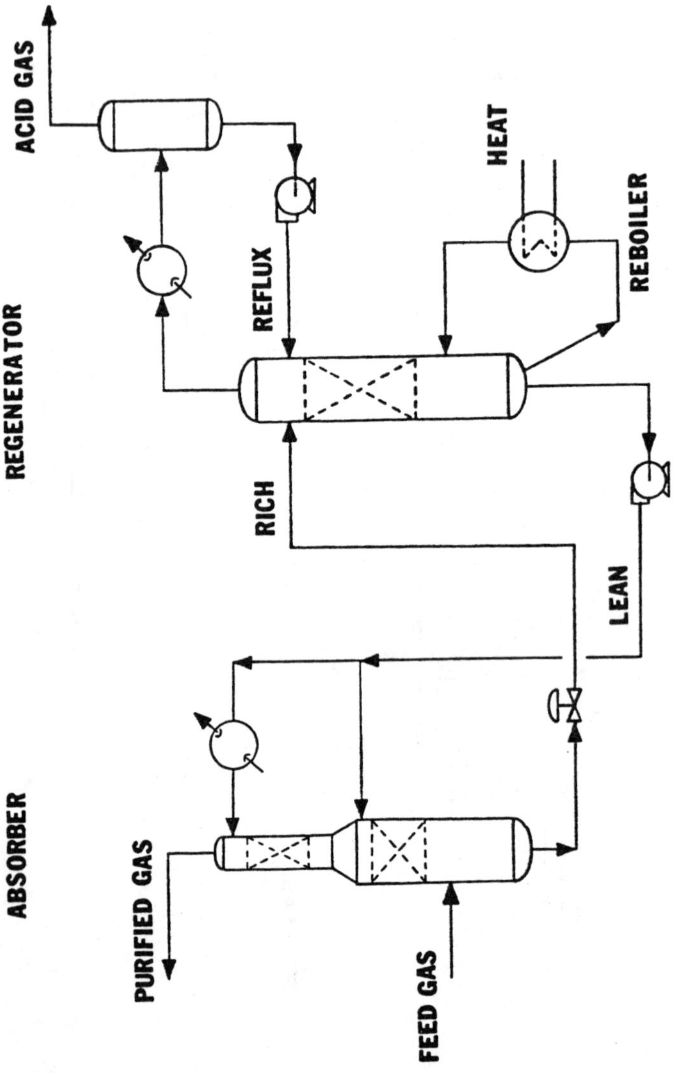

Figure 3. CATACARB split cooled design.

and sent to the top of the absorber permitting a greater purity of the treated gas. The bulk of the lean solution is sent hot, to the middle of the absorber thereby improving the heat efficiency of the system. The purity of the treated gas can be improved over the single stage design in this configuration. CO_2 partial pressure as low as about 0.3 PSI are achievable in the treated gas.

Two Stage Design

This is the most heat efficient configuration when a high purity of treated gas is required, such as 4 PPM H_2S and/or 50-1500 PPM CO_2. Two stages of absorption and regeneration are used as shown in Figure 4. The bulk of the solution is circulated as the semi-lean solution. A minor portion (about 15-20%) of the total solution is then thoroughly regenerated and cooled to be circulated as the lean solution.

In addition to these, other configurations are also available. These are used in specialized cases and are therefore not discussed in this paper. All of the above are conventional designs which require much less heat than the amine processes. The heat requirements of the process in any of these configurations can be dramatically improved by incorporating them with LOW-HEAT technology.

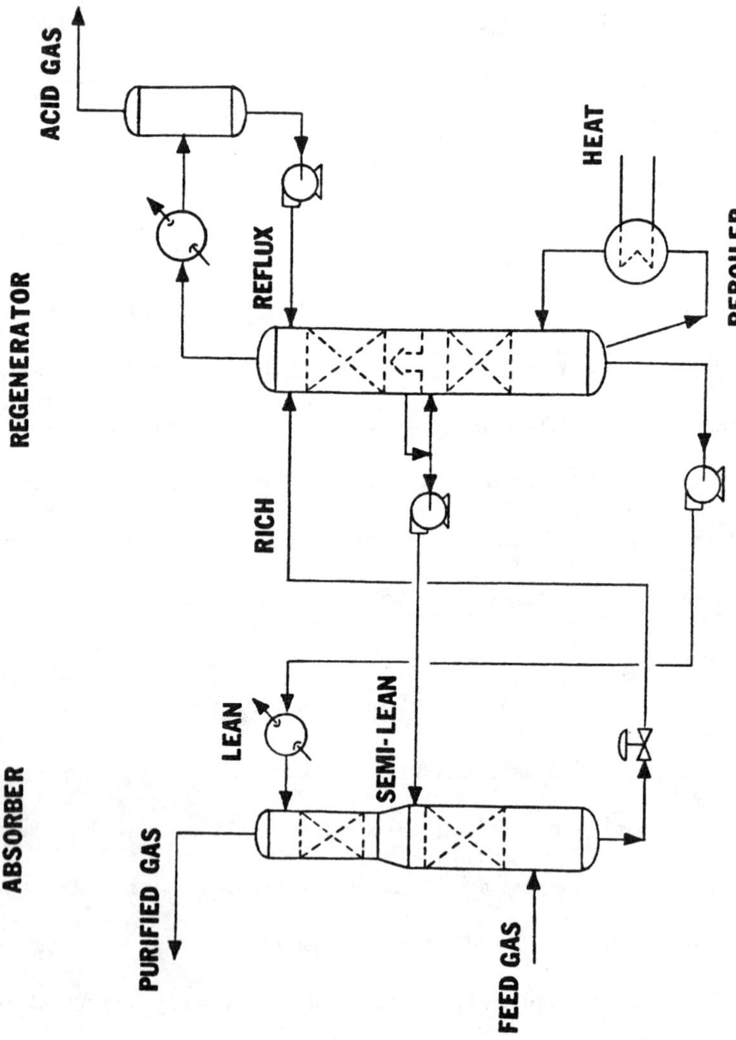

Figure 4. CATACARB two-stage design.

LOW-HEAT TECHNOLOGY

The heat requirements of the CATACARB Process used in the earlier configurations can be significantly reduced by incorporating LOW-HEAT technology. Reductions as much as 50-60% of the conventional heat requirements are possible. The most frequently used LOW-HEAT concept is that of solution flash and vapor recompression (5). The lean and/or semi-lean solution from the regenerator, which is at its boiling point, is flashed to a lower pressure. The flashed vapors are mostly steam with some acid gas. These flashed vapors are compressed to the regenerator and used to reduce the reboiler heat duty. The vapors can be compressed using either a steam jet (thermo-compressor) or a mechanical centrifugal steam compressor.

The principle behind this LOW-HEAT version is quite simple. The heat recovered from the solution flash represents heat transferred from the top of the regenerator to the bottom, since this heat would otherwise have entered the absorber and then the top of the regenerator via the rich solution. This steam which enters the bottom of the regenerator is used to strip the solution as it passes up the regenerator and, therefore, is more useful than the heat that enters the top.

The concept of LOW-HEAT design is not new to Eickmeyer and Associates. The first LOW-HEAT CATACARB unit was built in the 1960's and used solution flash and a thermo-compressor to recover the flashed vapors. It is only recently that this technique has gained popularity in view of the high energy costs and the greater emphasis on energy conservation.

In addition to the above commonly used LOW-HEAT concept, Eickmeyer and Associates have developed several innovative techniques to lower the heat requirements of the CATACARB Process. Since these concepts are not commonly published or known we are not including them in this paper, but will be happy to discuss them with potential clients. Any of the LOW-HEAT technology concepts can be used with all the configurations mentioned earlier, namely, single stage, split-cooled lean, two-stage and others.

The use of LOW-HEAT technology reduces the reboiler heat requirements and, therefore, the heat transfer surface in the reboiler. Since it reduces the overall heat requirements it also reduces the heat transfer surface in the acid gas cooler. Additionally, it can lower the circulation rate and therefore the pump cost and the diameter of the absorber and sometimes the regenerator. The above savings generally compensate for the additional equipment cost required for

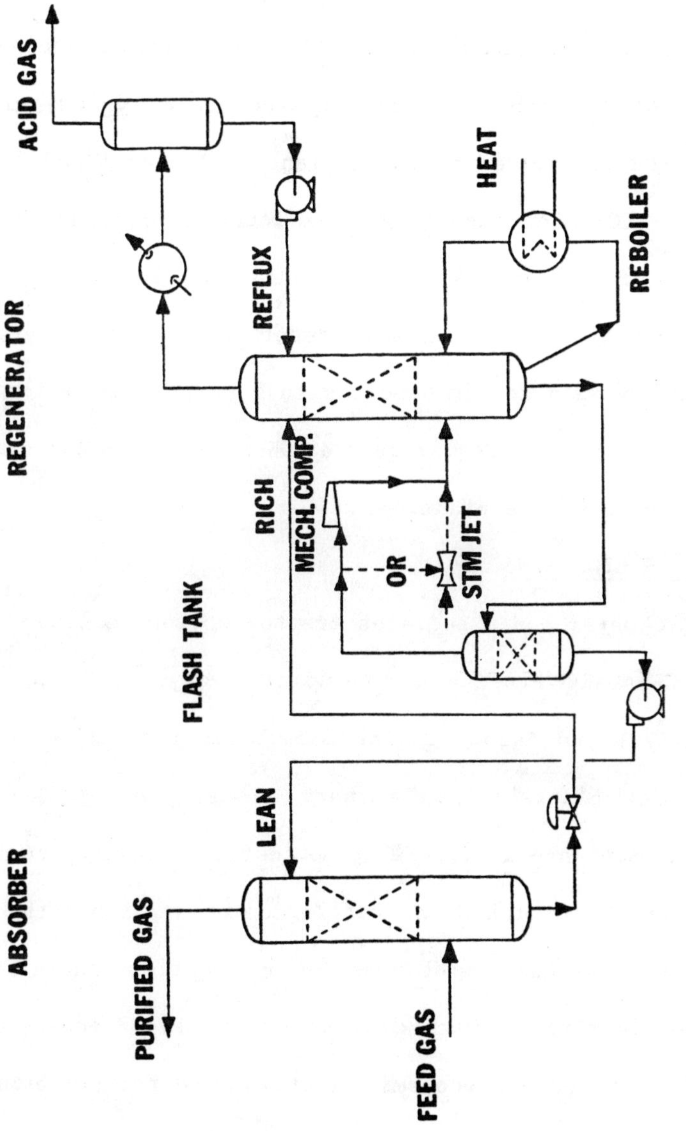

Figure 5. CATACARB single-stage LOW-HEAT design.

the LOW-HEAT technology such that the cost of a new installation using LOW-HEAT design is generally the same as a conventional CATACARB plant. The heat savings achieved by LOW-HEAT technology is illustrated in a typical ammonia plant and a typical hydrogen plant in Tables 1 and 2.

The LOW-HEAT technology has found wide application in revamp of existing CATACARB and other Hot Potassium Carbonate installations. The investment required in a revamp is typically paid for in 6 months to 2 years. In addition to lowering the heat requirement a LOW-HEAT revamp can increase the capacity of a plant.

PROCESS LICENSING

Eickmeyer and Associates are the worldwide licensors of the CATACARB Process. In addition to process licensing Eickmeyer and Associates perform basic process design for CATACARB plants. Preliminary process design and cost estimations are available at no cost to potential customers and generally without secrecy agreements. Since the CATACARB Process has a wide range of applications and a variety of configurations, including LOW-HEAT, which require optimization based on economic factors (6) for a particular application, we are not presenting generalized economics for they can be misleading. However, this information can be

Table 1
Comparison of CATACARB designs - 1000 MT/D Ammonia Plant

	Conventional	LOW-HEAT
1981 Capital Investment,MM$	6.7	7.2
Waste Heat from Gas,MMBTU/hr	54.5	54.5
STM,(2nd or 3rd use),MLB/hr	45	3.2
Power,KWH	1024	2060
C.W. (25•F Rise),MGPH	302	258
Heat Efficiency,MBTU/LB.MOLE of CO_2 Removed.	32.1	19.3

Table 2
Comparison of CATACARB designs - 90 MMSCFD Hydrogen Plant

	Conven-tional	LOW-HEAT Case 1	LOW-HEAT Case 2
1981 Capital Investment,MM$	5.8	5.8	6.2
Waste Heat from Gas,MMBTU/hr	101	101	74
STM,(2nd or 3rd use),MLB/hr	34	0	0
Power,KWH	750	750	1640
C.W. (25•F Rise,)MGPH	590	343	317
50 PSIG STM Generated for Export,(Boiler Cost not incl.)	0	0	38000 lbs/hr
Heat Efficiency, MBTU/LB.MOLE of CO_2 Removed.	48.5	37.2	27.0

made available to potential customers for their specific application within a day or two upon request at no cost.

CONCLUSION

The CATACARB Process has been successfully used in over 100 installations worldwide for acid gas removal. The LOW-HEAT technology and inorganic catalyst have increased the versatility of the Process.

REFERENCES

1. Eickmeyer, A.G., E.A. Wiberg and H.A. Gangriwala, "Energy Savings in the Gas Patch Using the Cata Processes," Paper presented at Petroenergy 81, Houston, Texas.

2. Eickmeyer, A.G., "The Catalysis of Potassium Carbonate Solutions in Carbon Dioxide Absorption," Chemical Engineering Progress, 58(4), April 1962, p.89.

3. Eickmeyer,A.G., U.S. Patent No.4,439,312, issued Feb.7,1984.

4. Eickmeyer,A.G., U.S.Patent No.3,932,582, issued Jan.13,1976.

5. Eickmeyer, A.G. and H.A. Gangriwala, "Acid Gas Removal in Synfuels Production," Energy Progress, Vol.1, No. 1-4, p.9.

6. Wiberg, E.A., H.A. Gangriwala and I-Meen Chao, "Process Design - Heart of the Estimate," presented at Kansas City Section AIChE 1982 Meeting.

15

Development of a New Hindered Amine-Promoted Hot Carbonate Scrubbing Solution for Simultaneous CO_2 and H_2S Removal from Gases

D. W. Savage G. R. Say
A. Bisio F. J. Heinzelmann
G. Sartori J. N. Iyengar
Exxon Research and Engineering Company

ABSTRACT

A new concept in CO_2- amine reaction chemistry has been applied to the simultaneous removal of CO_2 and H_2S from gases by reversible absorption with reaction in aqueous potassium carbonate solutions promoted with primary or secondary amines. Introduction of steric hindrance at the amino nitrogen site lowers the stability of the carbamate reaction product with CO_2 and leads to higher solution capacity for CO_2 and faster rates of CO_2 absorption compared to conventional promoters.

The feasibility of using carbonate solutions containing a sterically-hindered amine promoter has been tested in a large, gas treating pilot plant. The results established the basic operability of the process and confirmed performance advantages of 30 percent higher CO_2 equilibrium cyclic capacity and 100 percent faster mass transfer rates as compared with the conventional DEA (diethanolamine) promoted systems.

A commercial plant trial of the new treating process has been completed successfully. In this trial the scrubbing solution in an existing natural gas plant was changed over to the hindered amine-promoted solution, and the hindered amine solution permitted a 50% increase in acid gas removal capacity over conventional promoted carbonate treating, without any equipment modification.

INTRODUCTION

The promoted hot carbonate process is used widely for CO_2 removal in the manufacture of hydrogen and ammonia synthesis gas by the steam reforming process and in the cleanup of natural gas (1). Over 400 commercial hot carbonate treating plants have been built.

The basic hot carbonate process was developed in 1950 by Benson, Field, and coworkers (2). The use of promoters such as ethanolamines, arsenite, and amino acids was introduced to speed up the CO_2 mass transfer rates in the carbonate-bicarbonate buffer solution (3,4).

HINDERED AMINE CONCEPT

No detailed theory of promoter action in carbonate buffers has been published although certain features of organic and inorganic promoter chemistry have been discussed recently (5). A general theory of capacity and rate enhancement in amine-promoted buffers has been developed (10). The main features of the theory are discussed briefly in the following paragraphs in qualitative terms.

The common alkanolamines used as promoters for hot carbonate CO_2 removal increase the rates of CO_2 mass transfer without having any significant influence on CO_2 capacity, i.e., they do not affect the shape of the vapor-liquid equilibrium curve for the solution. Conventional alkanolamines such as diethanolamine form relatively stable carbamates with CO_2, as shown by the following equation:

$$CO_2 + 2RR'NH \rightleftharpoons RR'NCOO^- + RR'^+NH_2 \qquad (1)$$

RR'NH represents a secondary alkanolamine in which R and R' are linear substituents.

The amine is usually added as a minor constituent in a potassium carbonate solution (typically 25 percent K_2CO_3 in water). CO_2 absorption rate enhancement may be attributed to the presence of a significant concentration of free amine, RR'NH, in the solution interface. The total rate of chemical reaction in solution is given by

$$r = ([CO_2] - [CO_{2(e)}])(k_{OH^-}[OH^-] + k_{AM}[RR'NH]^n) \quad (2)$$

where:

$[CO_2]$	is the concentration of CO_2 dissolved in the buffer solution
$[CO_{2(e)}]$	is the concentration of CO_2 that would equilibrate the local liquid composition
k_{OH^-}	is the rate constant for CO_2 reaction with hydroxyl ion
$[OH^-]$	is the hydroxyl ion concentration
k_{AM}	is the rate constant for CO_2 reaction with amine
$[RR'NH]$	is the free amine concentration
n	is a constant having a value between 1 and 2

The values for k_{OH^-} and k_{AM} are of the same order of magnitude. Even at low concentration of the amine promoter, the concentration of free amine is significantly higher than the concentration of hydroxyl ion, thereby leading to a significant rate enhancement by the amine. The rate of CO_2 absorption from the gas phase is related to the chemical reaction rate in solution in a way which depends upon the regime of diffusion-reaction in the absorbent solution (6).

The absence of capacity enhancement with the common ethanolamine promoters in the usual range of hot carbonate application is believed to be primarily due to their strong carbamate-forming tendency. The latter ties up the amine in an essentially non-reactive form.

We have discovered that sterically-hindered amines can exhibit both a significant reversible capacity increment for CO_2 and superior CO_2 rate enhancement. Hindered amines are formed by placing a bulky substituent group close to the amino nitrogen site to lower the stability of the carbamate ion. Steric hindrance plays a role in reactions with CO_2 owing to the bulky nature of the CO_2 molecule. Hindered amines are defined as amino molecules containing at least one primary amino group attached to a secondary or a tertiary carbon atom (7). Two examples of sterically hindered amines are shown in Table 1. The overall chemistry of the CO_2 reaction with hindered amines may be written

$$CO_2 + RR'_BNH + H_2O \rightleftarrows HCO_3^- + RR'_BNH_2^+ \qquad (3)$$

where R'_B represents the bulky substituent (as contrasted with the linear substituents of conventional amines). In the theoretical limit, hindered amines will pick up one mole of CO_2 per mole of amine whereas conventional amines forming stable carbamates are limited to 0.5 moles of CO_2 per amine (see Equation 1). The superior rate enhancement may be attributed to the high kinetic constant for these amines, and the large fraction of the hindered amine present in the free form.

Table 1
Examples of Sterically Hindered Amines

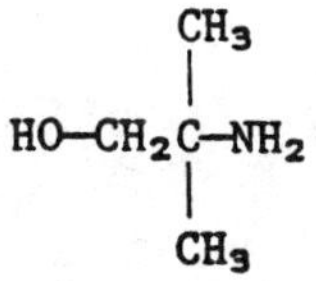

2-amino-2-methyl-1-propanol

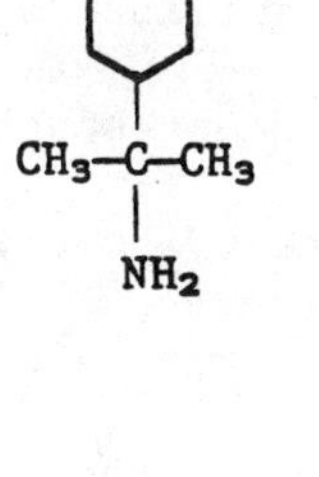

1,8-p-menthanediamine

CONCEPT VERIFICATION

The hindered amine promoter concept was verified in bench-scale equipment. The increase in CO_2 cyclic capacity with the hindered amine promoter is illustrated by the vapor-liquid equilibrium (VLE) diagram shown in Figure 1. The data were measure in a gas-liquid VLE cell in which a batch of solution was equilibrated with a gas of known CO_2 content by flowing the gas continuously through the solution for an extended period. Equilibrium partial pressures of CO_2 were calculated from the mole percent CO_2 in the gas, with a correction for the equilibrium water vapor pressure over the solution. Water vapor pressures, P_{H_2O}, for promoted solutions were estimated from the available data for unpromoted solutions (8), assuming P_{H_2O} to be directly proportional to the mole fraction of water in the solution. The loading of CO_2 in solution at equilibrium was determined by analyzing liquid samples for CO_2 content (by a method involving acidification and weighing the CO_2 evolved) and potassium content (by an x-ray fluorescence method using a Princeton Gamma-Tech., Inc. instrument, in which the source of radiation is ^{55}Fe).

Comparison of the VLE curves for hindered amine and conventional promoters shows a 20-30 percent equilibrium cyclic capacity enhancement with the hindered amine for

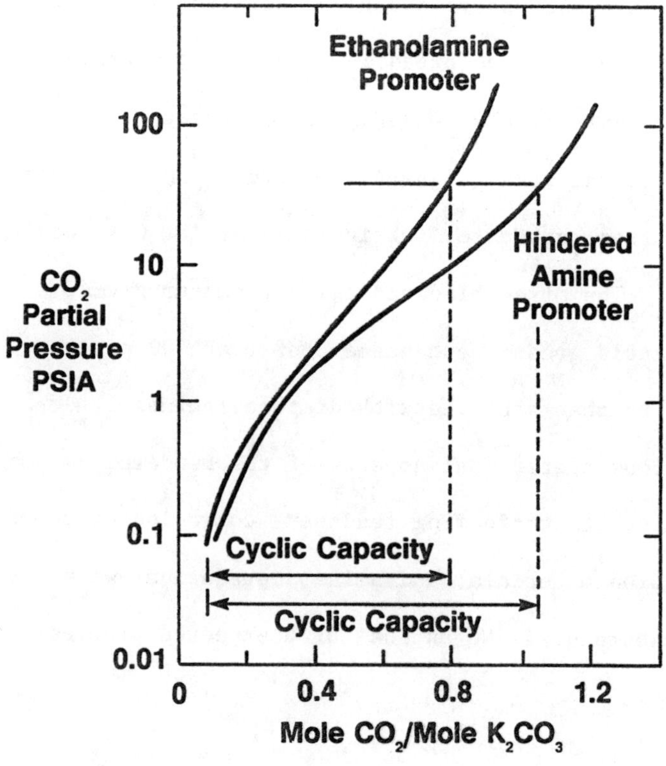

Figure 1. Hindered amine promoter increases cyclic capacity.

delta pressure cycles at 120°C over the pressure range from 50 to 0.1 psia CO_2 which simulates a rich (i.e. CO_2 saturated) and lean (i.e. regenerated) solution.

The superior CO_2 absorption rate with the hindered amine promoter is demonstrated by data, shown in Figure 2, obtained in a one-sphere absorber operating at 90°C. The apparatus has been described in detail elsewhere (9). As shown in Figure 2, the conventional promoter, diethanolamine, gives significant rate enhancement compared to unpromoted carbonate, particularly at low CO_2 loadings in solution. The mixed hindered amine promoter gives significantly greater enhancement of over 100 percent relative to that achieved with diethanolamine.

Various operational aspects of the hindered amine-promoted solution--foaming tendency, corrosion of common construction materials, and amine degradation--were checked in the laboratory. No unusual or unexpected problems were uncovered.

PILOT PLANT DEMONSTRATION

To apply the hindered amine gas treating solution in a commercial unit, it was necessary to carry out an integrated, continuous pilot plant operation in order to:

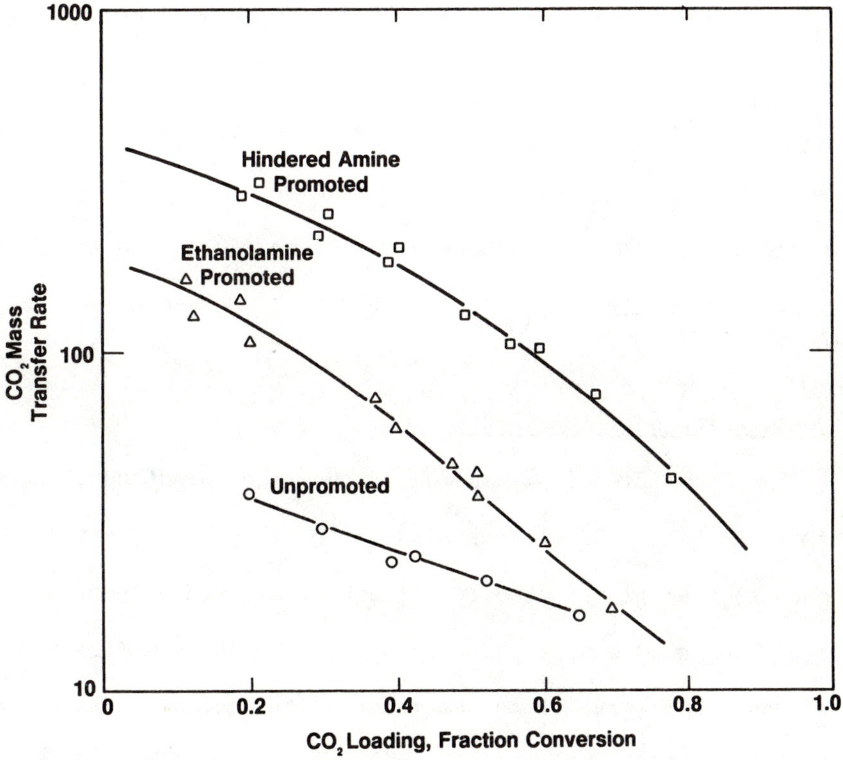

Figure 2. Hindered amine promoter significantly increases mass transfer rate.

(1) Verify batch absorption results regarding mass transfer coefficients and solution capacity benefits.

(2) Evaluate overall operability regarding foaming, corrosion, amine volatility, and amine degradation.

The approach used was to carry out comparative-type tests in which the hindered amine-promoted hot carbonate solution and the DEA-promoted hot carbonate solutions would both be tested under similar conditions.

Description of Pilot Plant

A simplified flow diagram of the pilot plant used is shown in Figure 3. It contains the basic parts of a typical gas treating plant: absorber, flash drum, regenerator, reboiler, and lean solution pump. All condensate from the overhead knock-out drums was returned to the unit. Water balance was maintained by controlling make-up water addition as a function of the solution specific gravity.

The inlet gas manifold contains separate flow controllers for CO_2, H_2S, and N_2 which was the main carrier gas. CO_2 is supplied by a liquid tank and is vaporized as required. H_2S is supplied in liquid cylinders and is

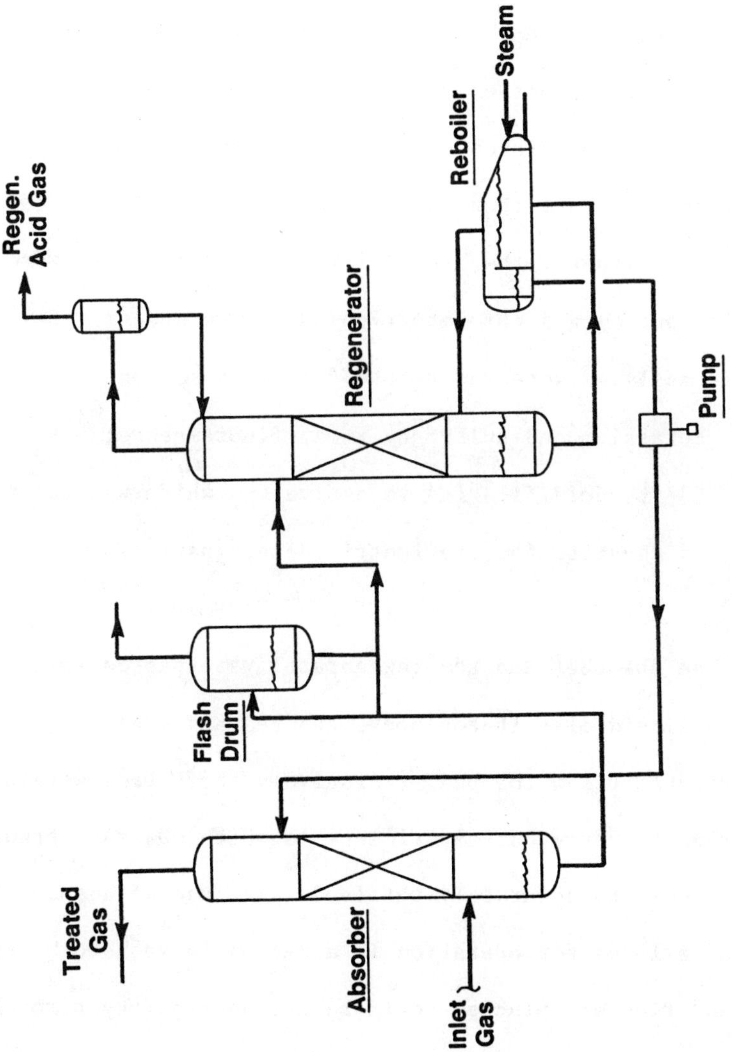

Figure 3. Pilot plant is a complete gas treating unit.

vaporized as needed. In order to obtain good acid gas material balances, the inlet gas, the treated gas from the absorber overhead, the flash drum acid gas, and the regenerator acid gas were all monitored for flow rate and composition. For each set of conditions to be studied, a material balance period was run in which a lined-out steady-state operation existed. Special precautions were followed in order to insure that CO_2 and H_2S flashing did not occur during sampling of the rich solution. Gas analyses were carried out by gas chromatography and mass spectrometry. Liquid analyses were run by the following methods:

Potassium and sulfur by x-ray fluorescence;

CO_2 by acidification to evolve CO_2 which was absorbed on Ascarite for gravimetric determination.

The absorber and the regenerator were packed columns with a liquid distributor above the bed and a gas distributor below the bed. Relatively short beds were used in order to carry out experiments in which CO_2 mass transfer rates could be accurately obtained. The use of a short bed height allowed for operation in a region in whch both lean end and rich end pinches could be avoided thereby permitting accurate measurements of the driving forces at the inlet and outlet. By varying the inlet gas composition and the lean solution composition, different portions of a commercial absorber tower could be simulated with the short bed. In the pilot plant studies described in this paper, no attempt was made to simulate an overall gas treating plant. Various calculations were carried out based on the pilot plant results in order to predict performance in a commercial gas treating plant.

MASS TRANSFER COEFFICIENTS

Promoted hot carbonate solutions provide considerable rate enhancement. An unpromoted K_2CO_3 solution having low normality of about 2N and operating at around 80°F with a lean solution conversion of around 0.50 would have negligible rate enhancement and would have an observed mass transfer coefficient approximately equal to the physical mass transfer coefficient k_L^o, for the particular packing and operating conditions for the experiments.

The mass transfer coefficient for a rate enhanced system is expressed as follows:

$$K_L = [k_L^{o\,2} + D_{CO_2} k_{\bar{OH}} [\bar{OH}] + D_{CO_2} k_{AM} [Am]]^{1/2}$$

where:

K_L = overall mass transfer coefficient

k_L^o = physical mass transfer coefficient

D_{CO_2} = diffusivity of CO_2 in liquid

With temperatures around 200°F, the values for $k_{\bar{OH}}$ and k_{AM} are large, and the resulting enhancement factor can become sizeable. At very low lean conversion levels, the $[OH^-]$ and [Am] concentrations will be high, thus further increasing the enhancement factor. Generally, for an amine promoted system, the first two terms under the square root are low compared to the last. With a very high enhancement, the absorption may become gas film controlled.

Experiments were carried out with a short absorber bed in order to avoid a pinched operation. In these tests there was considerable driving force across the entire absorber, so that the determination of mass transfer coefficients could be done with reasonable accuracy. Whenever a pinched condition is approached, the driving forces are not known accurately and, therefore, reliable mass transfer coefficients cannot be obtained.

Mass transfer coefficients were calculated by using log-mean driving forces at the top and bottom and were expressed in terms of the gas phase partial pressures. Figure 4 is a plot of the relative coefficients over a range of fractional conversions for the DEA and hindered amine promoted solutions. As can be seen, the hindered amine solution maintains a 2.5 to 3 fold higher mass transfer coefficient. This can be related to a higher effective amine concentration (since very little of the hindered amine is present as the carbamate as compared to the case with

DEA) and to a higher amine rate constant for reaction with CO_2. In all of these experiments, the critical operating conditions such as absorber temperature and pressure, solution rate, and inlet gas rate were all held constant. The percent CO_2 was adjusted to three or four different levels in order to achieve different average solution conversion levels. Another run was carried out in the pilot plant with a 25 wt. % K_2CO_3 solution with no promoters present at roughly the same operating conditions as shown for Figure 4. For a 0.44 average fractional conversion in the absorber a relative K_{Ga} of 0.3 was obtained. Based on this data point plus the pilot plant data in Figure 4, the DEA appears to provide an enhancement of 2.8 over the unpromoted K_2CO_3. Table 2 lists some of the relative mass transfer coefficient ratios for the unpromoted/DEA promoted/hindered amine promoted solutions for the single sphere and pilot plant experiments. The data show fairly close agreement.

Similar measurements were also obtained for the regenerator bed. In this case any flashing at the top of the regenerator and the effects of the reboiler had to be subtracted from the overall desorption occurring in the regenerator. The rich solution was flashed in the flash drum in each case to get a solution of its bubble point.

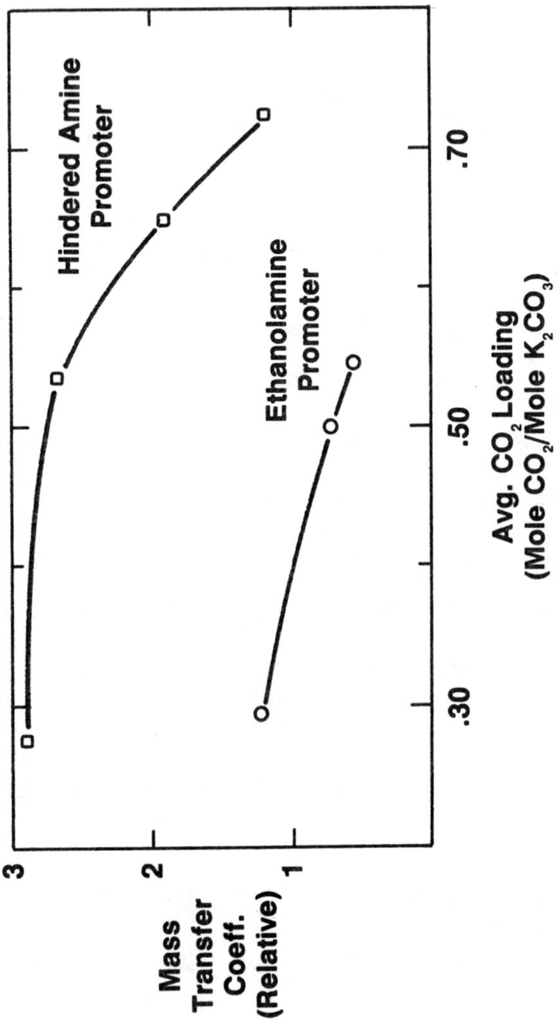

Figure 4. Pilot plant absorber results confirm mass transfer rate increase.

Table 2
Relative Mass Transfer Coefficients
at 0.44 Avg Fractional Conversion

	SINGLE SPHERE	PILOT PLANT
RELATIVE MASS TRANSFER COEFFICIENTS		
Unpromoted K_2CO_3	22	0.3
DEA Promoted K_2CO_3	53	0.85
Hindered Amine Promoted K_2CO_3	150	2.8
MASS TRANSFER COEFFICIENT RATIOS		
DEA Promoted K_2CO_3 / Unpromoted K_2CO_3	2.3	2.8
Hindered Amine Promoted K_2CO_3 / Unpromoted K_2O_3	6.8	9.3

In this manner, the mass transfer occurring across the regenerator packing could be isolated. Again, the hindered amine solution had about a two-fold higher mass transfer coefficient than the DEA promoted solution.

SOLUTION CAPACITY ADVANTAGE

Perhaps the best measurement of the capacity advantage for one gas treating solution versus that for another is to examine the equilibrium curves. The equilibrium data for the hindered amine promoted and the DEA promoted hot carbonate solutions are shown in Figure 1. From these curves a 20-30% cyclic advantage was predicted for the hindered amine solution.

The capacity advantage could be demonstrated in pilot plant runs by operating the absorber at the rich end pinch. The minimum solution rate can be determined by holding the inlet gas rate and composition constant and by gradually decreasing the solution rate until the absorber leak rate rises dramatically. In these types of tests, the steam rate would be set at a level to insure adequate solution regeneration.

Experiments similar to those described in the preceding paragraph were run except that a minimum packing height was used. While the short bed was insufficient to define a classic rich end pinch the experiments did give an indication of the capacity advantage.

These tests were run with approximately the same lean loading for both the DEA and hindered amine promoted solutions. These runs were carried out at the same lean solution temperature, inlet gas rate, and gas temperature, and absorber pressure. These runs indicated an approximate 30% capacity advantage for the hindered amine promoter.

Other runs were carried out with no regenerator steam, only CO_2 flashing. In these runs maximum rich solution loadings were obtained and again the hindered amine solution achieved an approximate 30% greater capacity for CO_2. The rich loadings were very close to the predicted CO_2 loadings that would be in equilibrium with the inlet gas at the absorber rich solution temperature.

Regeneration of the Solution

Another important aspect of evaluating a gas treating solution is the amount of steam for regeneration. In the pilot plant tests using a short regenerator bed, the steam rates required to reach a certain lean solution loading are higher than what would be required for a commercial unit that has many more mass transfer stages. The relative steam rates for two different solutions, however, offer a basis for comparing the ease of regeneration for the two solutions. Overall regeneration in these tests carried out in the same equipment are a direct result of the interactions of the three main factors: (1) regenerator mass transfer coefficients, (2) equilibrium data at regenerator temperatures, and (3) the heat of desorption.

Figure 5 shows the results of varying the steam rate for a set of conditions in which very low CO_2 pickup is occurring on the absorber side and, therefore, very little desorption was occurring in the regenerator. The low absorption/desorption was set by lowering the CO_2 in the inlet gas to a level of only 1.0 vol. %. As shown in the figure, the lean solution conversion decreases as the steam rate increases for a constant solution rate. The hindered amine solution was regenerated to a lower fraction conversion over the range studied.

During these tests the absorber was effectively lean end controlled, so another indication of the degree of regeneration is to examine the CO_2 content in the treated gas as a function of the steam rate. The hindered amine promoted solution produced a CO_2 leak rate less than half that for the DEA promoted solution as shown in Figure 6.

Other Development Issues

Aside from the process characteristics of a new gas treating solution, other important development issues include volatility, foaming, corrosion, and solution

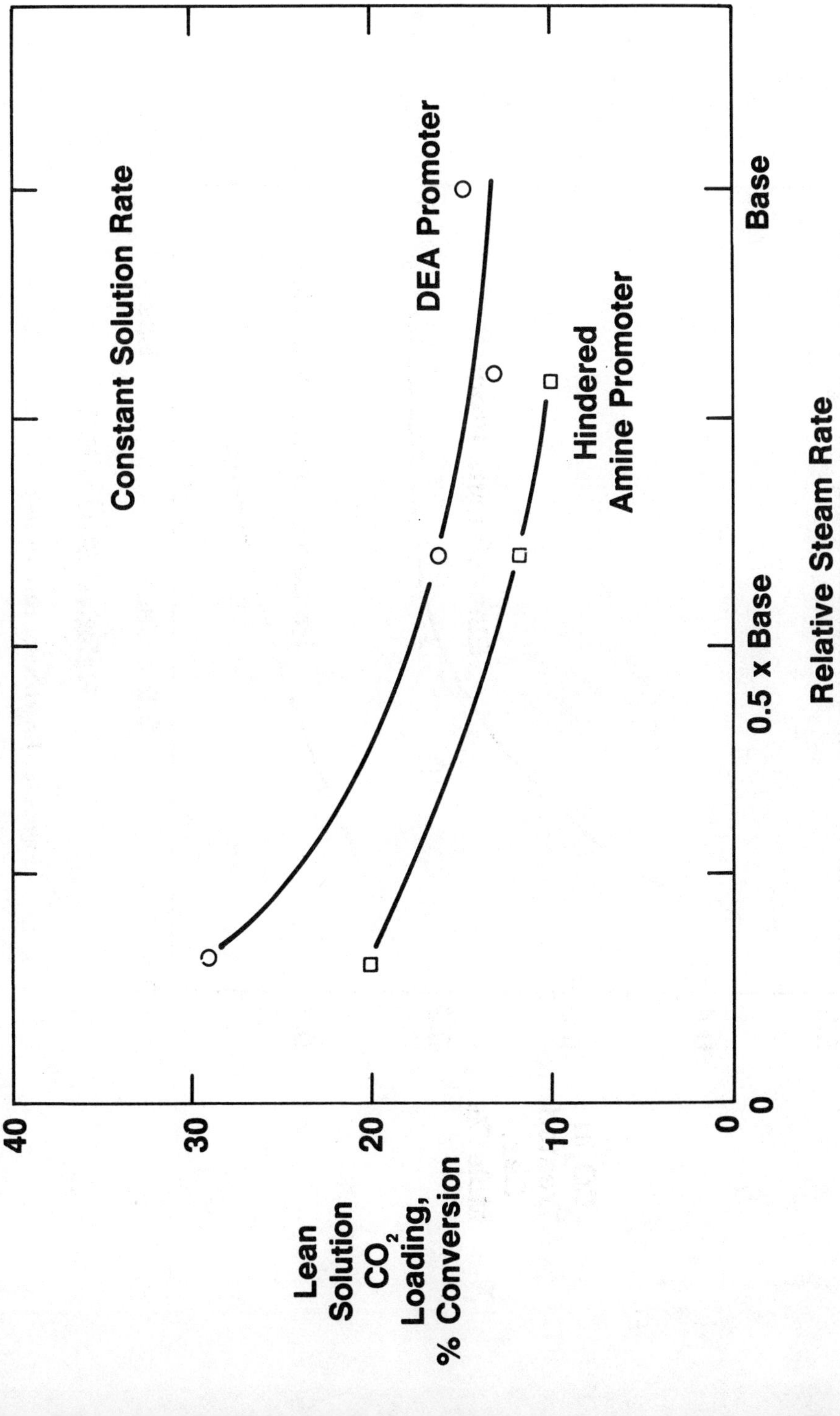

Figure 5. Lean solution loading vs. steam rate.

0484-0-137

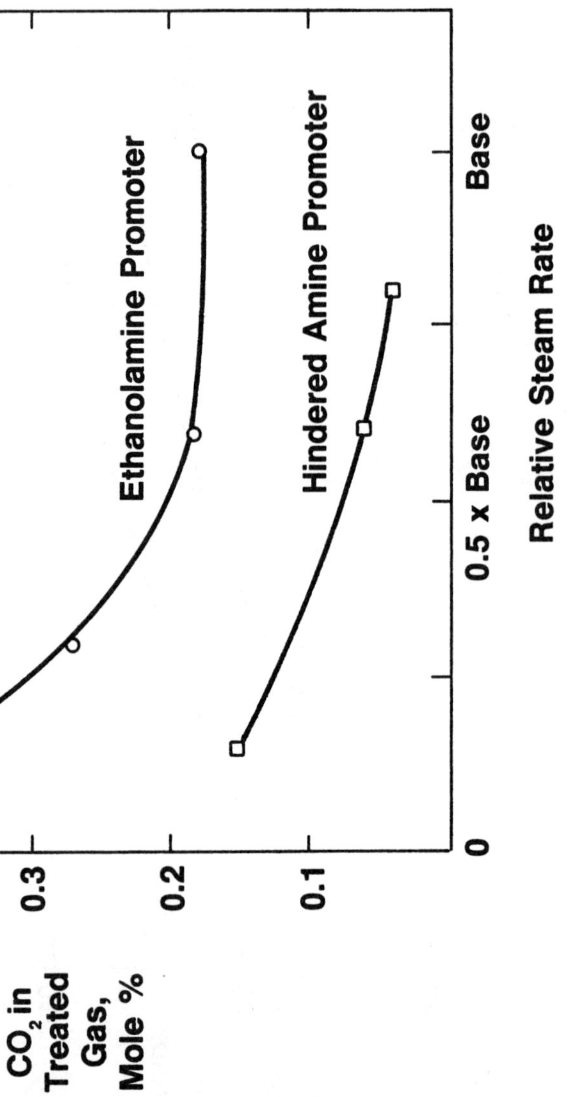

Figure 6. Regenerator performance also improved.

degradation.

Trace measurements of solution volatility were made by analyzing the treated gases for amine content. Solution foaming tendency was monitored by use of DP cells across the absorber and regenerator. As operations were moved closer to the flooding point, small amounts of foaming were detected quite clearly on the DP cells. At this point a small amount of anti-foam was added. In all cases, any foaming that occurred with either solution was easily brought under control with anti-foam addition. It was felt that any foaming observed in the pilot plant columns for both solutions would probably represent a worse case. With small diameter packing a small amount of foaming can cause more liquid bridging than for a larger diameter packing. The conclusion of these studies was that foaming would not be a problem in a commercial unit.

Corrosion was monitored in the unit by use of coupons located in the absorber bottom section and by Corrosometer probes which were located in the lean solution line and in the rich solution line ahead of the flash drum but after the rich solution heater. In all of the pertinent tests described in this paper, a small amount of H_2S was present, ranging from 0.1 to 0.5 mole percent H_2S in the inlet gas. With this amount of H_2S present, a protective sulfide coating was achieved in the entire unit which inhibits carbon steel corrosion. Promoted hot carbonate solutions can be very corrosive in an H_2S free environment without a corrosion inhibitor present. Some of the pilot unit data are shown in Figure 7 for an operation with and without H_2S present for the DEA promoted solution. The result of these tests was that with the H_2S present, corrosion rates were practically nil for both the DEA and hindered amine promoted solutions.

COMMERCIAL PLANT TEST

Following the successful pilot plant demonstration of the process, a trial in a commercial unit was undertaken. The plant chosen for the test was an existing promoted carbonate unit treating natural gas.

Natural gas feed to the unit normally contained approximately 3.5% acid gas (H_2S + CO_2). However, a shift in feed gas sources resulted in a need to markedly increase the acid gas treating capacity, with the acid gas content of the feed gas expected to increase from 3.5% to greater than 7.5%. With the conventional promoted carbonate system and the existing facilities, the maximum acid gas content that could be handled while meeting product gas specifications was determined to be 4.5-5.0%. This limit was set by the solution acid gas capacity and mass transfer

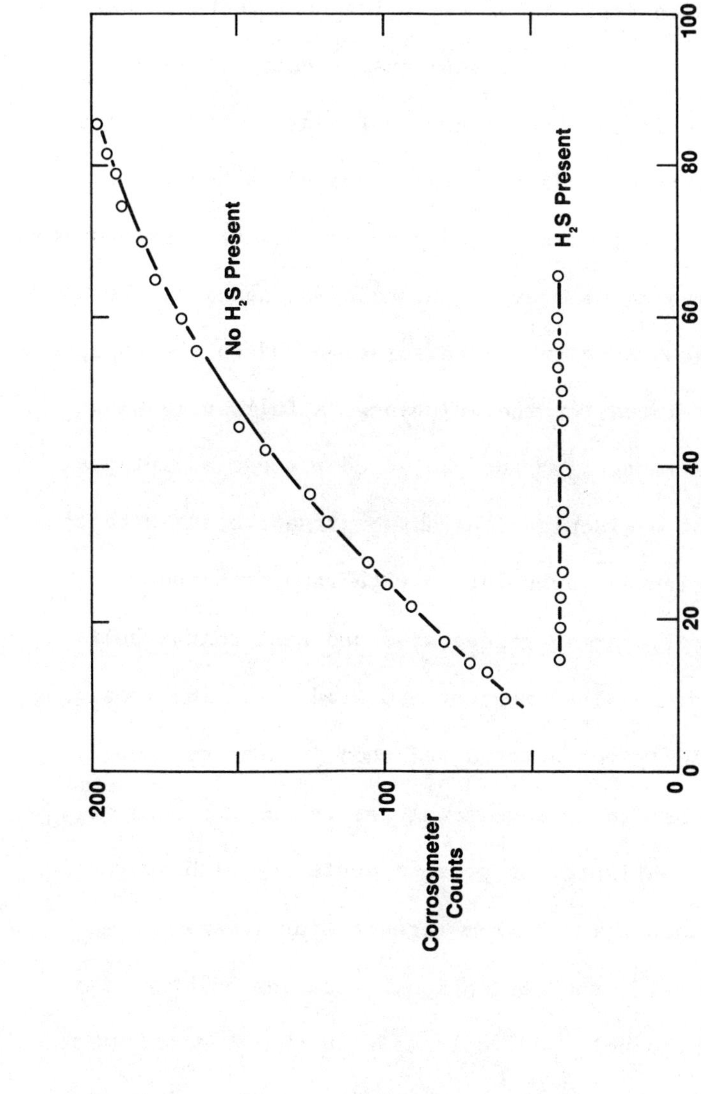

Figure 7. Corrosion with and without H_2S present.

characteristics. To achieve the desired acid gas capacity with the conventional process would have required extensive plant modifications to permit increased circulation rate and regeneration capacity.

Based on an analysis of the pilot unit tests it was determined that the desired acid gas capacity could, however, be achieved without any equipment modifictions by using the improved hindered amine promoted carbonate process. The prediction of the greatly improved performance of the hindered amine promoted process was based on the higher acid gas capacity of the solution and on the enhanced mass transfer rates for the solution. A full scale plant test was made to confirm the projected process advantages.

The plant was designed as a two stage process with the bulk of the circulating solution withdrawn from the regenerator only partly regenerated and sent to the bulk absorber for removal of most of the acid gas. The remaining solution is fully regenerated and then is sent to the cleanup absorber to achieve final removal of the acid gas. This process configuration permits achieving high levels of acid gas cleanup with minimum regeneration steam.

For the test, the conventional treating solution was removed from the unit and replaced with the hindered amine promoted carbonate solution. The unit was started up with the hindered amine solution, and a full scale capacity test was made to confirm the process performance. The tests demonstrated the superior performance of the hindered amine process by obtaining a 50% increase in acid gas removal capacity over that achievable with the conventional process at regeneration steam rates equivalent to that of the conventional solution. Results are summarized in Table 3.

In addition to superior process performance, other operability characteristics were closely monitored during the plant test. Foaming was controlled completely with commercially available antifoams, even at higher than design gas rates. Corrosion was carefully monitored and the unit exhibited no evidence of corrosion even at the higher acid gas loadings for the hindered amine process over a two year period of operation. Overall solution composition was monitored, and no unusual or unexpected problems were observed.

SUMMARY

A new hindered amine concept has been successfully applied in the formulation of a gas treating solution for the simultaneous removal of CO_2 and H_2S from gas streams. The development has progressed from the initial laboratory fundamental studies, through an extensive pilot plant

Table 3
Commercial Plant Test Demonstrates Advantages of Hindered Amine Promoted Carbonate Process

	CONVENTIONAL AMINE PROMOTER	HINDERED AMINE PROMOTER
Treating Capacity, % Acid Gas	5.0	8.0
Solution Circulation Rate Per MSCF Acid Gas	Base	0.7X Base
Regeneration Steam Per MSCF Acid Gas	Base	0.7X Base

development program and was applied in a commercial unit. The hindered amine promoted hot carbonate process offers clear advantages over conventional promoters in applications such as debottlenecking, energy conservation, and new unit construction. The use of the hindered amine process leads to a reduction in solution rate and a reduction in regeneration steam requirements as well as higher mass transfer coefficients. Lower equipment costs for new units provide further credits.

Hindered amine promoted hot carbonate solutions will provide similar advantages in H_2S-free CO_2 removal units, such as in ammonia or hydrogen plant applications. However, all hot carbonate processes require a corrosion inhibitor, if H_2S is not present in the gas stream. We currently are testing the hindered amine promoted hot carbonate solution, with corrosion inhibitor, and plan to demonstrate this technology in a commercial CO_2 removal unit later this year.

REFERENCES

1. Tennyson, R. S. and Schaaf, R. P., Oil and Gas Journal, Vol. 75, No. 2, p. 78, 1977.

2. Benson, H. E. and Field, J. H., U. S. Patent 2,886,405, 1959.

3. Mayland, B. J., U. S. Patent 3,144,301, 1964.

4. Riesenfeld, F. C. and Mullouney, J. F., Petroleum Refiner, Vol. 38, No. 5, p. 161, 1959.

5. Astarita, G., Savage, D. W., Longo, J. M., Chemical Engineering Science, Vol. 36, p. 581, 1981.

6. Astarita, G., Mass Transfer with Chemical Reaction, Elsevier, 1967.

7. Sartori, G., Savage, D. W., U. S. Patent No. 4,094,957, 1978.

8. Tosh, J. S., Field, J. H., Benson, H. E., and Haynes, W. P., U. S. Bureau of Mines Report of Investigations, No. 5484, 1959.

9. Savage, D. W., Astarita, G., Joshi, S., Chemical Engineering Science, Vol. 35, p. 1513, 1980.

10. Sartori, G., Savage, D. W., Industrial and Engineering Chemistry Fundamentals, Vol. 22, p. 239, 1983.

16

Desulfurization of Hot Coal by Zinc Ferrite

T. Grindley
U.S. Department of Energy
Morgantown Energy Technology Center

G. Steinfeld
Science Applications, Inc.

INTRODUCTION

Morgantown Energy Technology Center (METC), the Department of Energy's lead center for coal gasification and gas stream cleanup technology, has been engaged for many years in the investigation of technology for the purification of coal-derived gas for various applications, particularly gas turbine and fuel cell combined-cycle power generation (1). Sulfur compounds, which are major contaminants, must be removed to satisfy environmental and process reasons.

The current program is directed primarily at gas purification for molten carbonate fuel cell application, which has especially stringent requirements (2,3).

Another potential application is the provision of fuels for gas turbine combined-cycle plants.

High-temperature desulfurization, which would obviate the water quenching and scrubbing of conventional technology, potentially has significant thermal efficiency advantages and has been a principal goal of METC's developmental efforts. Through bench-scale evaluation of promising sorbents in simulated and actual coal gas streams, we have demonstrated that zinc oxide-containing sorbents can remove sulfur compounds to less than 10 ppm in the temperature range 1,000° to 1,200°F (~ 540 to 650°C) and have the potential to be regenerated for repeated use.

A zinc ferrite sorbent formulated at METC was extensively tested in a side stream of a pressurized, fixed-bed gasifier consuming a range of coals during about 500 hours of operating time in desulfurization and regeneration modes.

The principal purpose of this paper is to describe the developmental activities at METC since 1980, including laboratory and gasifier side stream testing, which have been directed at evaluating regenerable solid sorbents capable of removing sulfur compounds from hot coal-derived gas down to about 1 ppm.

PROCESS BACKGROUND

The METC project to investigate zinc oxide-containing sorbents for hot coal gas desulfurization is an outgrowth of previous work at METC using iron oxide as a sorbent (4,5,6,7).

Iron oxide is somewhat limited in the extent to which it can remove sulfur compounds, specifically hydrogen sulfide, at high temperatures. Zinc oxide has the capability of removing H_2S to a level about two orders of magnitude less (8,9,10). The H_2S concentrations over a range of temperatures and steam concentrations in equilibrium with the two oxides, as computed from thermodynamic properties, are shown in Table 1. Thermodynamic analysis by both Westmoreland, et al., and Jalan, et al., based on NBS data, indicates that vanadium oxide may be superior to zinc oxide in its degree of H_2S removal, but it appears that the NBS tabulated heat of formation of vanadium sulfide, as employed by them, is inaccurate (11), leading to an erroneous conclusion. Thus, of the metal oxides evaluated, zinc oxide possesses the highest affinity for sulfur, according to the thermodynamic computations.

Table 1
Hydrogen Sulfide in Equilibrium with Zinc Oxide
Calculated from Thermodynamic Properties
$ZnO + H_2S \rightarrow ZnS + H_2O$

Volume % Steam in Gas	H_2S Concentration (ppm) in Wet Exit Gas					
	450°C (842°F)	500°C (932°F)	550°C (1022°F)	600°C (1112°F)	650°C (1202°F)	700°C (1292°F)
30	1.0	2.2	4.4	7.9	13.2	20.7
20	0.69	1.5	2.9	5.3	8.8	13.8
10	0.35	0.75	1.5	2.6	4.4	6.9
5	0.17	0.37	0.7	1.3	2.2	3.5

HYDROGEN SULFIDE IN EQUILIBRIUM WITH IRON OXIDE
CALCULATED FROM THERMODYNAMIC PROPERTIES

$$Fe_3O_4 + 3H_2S + H_2 \longrightarrow 3\ FeS + 4\ H_2O$$

GAS CONTAINS 20 MOLE % HYDROGEN

Volume % Steam in Gas	H_2S Concentration (ppm) in Wet Exit Gas					
	450°C (842°F)	500°C (932°F)	550°C (1022°F)	600°C (1112°F)	650°C (1202°F)	700°C (1292°F)
30	337	463	620	813	1040	1310
20	196	270	361	473	606	761
10	78	107	143	188	241	302
5	31	42	57	75	96	120

Both iron oxide and zinc oxide, in their sulfidized form, can be regenerated by roasting with air or oxygen, diluted with inert gases such as steam and nitrogen to moderate the high temperatures which can result from the exothermic oxidation reaction.

Sulfidized iron oxide is somewhat easier to regenerate than zinc oxide because of the lower thermal stability of iron sulfate as compared with that of zinc sulfate; therefore, the regeneration reaction temperature need not be so high.

The compound zinc ferrite, $ZnFe_2O_4$, which is formed by heating a mixture of zinc and iron oxides, tends to combine the high sulfur affinity of zinc oxide with the good regenerability of iron oxide; the formation of zinc vapor in a reducing atmosphere is also thought to be inhibited in comparison with that of zinc oxide. The chemical changes taking place during sulfidation and regeneration may be represented as follows:

Sulfidation

$$3ZnFe_2O_4 + H_2 \longrightarrow 3ZnO + 2Fe_3O_4 + H_2O$$

$$Fe_3O_4 + 3H_2S + H_2 \longrightarrow 3FeS + 4H_2O$$

$$ZnO + H_2S \longrightarrow ZnS + H_2O$$

Regeneration $2FeS + 7/2O_2 \longrightarrow Fe_2O_3 + 2SO_2$

$ZnS + 3/2O_2 \longrightarrow ZnO + SO_2$

$ZnO + Fe_2O_3 \longrightarrow ZnFe_2O_4$

The last reaction is a solid-solid reaction.

Zinc ferrite normally operates at gas conditions where magnetite, Fe_3O_4, which is formed from zinc ferrite, is stable. This stability region is shown in Figure 1, which is an iron-carbon-oxygen equilibrium diagram (computed for atmospheric pressure) plotted in terms of temperature and the ratio $CO/(CO + CO_2)$. Operating lines for typical gases from the Westinghouse (fluid-bed) and Lurgi (fixed-bed) gasifiers are marked (12).

The sulfur dioxide in the regeneration tail gas can be recovered or fixed as calcium sulfate, etc., by various techniques.

LABORATORY EVALUATIONS

Sorbent Comparison

The principal sorbent to be evaluated, zinc ferrite, was initially made from an equimolar mixture of commercially available ferric oxide and zinc oxide powders. These were extruded into 3/16-inch diameter extrusions

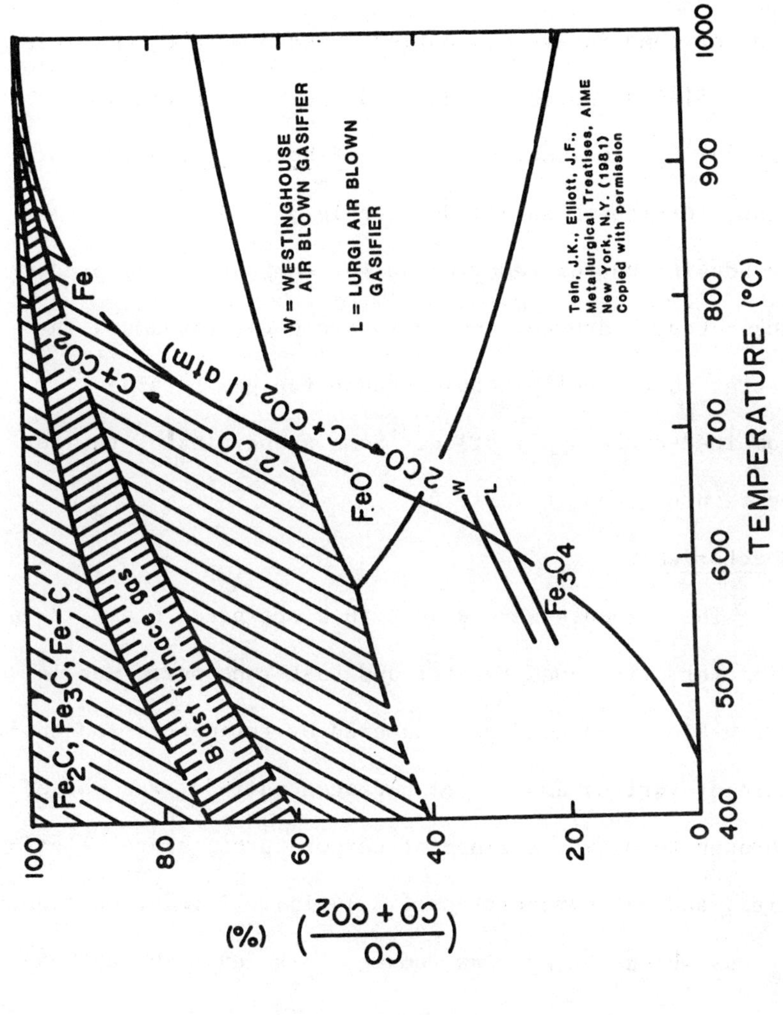

Figure 1. Fe-C-O Equilibrium diagram.

after the addition of a small quantity of bentonite and water, dried at 200°F for several hours, and then calcined at 1,500°F.

Two other zinc oxide-containing sorbents were selected for comparative evaluation. These were a commercial, nominally 100 percent zinc oxide sorbent, Harshaw Zn-0401 E 3/16-inch, and a cement-bound zinc oxide sorbent, Institut Francais du Petrole MEP 682. The former is advertised for removing sulfur compounds from producer, coke-oven, hydrogen, and hydrocarbon gas streams. The latter is a developmental sorbent for gas turbine combined-cycle application. Some properties of the sorbents are given in Table 2.

Reactor Unit

The sorbents were placed in a stainless steel, 2-inch diameter, fixed-bed reactor and test runs performed in which simulated coal gas, metered by rotameters and containing various amounts of hydrogen sulfide, was passed through them over a range of temperatures, space velocities, and gas compositions (13,14,15). A reaction zone, across which the H_2S was reduced to a level approaching

Table 2
Sorbent Characterization

	Harshaw Zinc Oxide Zn - 0401E[6]	IFP Zinc Oxide MEP - 682[7]	METC Zinc Ferrite[8]
Extrusion Diameter (mm)	4.75	3	4.75
Extrusion Length (mm)		6-10	6-19
Zinc Oxide Content (wt%)	100	70	33[1]
Iron Oxide Content (wt%)	0	0	66[2]
BET Surface Area (m^2/g)	3 (3.2)[3]	12 (11.7)[3]	4[3]
Apparent Bulk Density (g/cm^3)	1.2	1	
Helium Density (g/cm^3)	6.39[4]	4.74[4]	5.72[4]
Extrusion Density (g/cm^3)		1.87	
Cumulative Pore Volume (cm^3/g) by Mercury Porosimetry	0.26	0.30	
Cumulative Pore Volume (cm^3/g) by Nitrogen Adsorption	0.0065[5]	0.0381[5]	0.0085[5]
Crush Strength (Kg/Extrusion)	3.6		

NOTES:
1. Sherwin Williams HSA Zinc Oxide
2. Fisher Certified Iron Oxide (Fe_2O_3).
3. BET Surface Area by METC.
4. Helium Density by METC.
5. Nitrogen Adsorption by METC. Pore diameters up to 600 Angstrom.
6. Data from Harshaw except where noted.
7. Data from IFP except where noted.
8. Data from METC

chemical equilibrium, passed through the bed until breakthrough occurred. A schematic of the reactor system is shown in Figure 2. Hydrogen sulfide in the reactor exit gas after condensation of steam was measured by detector tubes capable of measuring concentrations down to about 1 ppm. Supplementary measurements were made with an on-line gas chromatograph. Sorbent temperatures were measured by means of thermocouples projecting into the sorbent bed.

Regeneration of the sorbents, in which a hot reaction zone transited the sorbent bed, was carried out by passing a steam-air mixture through them at about 1,000° F (~ 540°C). The sulfur dioxide formed was measured by detector tubes after condensation of the steam. The SO_2 concentration in the exit gas on a dry basis was about 12.5 volume percent, except at the ends of the regenerations when the level tailed off. The SO_2 level would be commensurately less if inert gas were substituted for steam.

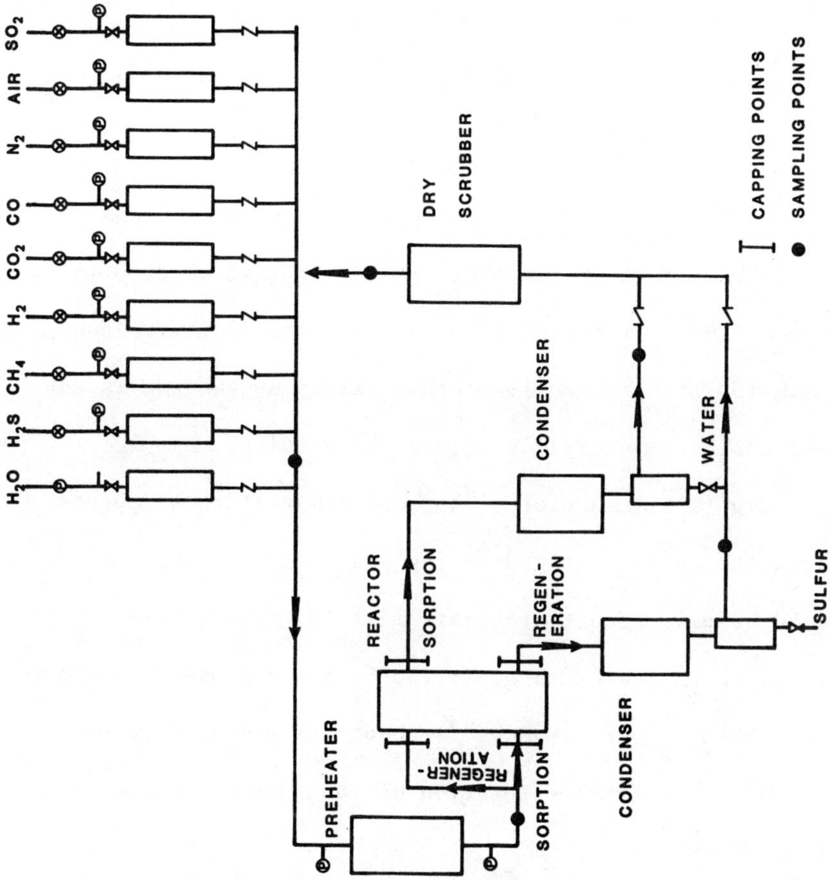

Figure 2. Process flow schematic for lab scale unit.

Results

Figure 3 is a plot of the exit hydrogen sulfide level attained over three cycles of absorption and regeneration, the inlet gas composition being:

	H_2S	H_2O	H_2	N_2
Volume Percent	2.7	6.9	14.6	75.8

It can be seen that the H_2S level is about 1 to 5 ppm before breakthrough and there is no significant drop in sorbent sulfur loading following regeneration. The substitution of nitrogen for carbon oxides in the gas is not considered to materially affect the results.

Figure 4 is a plot of sorbent temperature at points along the reactor length during a regeneration with a 50/50 volume percent steam-air mixture. These temperatures, which reached about 1,700°F, did not appear to harm the sorbent's performance. Sorbent analysis indicated virtually complete restoration of the sorbent to zinc ferrite.

Figure 5 shows how the amount of sulfur absorbed before H_2S breakthrough by the three sorbents under

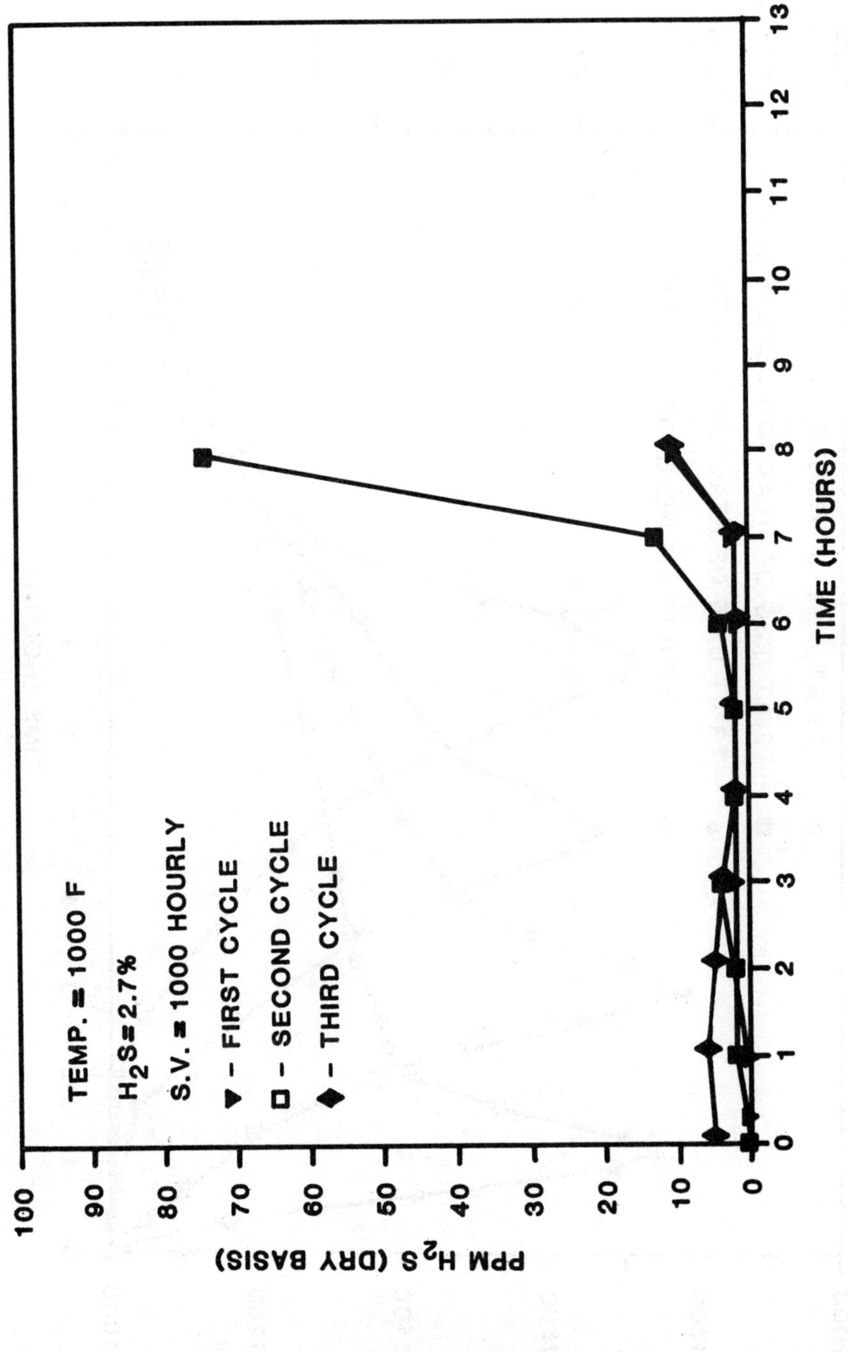

Figure 3. H_2S breakthrough curves for three consecutive cycles.

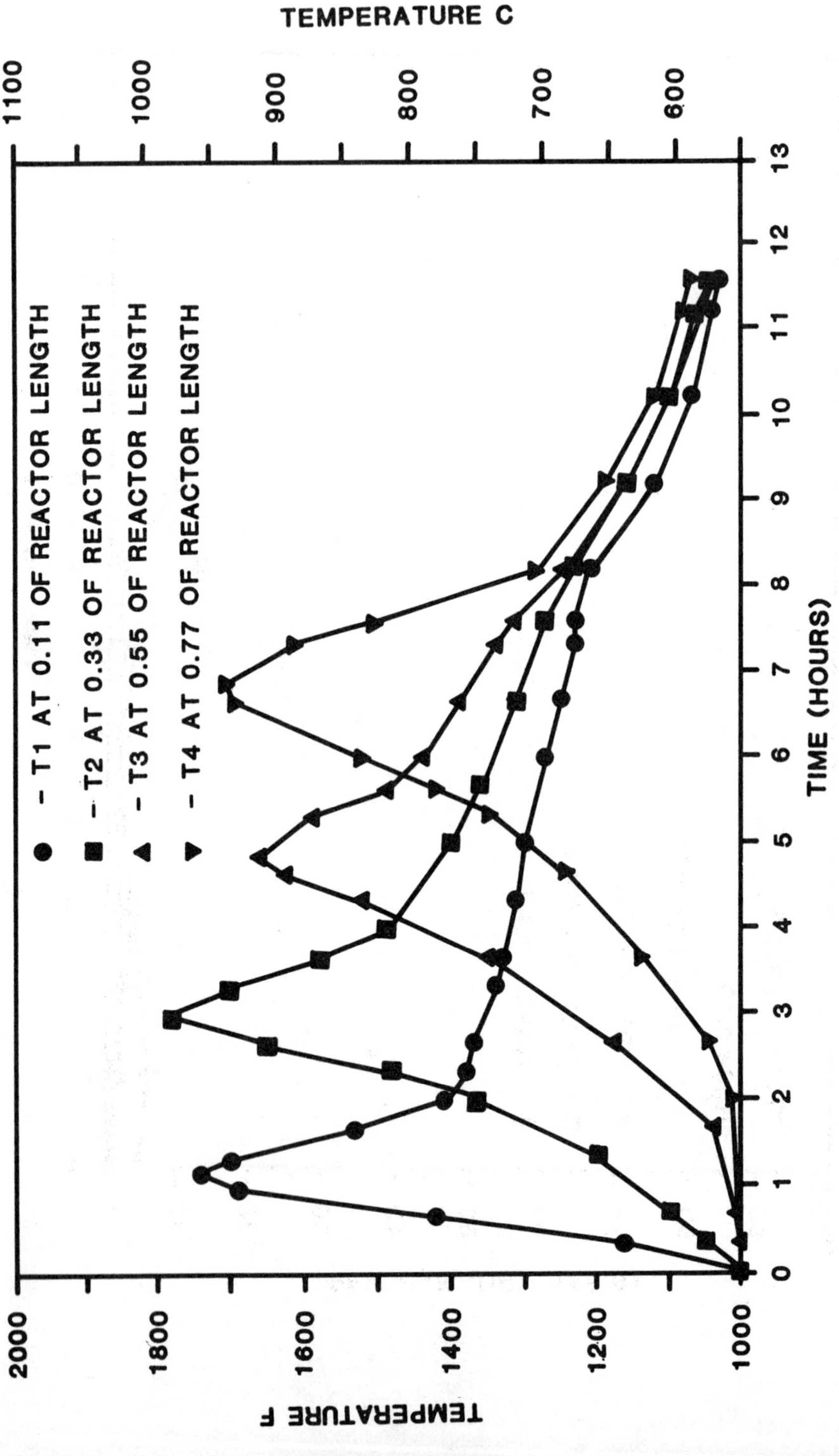

Figure 4. Regeneration exotherm.

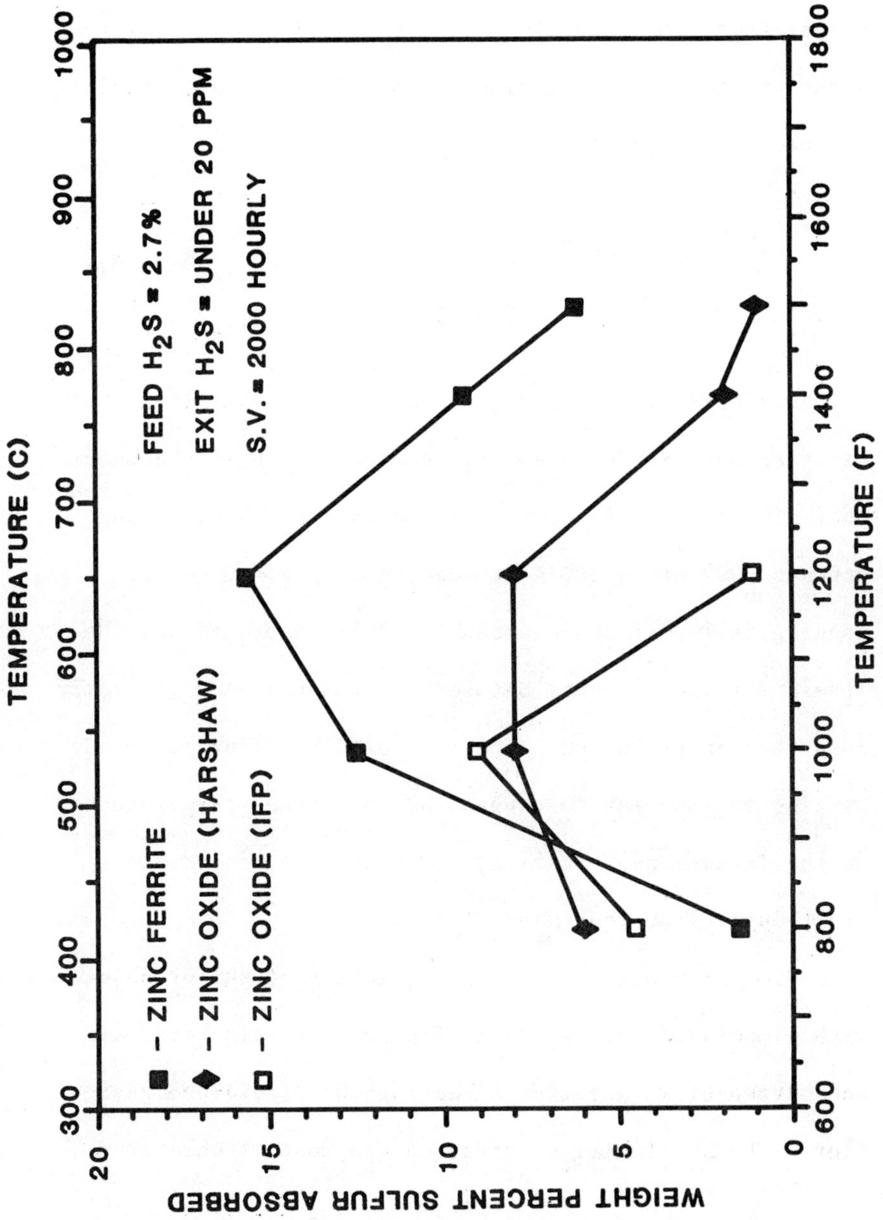

Figure 5. Sulfur absorption capacity.

evaluation varies with temperature. Except at the lowest temperature, the sorbent is in a successively regenerated state as the temperature increases. The simulated coal gas had the composition:

	H_2S	H_2O	CO_2	CO	H_2	CH_4	N_2
Volume Percent	2.7	27.6	9.8	11.7	17.1	2.9	28.2

which is similar to gas from a Lurgi, dry-ash, air-blown gasifier but with a larger H_2S concentration. It appears that zinc ferrite is most efficient at absorbing sulfur between 900° and 1,300°F, though in all cases, sulfur loading falls off above about 1,200°F. Hydrogen sulfide levels for the sorbents before breakthrough were generally less than about 10 ppm. The fall-off in sulfur absorbed by zinc ferrite may be caused, at the higher temperatures, by the formation of wustite, FeO, which is poorer at absorbing sulfur than magnetite (9).

Similar plots showing the variation of sulfur absorbed with temperature are shown in Figure 6 for simulated air- and oxygen-blown gas from a Westinghouse fluid-bed gasifier. The artificially increased H_2S content permitted

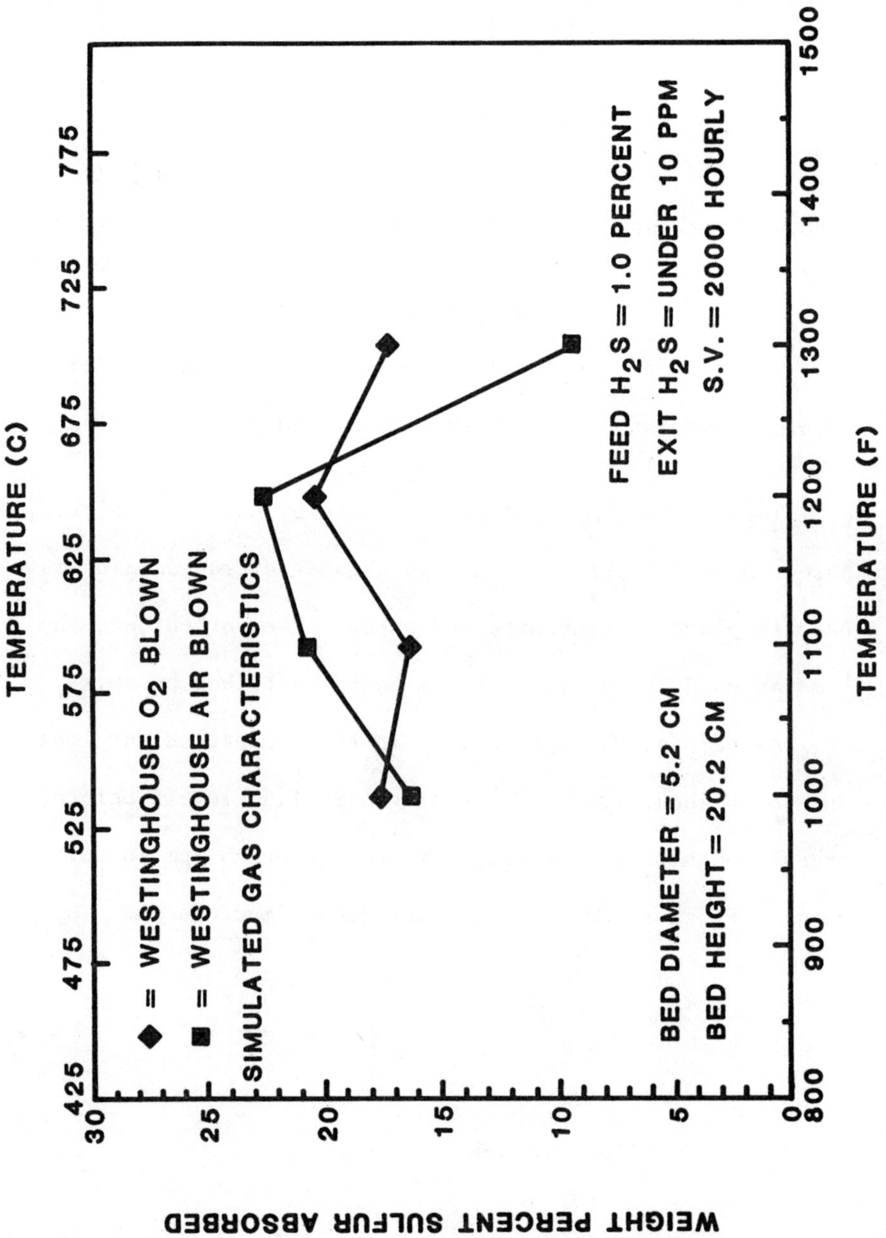

Figure 6. Sulfur absorption capacity of zinc ferrite.

experimental runs to be made in a reasonable time. The gas compositions are:

Air Blown

	H_2S	H_2O	CO_2	CO	H_2	CH_4	N_2
Volume Percent	1.0	19.9	10.9	12.4	13.7	0.6	41.5

Oxygen Blown

	H_2S	H_2O	CO_2	CO	H_2	CH_4
Volume Percent	1.0	32.0	25.0	30.0	11.0	1.0

The sorbent employed was prepared by United Catalysts, based on METC's formulation. An organic extrusion aid was used in place of bentonite and water. The sorbent was the same as used in the side stream tests described below.

As before, there is a drop in the sorbent sulfur loading above about 1,200° F. Hydrogen sulfide levels before breakthrough were generally less than about 5 ppm in both cases. Fresh sorbent was used at each temperature.

Supplementary Studies

The evaluations were supported, through contracted efforts, by similar laboratory-scale work to investigate the behavior of zinc ferrite in simulated gas containing tars and hydrogen chloride (16), sulfate formation during regeneration (17), and trace sulfur removal by chemisorption (18). A fixed-bed reactor model of the sulfidation and regeneration process was generated also by an outside contract (19). It was found that tars are deoxygenated and dealkylated to a large extent with little cracking to coke occurring. A small amount of hydrogen chloride was absorbed. An increase in the regeneration temperature from 1100 to 1200°F (~ 593 to 650°C) was accompanied by a rapid decrease in the residual sulfate content of the regenerated sorbent.

COAL GASIFIER SIDE STREAM TESTING

Side Stream Unit

To gain a more realistic assessment of the desulfurization sorbents tested at laboratory-scale, a larger reactor unit was constructed on a skid and placed in a side stream of METC's 24 t/d gasifier (14,15,20). Figure 7 is a drawing of the unit operating in absorption mode. The

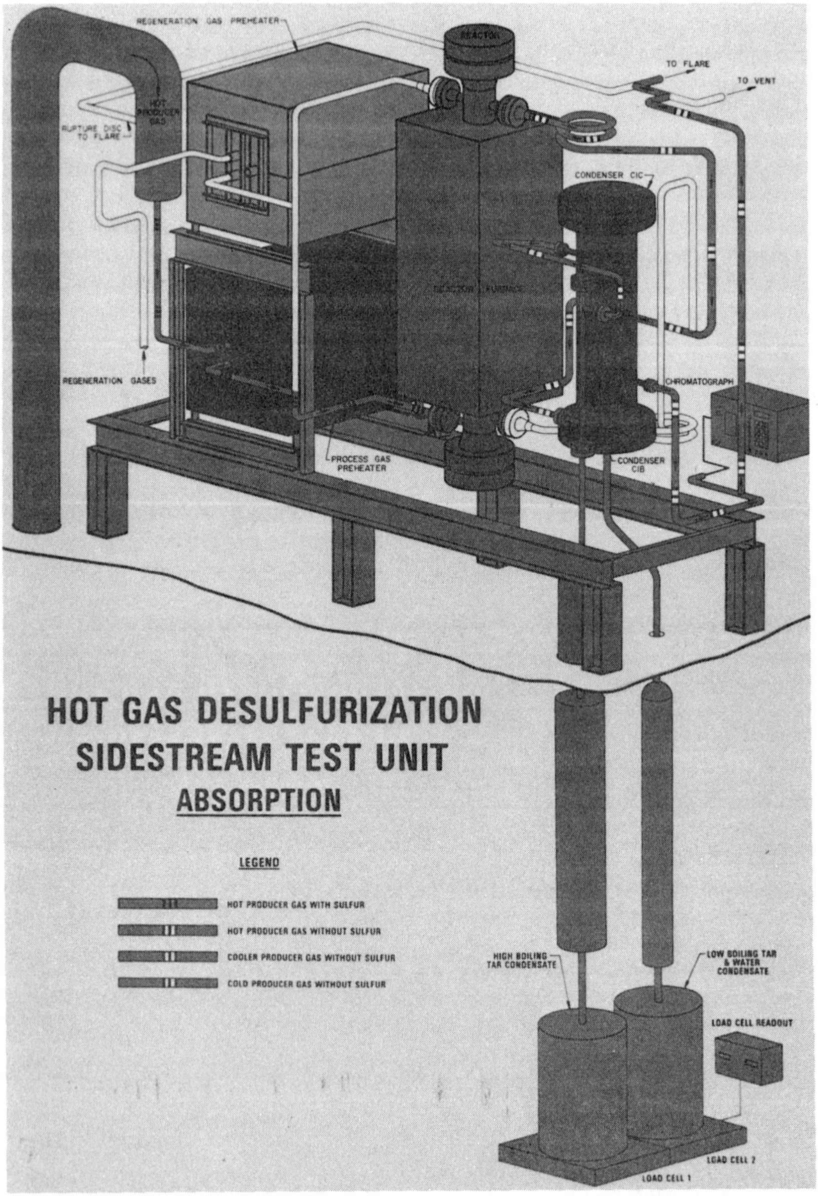

Figure 7. Hot gas desulfurization sidestream test unit—absorption

principal component is a 6-inch diameter, 6-foot long reactor of Incoloy 800H heated by a clamshell type, electrically heated furnace. It is aluminized internally for increased corrosion resistance and is designed for operation up to a pressure of 300 psig (~ 2MPa(ga)) and a temperature of 1,800°F (~ 980°C).

In operation, hot coal gases or regeneration gases pass through a fixed bed of about half a cubic foot of desulfurization sorbent in the reactor. Water and tars are condensed from the gases before discharge. The gases entering and leaving the unit are analyzed by various techniques, as also are the condensates. The sorbent is characterized at the conclusion of a test. Important parameters which are monitored are reactor pressure drop, pressure and temperature, and gas flow rate.

Results

The side stream unit was operated for about 500 hours during 4 gasifier runs of about 2 weeks each. During this time, various bituminous coals and lignite were fed to the gasifier, which was operated in both air- and oxygen-blown modes.

Hot gas was taken from the exit of the gasifier cyclone through a small secondary cyclone in a heated stainless steel pipeline about 100 feet long. Analyses of the coals fed to the gasifier are shown in Table 3. A typical gas composition from one of the bituminous coals is:

	H_2O	CO_2	CO	H_2	CH_4	N_2
Volume Percent	5	10	17	16	3	49

Sulfur compound levels are given in test results below. The gas contains additionally about 0.5 weight percent of tar and other minor constituents.

Five separate batches of zinc ferrite sorbent underwent 13 sulfidations and 9 regenerations. Reactor temperature was generally 1,000° to 1,300°F (~ 540 to 705°C), pressure 120 to 240 psig (~ 0.8 to 1.6 M Pa [ga]), and space velocity, 1,000 to 8,000 h^{-1}. Figure 8 shows the H_2S breakthrough plots for two consecutive sulfidation runs with an intermediate regeneration. In the first sulfidation, with fresh sorbent, the H_2S level is less than 5 ppm until breakthrough. Except for a short initial time

Table 3
Ultimate Analysis of Various Coals

COMPONENT	WT. %				
	PITTSBURGH SEAM COAL			KITTANNING SEAM COAL	NORTH DAKOTA LIGNITE
	ARKWRIGHT	BLACKSVILLE	BLACKSVILLE BRIQUETTES	KITTANNING	INDIAN HEAD
MOISTURE	0.39	1.1	0.96	0.82	0.52
ASH	7.15	10.55	15.33	6.43	10.28
S	1.82	3.20	2.39	0.98	1.66
H	5.12	5.58	1.72	5.55	4.76
N	3.05	1.30	1.29	1.61	1.27
TOTAL CARBON	77.10	73.17	70.49	78.23	67.54
VOLATILES	38.35	38.25	34.14	34.79	40.33
Btu/lb	13715	13308	12615	14008	11956

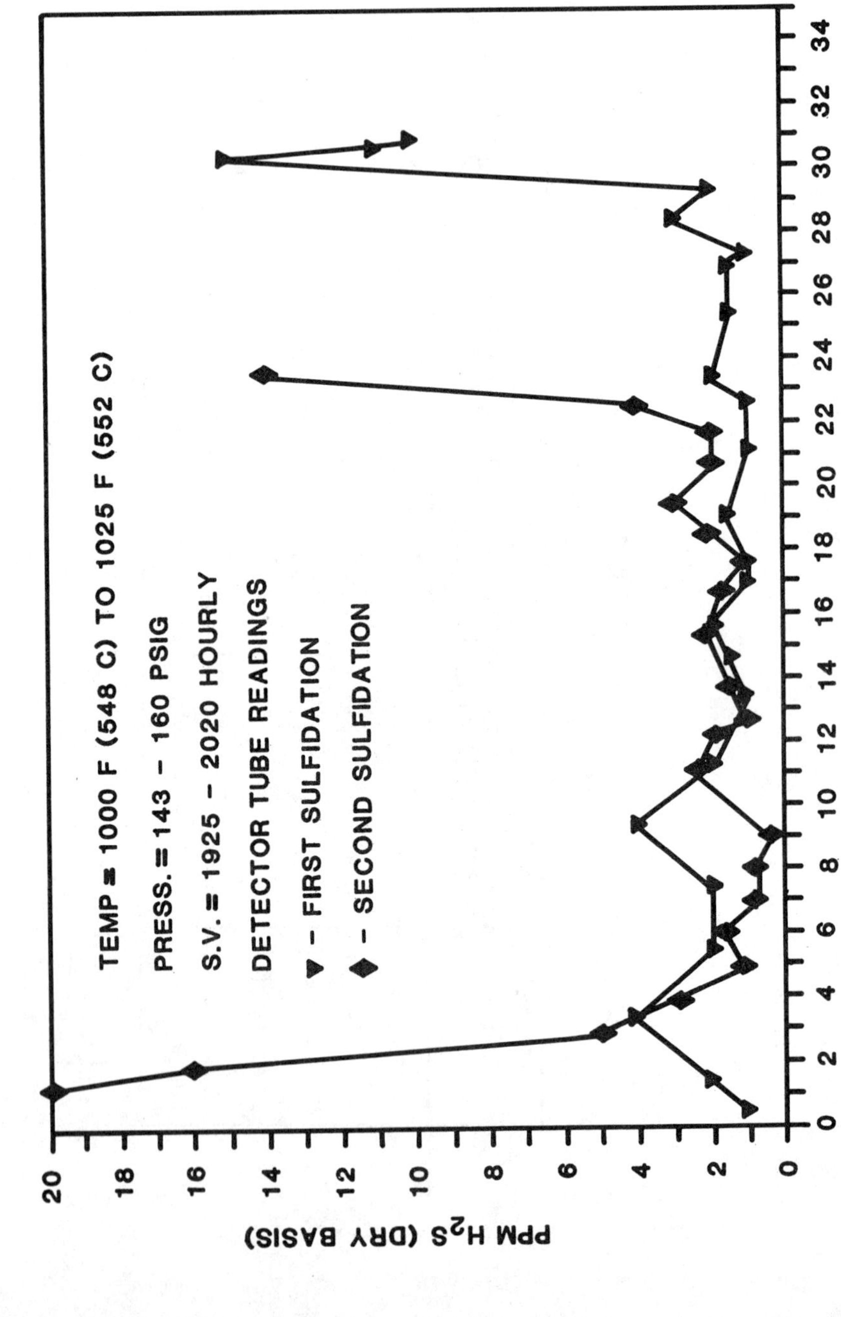

Figure 8. H_2S breakthrough curves for zinc ferrite.

when H_2S is high, presumably because of residual sulfate in the sorbent, the second sulfidation is similar. The amount of sulfur absorbed is about the same in both cases.

Figures 9, 10, 11, and 12 are semilog plots showing inlet and outlet H_2S, COS, CS_2, and thiophene, C_4H_4S, levels, respectively in gas produced from Pittsburgh (Tests 13, 15, and 19) and Kittanning (Test 17) coals. Each species is generally removed to a level between 1 and 10 ppm, giving a total sulfur content less than about 10 ppm. Measurements were made on grab samples by laboratory gas chromatograph.

Particulate analysis, loading, and size distribution in the hot gas entering and leaving the side stream reactor were measured in samples taken with a ceramic filtering system. The measurements are shown in Table 4 for Pittsburgh (Tests 13 and 15) and Kittanning (Test 17) coal gases. Particulate loading in each case is of the order of $1g/Nm^3$. The significant reduction in loading across the reactor is approximately the same in each test. Larger particles are removed preferentially. Exiting particles have an ultimate analysis similar to that for entering particles. Figure 13 plots the reactor pressure drop with

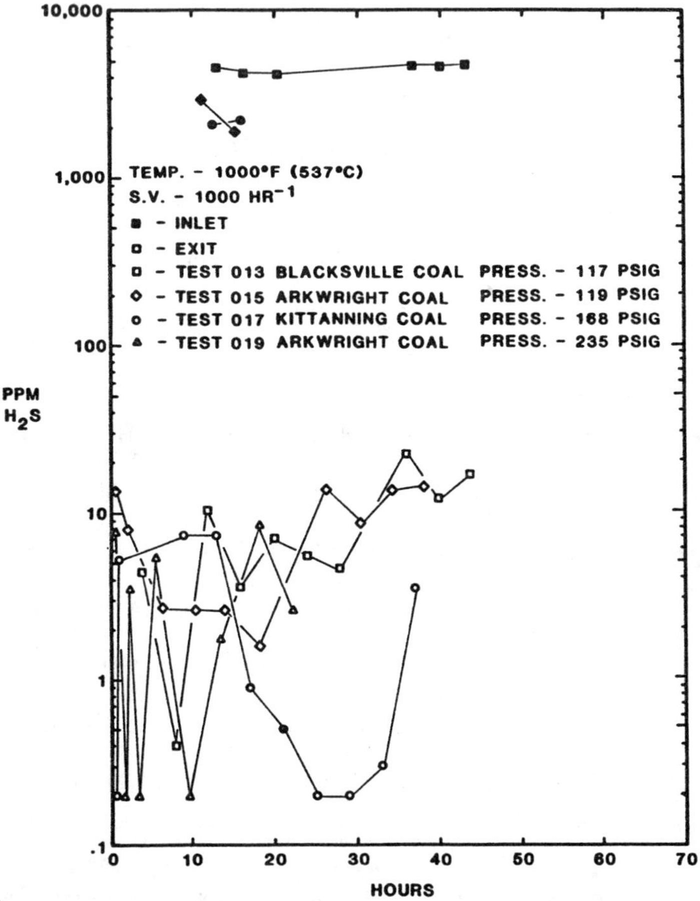

Figure 9. Removal of hydrogen sulfide from coal gas by zinc ferrite.

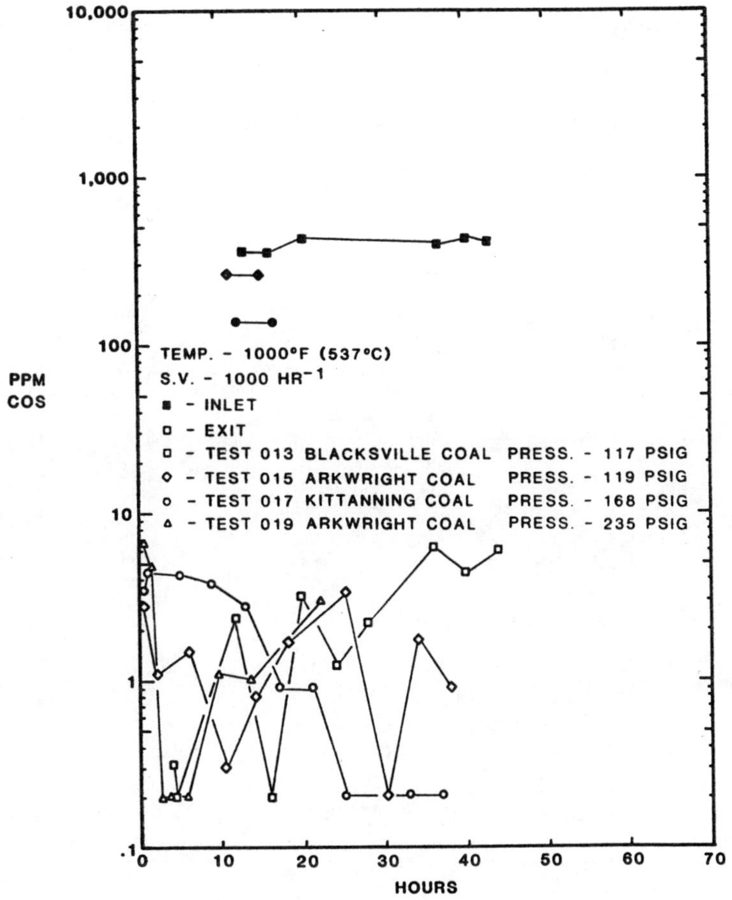

Figure 10. Removal of carbonyl sulfide from coal gas by zinc ferrite.

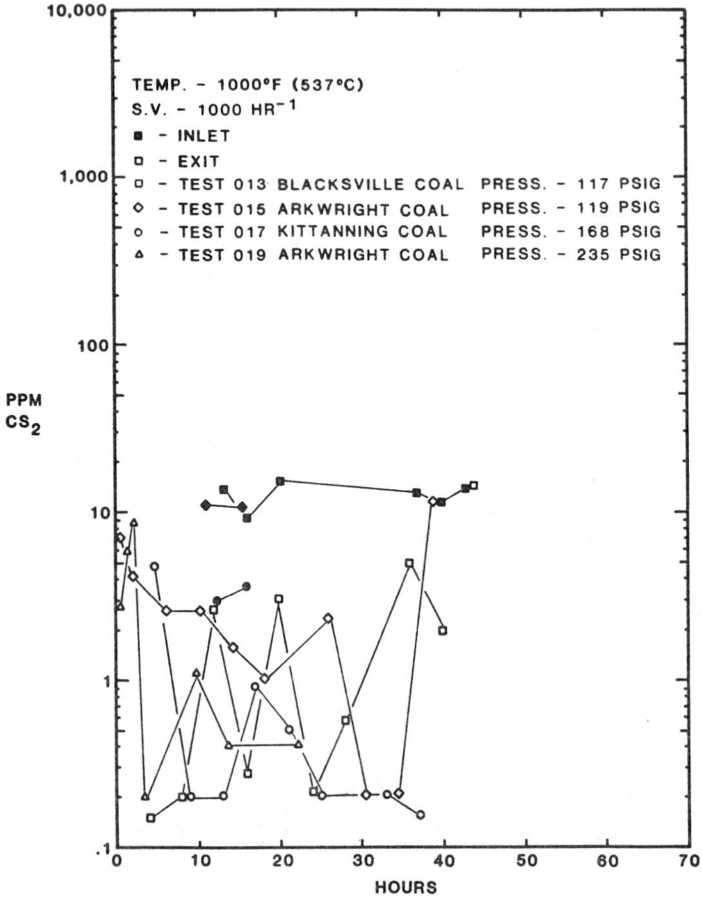

Figure 11. Removal of carbon disulfide from coal gas by zinc ferrite.

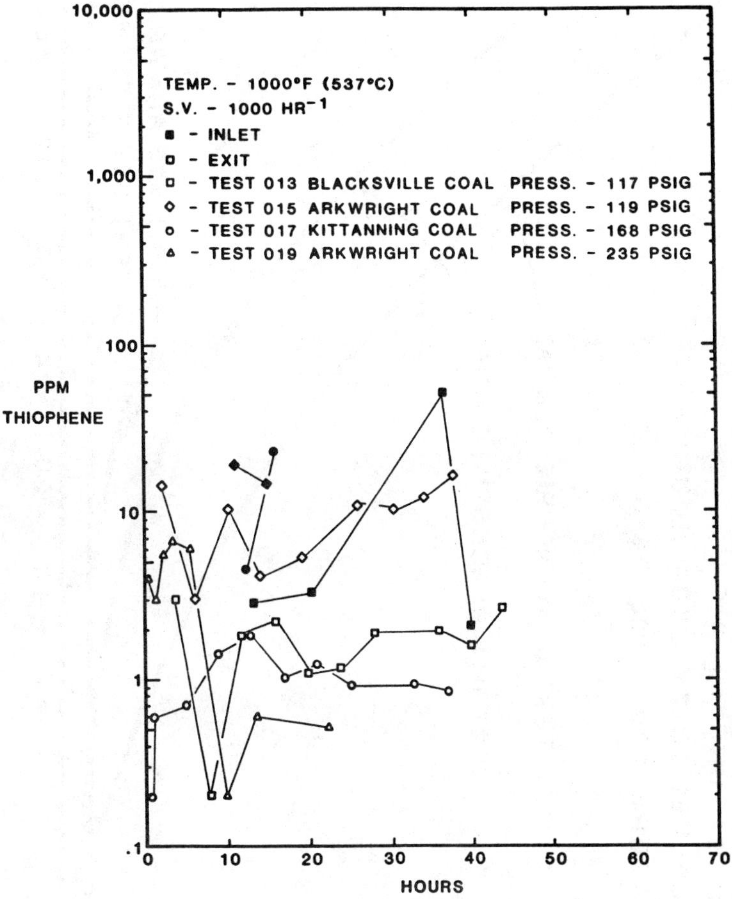

Figure 12. Removal of thiophene from coal gas by zinc ferrite.

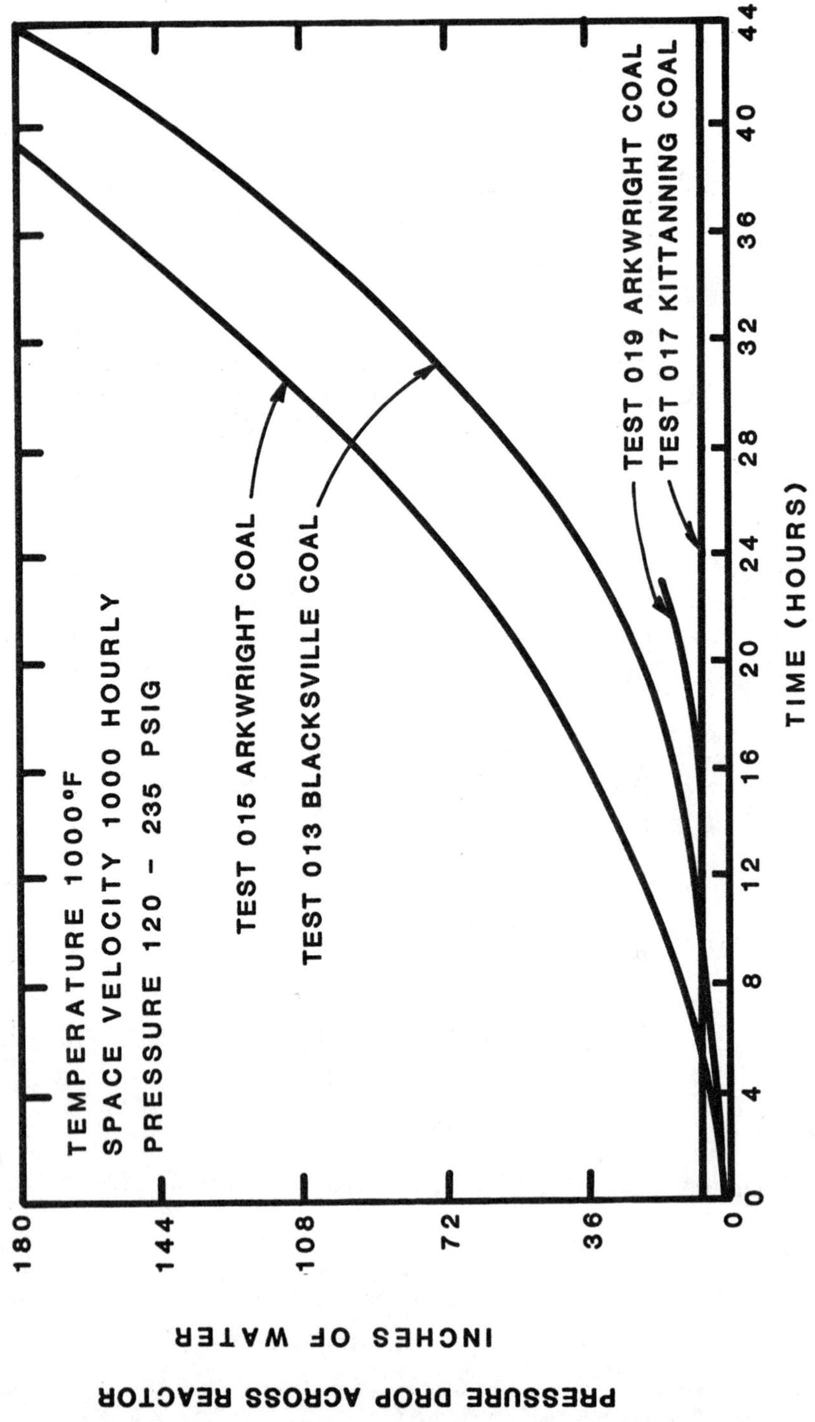

Figure 13. Reactor pressure drop during operation of gases derived from various coals.

Table 4
Particulate Sampling Data
Hot Gas Desulfurization Test Unit
Gasifier Run 103

TEST #	FROM START OF TEST	PARTICLE LOADING g/Nm^3		MEDIAN PARTICLE SIZE μm		CHEMICAL ANALYSIS WT PERCENT				
		INLET	EXIT	INLET	EXIT	C	H	N	S	Ash
013	16.6 HRS	2.12		21.3		81.1	3.8	2.4	0.1	4.4
013	21.1 HRS		1.36		14.5	67.9	4.0	3.5	NM	10.2
013	42.5 HRS		0.469		13.1	73.8	4.9	4.3	NM	3.3
015	18.3 HRS	1.52		26.0		79.1	1.4	1.7	0.3	9.5
015	15.5 HRS		0.381		21.6	72.4	4.8	2.8	NM	7.4
017	16.6 HRS	2.09		13.0		75.9	4.6	3.2	0.9	2.9
017	27.7 HRS		0.318		5.3	76.5	2.1	1.4	1.1	12.9

NM = Not Measured due to insufficient sample size.

time. There appears to be a positive correlation between median particle size and rate of increase of pressure drop. For Test 17, where the median particle size entering the reactor is 13 μm, virtually no increase in pressure drop occurred during 50 hours on-stream time. In Tests 13 and 15, the pressure drop became intolerably high before H_2S breakthrough. Test 19 was ended prematurely because of gasifier shutdown, but in this case also pressure drop has an upward trend. After the intermediate regenerations, this pressure drop was restored to a low value, presumably because of particulate elutriation and coke burn-off.

Based on the analysis of the composite water condensate downstream of reactor, Tables 5 and 6 give the concentrations of other contaminants in Pittsburgh coal gas and regeneration gas, respectively. Halides and alkali metals appear to be removed by the sorbent by as much as 70 percent, based on their subsequent release from the sorbent during regeneration. The regeneration gas is enriched in vanadium, calcium, aluminum, mercury, and ammonium, indicating these may also have been removed to some extent. Very little zinc is found in either the coal or regeneration-exit gases.

Table 5
Trace Components in Exit Gas During Sulfidation

COMPONENT	CONCENTRATION (ppmw)		
	AIR BLOWN GASIFIER RUN 101 (TESTS 001, 003, and 005)		OXYGEN BLOWN GASIFIER RUN 101 (TEST 005)
	Sample 1	Sample 2	Sample 3
Chloride, (Cl)	45	62	289
Cyanide, total (Cn)	<1	NM	NM
Nitrogen, Ammonia, total (N)	758	478	329
Phenolics, total (phenol)	1.5	NM	NM
Sulfate (SO_4)	8.9	12	43
Sulfide (S)	4	NM	NM
Sulfite (SO_3)	10	NM	NM
Fluoride (F)	2.2	2	16
Thiocyanate (SCN)	8	NM	NM
Antimony, total (Sb)	NM	NM	NM
Arsenic, total (As)	.3	.06	0.10
Iron, total (Fe)	.7	1	3.5
Lead, total (Pb)	<.002	NM	NM
Calcium, total (Ca)	.3	NM	NM
Potassium, total (K)	.2	.3	0.6
Vanadium, total (V)	<.02	<.002	0.03
Zinc, total (Zn)	.05	.08	0.02
Sodium, total (Na)	.13	.5	1.23
Chromium (Cr)	NM	.1	0.16
Nickel (Ni)	NM	.03	0.05
Phosphate (PO_4)	NM	<.1	<1.6
Bromine (Br)	NM	<.2	<2.1
Nitrite (NO_3)	NM	.8	<1.1
Aluminum (Al)	NM	.007	<0.011
Selenium (Se)	NM	.01	0.03
Mercury (Hg)	NM	.33	0.92

NOTES:

1. Sampling location is at condenser CIC knockout.
2. Sample 1 metals analysis obtained by atomic absorption
3. Sample 1 anions and cations determined by wet chemistry (Titrimetric, colorimetric, and specific ion electrode).
4. Sample 2&3 metals analysis obtained by atomic absorption/atomic emission spectroscopy.
5. Sample 2&3 anions and cations determined by ion chromatography.
6. Concentration in gas phase calculated based on analysis in aqueous phase.
7. NM = Not Measured.

Table 6
Trace Components in Exit Gas During Regeneration

COMPONENT	CONCENTRATION (ppmw)	
	TESTS 002 & 004	PLANT STEAM
Zinc (Zn)	0.72	0.582
Chromium (Cr)	0.24	<0.02
Iron (Fe)	5556	0.351
Nickel, (Ni)	0.70	<0.8
Potassium (K)	6.38	0.150
Sodium (Na)	8.79	0.293
Calcium (Ca)	28.2	0.873
Vanadium (V)	0.3	<0.05
Ammonia (NH_4)	373	0.54
Fluoride (F)	24	0.11
Chloride (Cl)	839	1.02
Phosphate (PO_4)	<2.4	<0.03
Bromine (Br)	17.5	<0.03
Nitrite (NO_3)	<1.6	1.12
Sulphate (SO_4)	675	2.38
Aluminum (Al)	1.45	0.169
Selenium (Se)	<0.008	<0.01
Mercury (Hg)	70.3	<0.2
Arsenic (As)	<0.008	<0.01

NOTES:

1. Sampling location is at condenser CIC knockout.
2. Metals analysis was by atomic adsorption/atomic emission spectroscopy.
3. Anions and cations determined by ion chromatography.
4. Concentration in gas phase calculated based on analysis in aqueous phase.

Ultimate analysis of tar condensate, which amounts to up to 1 weight percent of the gas, yields about 1 weight percent for its sulfur content, translating to up to 100 ppm carried through the reactor in the gas.

The sorbent, on completion of each test series, was removed from the reactor and characterized at a number of points along the reactor length. Table 7 gives the mineral analysis of the sorbent after it has been in contact with oxygen-blown Arkwright coal gas. Table 8 gives other properties. It can be seen that at the reactor inlet most of the sorbent has been converted to iron and zinc sulfides. At the outlet, only zinc sulfide is found, indicating a transition to an H_2S level too low to form iron sulfide. There is not a marked change in surface area and crush strength compared with fresh sorbent. The sorbent contains up to 4 weight percent of carbon. Mineral analysis of sorbent in the case of air-blown gas indicated some formation of iron carbide and wustite (above 1,200°F [~ 650°C]). This is ascribed (see Figure 1) to the operation of the gasifier at the low steam/air ratio of 0.2 weight/weight to attain a high cold gas thermal efficiency, which produced a gas of high CO/CO_2 ratio.

Table 7
Analysis of Minerals in Sulfided Zinc Ferrite

BED HEIGHT (INCHES)	HEMATITE Fe_2O_3	FRANKLINITE $ZnFe_2O_4$ OR MAGNETITE Fe_3O_4 (4)	ZINCITE ZnO	ALPHA ZnS	BETA ZnS	PYRRHOTITE FeS
	PERCENT TII (3)					
0 (1)	-	10	Trace	13	51	25
6	-	4	-	12	54	29
16	-	18	Trace	19	36	27
23	-	10	-	19	42	28
26	-	19	Trace	24	25	32
36	-	68	14	10	7	-
46 (2)	-	70	18	9	4	-
Fresh Sorbent	4	89	6	-	-	-

NOTES:

(1) Bottom/Inlet

(2) Top/Outlet

(3) Percent TII = Percent Total Integrated Intensity: Directly proportional to the concentration of the crystalline phase. Amorphous Phase is not accounted for.

(4) Franklinite and Magnetite have the same crystal structure and, therefore, cannot be readily distinguished by the diffraction pattern.

Table 8
Sorbent Analysis of Sulfided Zinc Ferrite

BED HEIGHT (INCHES)	TOTAL CARBON (%WT)	TOTAL SULFUR (%WT)	BET SURFACE AREA (SQ M/G)	DENSITY (G/CC) (3)	PORE VOLUME (CC/G) (5)	CRUSH STRENGTH (Kg DWL) ± S.D. (7)
0 (1)	1.97	29.88	1.4332	4.366	0.002152	7.0 ± 3.29
6	1.88	31.41	1.4903	4.548	0.002068	8.4 ± 3.55
16	2.58	27.78	2.0936	4.261	0.004003	7.3 ± 3.54
23	2.89	25.92	2.4263	4.326	0.004616	8.1 ± 3.09
26	3.48	28.97	2.2432	4.344	0.005324	4.4 ± 2.09
36	2.37	7.29	2.2978	4.863	0.003678	4.2 ± 2.57
46 (2)	3.27	8.60	3.1138	5.141	0.005498	2.4 ± 1.76
--METC (8)	0	0.13	3.1305	6.159	0.004316	6.3 ± 2.8
FRESH SORBENT --UCI (9)			3.2	85 (4)	0.25	7.2
ABSORBER FINES	6.93	26.29				

NOTES:

(1) Bottom/Inlet during sulfidation - Exit during regeneration.
(2) Top/Outlet during sulfidation - Inlet during regeneration.
(3) Helium Density
(4) Bulk Density (Lbs/Cu.Ft.)
(5) Nitrogen Adsorption
(6) Mercury Porosimetry
(7) Crush strength data are average and standard deviation of 15 measurements.
(8) Analysis of fresh sorbent done at METC.
(9) Analysis of fresh sorbent done at United Catalysts.

The corresponding mineral analysis and other properties after a regeneration are given in Tables 9 and 10. In this case, it can be seen that the sorbent has been restored to its original composition, predominantly zinc ferrite. There appears to be a small enrichment in zinc at the top of the reactor. Again there is not a big difference between the surface area and crush strength of the regenerated and fresh sorbent. The sorbent contains up to about 1 weight percent of residual sulfur as sulfate, mostly at the bottom of the reactor. Virtually all of the carbon has been burned out.

SULFUR RECOVERY

The regeneration offgas obtained upon regenerating the sorbent with air may contain from 2 to 12 volume percent sulfur dioxide on a dry basis, depending upon the dilution gas added for temperature control. Conventional technology widely used in the metallurgical industry (21) is available to recover or fix the sulfur dioxide content of this gas. Common processes are absorber-strippers for sulfur dioxide recovery and ultimate conversion to sulfur or sulfuric acid and limestone scrubbers, producing a gypsum product.

Table 9
Analysis of Minerals in Regenerated Zinc Ferrite

BED HEIGHT (INCHES)	HEMATITE Fe_2O_3	FRANKLINITE $ZnFe_2O_4$ OR MAGNETITE Fe_3O_4 (4)	ZINCITE ZnO	ALPHA ZnS	BETA ZnS	PYRRHOTITE FeS
	PERCENT TII (3)					
0 (1)	4.9	93.3	1.9			
6	4.4	93.0	2.6			
16	3.8	93.7	2.5			
23	3.9	93.7	2.5			
26	3.1	94.0	2.9			
36	2.7	93.4	3.9			
46 (2)	2.0	92.3	5.8			
FRESH SORBENT	4	89	6			

NOTES:

(1) Bottom/Inlet during sulfidation - exit during regeneration.

(2) Top/Outlet during sulfidation - inlet during regeneration.

(3) Percent TII = Percent Total Integrated Density: Directly proportional to the concentration of the crystalline phase. Amorphous Phase is not accounted for.

(4) Franklinite and Magnetite have the same crystal structure and, therefore, cannot be readily distinguished by the diffraction pattern.

Table 10
Sorbent Analysis of Regenerated Zinc Ferrite

BED HEIGHT (INCHES)	TOTAL CARBON (% WT)	TOTAL SULFUR (% WT)	% SULFUR AS SULFATE SO_4	% SULFUR AS SULFIDE S	BET SURFACE AREA (SQ M/G)	DENSITY (G/CC)(3)	PORE VOLUME (CC/G)(5)	CRUSH STRENGTH (Kg DWL) ± S.D.(7)
0(1)	0.01	1.99						
0-6			1.21	0.0006	1.9591	5.582	0.003666	7.45 ± 3.41
6	0.08	1.43						
6-16			0.66	0.0033	1.9756	5.645	0.002679	5.66 ± 2.99
16	0.02	1.33						
23	0.01	1.36	0.71	0.0012	2.0393	5.567	0.003835	5.42 ± 3.50
26	0.01	0.97						
26-36			0.42	0.0008	1.6614	5.572	0.002274	4.20 ± 4.21
36	0.02	0.59						
36-46			0.20	0.0096	1.5306	5.769	0.001548	3.88 ± 3.68
46(2)	0.03	0.14						
-- METC(8)	0	0.13			3.1305	6.159	0.004316	6.3 ± 2.8
FRESH SORBENT --UCI (9)					3.2	85(4)	0.25(6)	7.2

NOTES:

(1) Bottom/Inlet during sulfidation - Exit during regeneration.
(2) Top/Outlet during sulfidation - inlet during regeneration.
(3) Helium Density
(4) Bulk Density (Lbs/Cu ft).
(5) Nitrogen Adsorption.
(6) Mercury Porosimetry.
(7) Crush strength data are average and standard deviation of 15 measurements.
(8) Analysis of fresh sorbent done at METC.
(9) Analysis of fresh sorbent done at United Catalysts.

SYSTEMS STUDIES

The comparative advantages of high- and low-temperature coal gas desulfurization have been the subject of much controversy, leading to a number of studies of the topic (22,23,24,25,26). These are generally inconclusive because of the lack of detailed design data for the high-temperature case. Potential savings of up to 20 percent in the cost of electricity have been reported (26).

CONCLUSIONS/ACCOMPLISHMENTS

It has been demonstrated that, in the operating regime 1,000 to 1,200°F (~ 540 to 650°C) and 120 to 240 psig(~ 0.8 to 1.6 M Pa [ga]), space velocity 1,000 to 2,000 h^{-1}, a zinc ferrite sorbent can function in a hot gas stream from a fixed-bed coal gasifier, removing low molecular weight sulfur compounds to a level of about 10 ppm. High molecular weight sulfur compounds, which are constituents of the associated tar and translate to up to 100 ppm of sulfur in the gas, are not removed. The sorbent picks up about 30 percent of its weight of sulfur and can be regenerated by air at 1,400 to 1,500°F (~ 760 to 815°C), to near its original condition with a residual sulfur content (as sulfate) of less than 1 weight percent. Absorption performance after three regenerations is close to that of fresh sorbent.

A considerable amount of trace elements and ions, possibly deleterious to downstream processes, are absorbed and later released into the regeneration tail gas.

Based on very limited gas characterization data, it appears that reactor pressure drop build-up may be related to gas particle size distribution. For a median particle size of 13μm, pressure drop increase was very small. The gas contained up to 1 weight percent of tar.

Upon regeneration, the reactor pressure drop is restored to a low level, presumably as a result of particulate elutriation and coke burn-off. Long-term testing over many cycles will be required to establish the durability of the sorbent, and such a project is presently under consideration. Previous work with iron oxide sorbents gives grounds for optimism that a satisfactory durability can be attained.

ACKNOWLEDGMENTS

This paper is published by permission of Morgantown Energy Technology Center. We wish to thank colleagues for their cooperation in its preparation, particularly, L. C. Headley, K. Pater, R. V. Rahfuse, R. R. Romanosky, C. H. McCurry, R. J. Anderson, R. D. Bagnell, R. E. Carr, J. E. France, F. D. Slagle, J. R. Rotunda, G. R. Fizer, R. J. Corley, J. M. Clise, E. N. Eisentrout, P. J. Barnhouse, R. D. Barnes, and J. T. Cesa.

Other analytical and technical support was provided by EG&G, Inc., and Kemron Environmental Services.

REFERENCES

1. Grindley, T., V. P. Kothari, and K. Pater, "Coal Gas Desulfurization Activities at Morgantown Energy Technology Center." Proceedings of 12th Biennial Lignite Symposium, May 1983.

2. "Coal-Derived Fuel Gases for Molten Carbonate Fuel Cells." TRW for Morgantown Energy Technology Center, Report No. DOE/METC/8085/6, 1979.

3. "Monitoring Contaminants in Coal-Derived Gas for Molten Carbonate Fuel Cells." TRW for Argonne National Laboratory, Report No. DOE/METC/82-44, 1981.

4. Abel, W. T., F. G. Schultz, and P. F. Langdon, "Removal of Hydrogen Sulfide From Hot Producer Gas by Solid Absorbents." Bureau of Mines Report of Investigations 7947, 1974.

5. Oldaker, E. C., A. M. Poston, and W. L. Farrior, "Removal of Hydrogen Sulfide from Hot Low-Btu Gas with Iron Oxide-Fly Ash Sorbents." Report No. MERC/TPR-75, Vols. I and II, Morgantown Energy Technology Center, 1975.

6. Joshi, D. K. and E. L. Leuenberger, "Hot Low-Btu Producer Gas Desulfurization in Fixed Bed of Iron Oxide-Fly Ash." Air Products and Chemicals, Inc., for Energy Research and Development Administration, Report No. ERDA/FE-2033-19, Vols. I and II, 1977. Joshi, D. K., J. H. Olsen, M. L. Hayes, and V. Shah, "Hot Low-Btu Producer Gas Desulfurization in Fixed Bed of Iron Oxide-Fly Ash." Air Products and Chemicals, Inc., for Department of Energy, Report No. DOE/FE-2757-3, 1979.

7. Jenkins, R. J., "Hot Low-Btu Producer Gas Desulfurization in Fixed Bed of Iron Oxide-Fly Ash." Air Products and Chemicals, Inc., for Department of Energy, Report No. DOE/FE-2033-14, 1976.

8. Case, G. D., et al., "Chemistry of Hot Gas Cleanup in Coal Gasification and Combustion." Report No. MERC/SP-78/2, 1978.

9. Westmoreland, P. R. and D. P Harrison, "Evaluation of Candidate Solids for High-Temperature Desulfurization of Low-Btu Gases." Environmental Science and Technology, 10 (659), 1976.

10. Jalan, V. and D. Wu, "Molten Carbonate Fuel Cell Power Plant Desulfurization Systems." Giner, Inc., for U.S. Department of Energy, Report No. DOE/ET/15370-T1 & 2, 1980. Jalan, V., et al., "Studies

Involving High-Temperature Desulfurization/Regeneration Reactions of Metal Oxides for the Fuel Cell Program." Giner, Inc., for Argonne National Laboratory. Report No. ANL/J/ED3/12, 1981.

11. Correspondence with D. Garvin, Chemical Thermodynamics Data Center, National Bureau of Standards, July 16, 1982.

12. Parekh, R. D., "Handbook of Gasifiers and Gas Treatment Systems." UOP/SDC for Department of Energy, Report No. WD-TR-82/008-010, 1982.

13. Grindley, T. and G. Steinfeld, "Development and Testing of Regenerable Hot Coal Gas Desulfurization Sorbents." Presented at AIChE Spring National Meeting. Report No. DOE/MC/16545-1125, 1982.

14. Grindley, T. and G. Steinfeld, "Development and Testing of Regenerable Hot Coal Gas Desulfurization Sorbents." Second Annual Contractors' Meeting on Contaminant Control in Hot Coal-Derived Gas Streams. Report No. DOE/METC/82-47, 1982. Grindley, T. and G. Steinfeld, "Zinc Ferrite Hydrogen Sulfide Absorbent." Third Annual Contractors' Meeting on Contaminant Control in Hot Coal-Derived Gas Streams. Report No. DOE/METC/84-6, 1983.

15. Steinfeld, G., "Hot Gas Desulfurization." Science Applications, Inc., for Morgantown Energy Technology Center, Report No. 1-443-13-710-00, 1981. Steinfeld, G., "Hot Gas Desulfurization." Science Applications, Inc., for Morgantown Energy Technology Center, Report No. DOE/MC/16545-T1, 1982. Steinfeld, G., "Status of METC Investigations of Coal Gas Desulfurization at High Temperature." Science Applications, Inc., for Morgantown Energy Technology Center, Report No. DOE/MC/16545-1553, 1984.

16. Green, D. A., K. E. Leese, and W. J. McMichael, "Pollution Control in Coal Gasification." Research Triangle Institute for Morgantown Energy Technology Center. Report No. DOE/MC/16441-1406, 1983.

17. Krishnan, G. N. and H. Wise, "Study of Sulfate Formation in Regeneration of Zinc Ferrite Sorbents." SRI International for Morgantown Energy Technology Center. Contract No. DE-AC21-83MC20092, 1983.

18. Patel, P. S. and S. Rich, "Trace Contaminant Removal from Hot Coal Gas for Molten Carbonate Fuel Cell Application." Energy Research Corporation for Morgantown Energy Technology Center. Contract No. DE-AC21-83MC20094.

19. Schrodt, J. T., "Dynamic Simulation of a Fixed-Bed Zinc/Iron Oxide, High-Temperature, Desulfurization Process." Prepared for U.S. Department of Energy, Morgantown Energy Technology Center. Report No. DOE/MC/21582-1500, 1983.

20. Grindley, T. and G. Steinfeld, "Testing of Zinc Ferrite Hydrogen Sulfide Absorbent in a Coal Gasifier Side Stream." Fourth Annual Contractors' Meeting on Contaminant Control in Hot Coal-Derived Gas Streams, 1984.

21. Masarky, N. H., et al., "Control of Air Pollution Emissions from Molybdenum Roasting," Molycorp, Inc., for Environmental Protection Agency. Report No. EPA-600/2-83-093, 1983.

22. Lemmon, A. W., B. C. B. Hsieh, and A. E. Weller, "Evaluation of Low-Temperature Cleanup Systems for Low-Btu Gas." Battelle Columbus Laboratories for Tennessee Valley Authority, 1975.
Jenkins, D. M., and A. W. Lemmon, "Evaluation of High-Temperature Desulfurization Systems for Low- and Intermediate-Btu Gas." Battelle Columbus Laboratories for Tennessee Valley Authority, 1975.

23. Onursal, A. B., "Hot Gas Cleanup Process," Dynalectron, Inc., for Environmetal Protection Agency. Report No. EPA-600/7-79-169, 1979.

24. Jones, C. and J. M. Donohue, "Comparative Evaluation of High- and Low-Temperature Gas Cleaning for Coal Gasification, Combined-Cycle Power Systems." Stone and Webster for EPRI, Report No. AF-416, 1977.

25. Robson, F. L. and W. A. Blecher, "Assessment of Fuel Gas Cleanup Systems." United Technologies Research Center for Morgantown Energy Technology Center, Report No. DOE/METC/12050-149, 1981.

26. Jablonski, G., et al., "Evaluation of Gasification and Gas Cleanup Processes for Use in Molten Carbonate Fuel Cell Power Plants." Westinghouse R&D Center for Morgantown Energy Technology Center, Report No. DOE/MC/16220-1306, 1982.

17

High-Temperature Desulfurization of Coal Gases

Vinod Jalan
Giner, Inc.

Utilization of coal gases requires development of technically feasible and economical means for sulfur removal. The future of coal as a fuel or chemical raw material provides strong incentives (1) for improvements in coal cleaning and preparation, combustion processes and gas cleanup (fuel and flue). Sulfur removal for coal gas utilization involves a cleanup stage following gasification and prior to subsequent use in energy generation (combined cycles or fuel cells) or in synthesis of fuels (methanol, methane, hydrocarbons). There is substantial evidence that there is merit in using dry scrubbing with metal/metal oxide sorbents in this application.

The purpose of this review is to summarize the state of the art for hot desulfurization of coal gases following gasification. This is an area in which there are essentially no commercial processes, but for which there are strong incentives for technology development.

THE MERITS OF HOT DESULFURIZATION

Hot gas contaminant control provides substantial advantages for two energy conversion systems (2, 3, 4) namely, pressurized fluidized bed combustion (PFBC) and the molten carbonate fuel cell (MCFC). For PFBC (Table 1) hot gas contaminant control involves not only removal of sulfur but also particulates, alkalies, halogens and ammonia.

Table 1
Gasifier/Turbine Hot Gas Cleanup
Advantages (3,4)

CCG-Plant: (Fluidized Air Blown Gasifier--600 MWe--1800°F Cleaning Temp (Dolomite) 2200°F T.I.T.-2400 PSIG/1000°F Steam)

	Cold Cleanup	Hot Gas Cleanup
o Thermal Efficiency (%)	38	40
o Capital Cost ($/KW)	830	670
o Cost of Electricity (Mills/KWHR.)	74	67

For the MCFC (Table 2) in particular there are severe requirements for reduction of sulfur levels to below 1 ppm to meet the tolerance of the fuel cell (5, 6).

Table 2
MCFC High Temperature Cleanup
Advantages (3)
250 MWe Utility Plant

	Cold Cleanup (W O_2, SELEXOL)	Hot Cleanup (W O_2, METC, FeO)
Investment Costs (Total Costs Millions)	391	358
Cost of Electricity (Mills/KWHR.)	74.7	67.7
Heat Rate BTU/KWHR.	7,660	7,823

(Single 250 MW Fuel Cell Utility Plant Incurs a $10.7 Million Savings at 70% Utilization Factor at 7 Mills/KWHR.)

Hot gas cleanup of the fuel gases used in both PFBC and MCFC offers potential economics over available low temperature technology in providing simpler process hardware, minimizing impacts of secondary waste liquor treatment and capital cost savings of about 25 percent (2). Two assessments of these potential advantages are summarized here.

Jones (7) made a comparison for low and high temperature gas cleanup of a 1000 MW coal gasification-combined cycle power plant for two possible gas turbine inlet temperatures (1070°C and 1320°C). Four gasification options were considered: 1) air blown Lurgi (fixed bed), 2) oxygen blown Lurgi (fixed bed), 3) oxygen blown BGC (slagging fixed bed) and 4) the air blown Foster Wheeler (entrained bed). The high-temperature hot gas cleanup (HGC) used was the METC iron oxide fixed bed desulfurization process. The low-temperature gas cleanup (LGC) used was the Benfield process. The HGC alternative provided advantages in thermal efficiency, capital cost, electricity cost and purification system component costs. Savings of 34 percent in capital requirements and 30 percent for cost of electricity were obtained with HGC.

In a more recent assessment Robson et al (8) evaluated four gasifier configurations - three entrained flow with Texaco, Shell-Koppers (now Shell Gasification Process) and Foster Wheeler/Bituminous Coal Research (FW/BCR), and an IGT fluid bed type - twenty-four variations of air and oxygen blown gasification with three temperature regimes for desulfurization. The low temperature (<300°F) desulfurization was conducted with Selexol liquid scrubber which removed most of the H_2S but somewhat less of the COS. The medium temperature (1000°F) desulfurization was conducted with the METC iron oxide fixed sorbent bed. The high temperature (~ 1500°F) desulfurization used was the Conoco half calcined dolomite in a fluidized bed. The COS removal capability of the latter two desulfurization processes are largely undemonstrated but considered promising. All of these processes (Table 3) indicated a capability to reduce sulfur emissions to values suitable to meet current EPA emission standards. The medium and high temperature processes provided advantages in thermal efficiency, plant capital cost, and overall electricity cost. The greatest advantage provided by the high temperature processes was in reduced capital cost of a plant. The high temperature processes pose some still unresolved problems in handling alkalies and particulates which defer final judgement on the ultimate economics to be realized. Readers are referred to Lorton (9) for a recent evaluation of low temperature sulfur removal processes for advanced fuel cell systems.

HIGH TEMPERATURE PROCESSES

The cleanup of sulfur from fossil fuel gases and flue gases using dry scrubbing over solid sorbents has been studied extensively. In fuel gas desulfurization, principal attention has been given to mid-temperature (200°C) physical adsorption of H_2S, COS and organic sul-

Table 3
Summary of Integrated Systems Performance

	Low Temperature Selexol Slurry Feed	Low Temperature Selexol Dry Feed	Medium Temperature Iron Oxide Slurry Feed	Medium Temperature Iron Oxide Dry Feed	High Temperature Conoco Dry Feed
Thermal Efficiency - %	39.4	41.1	40.8	41.7	42.9
Capital Cost - $/kW	934.0	930.0	859.0	873.0	883.0
Cost of Electricity Mills/kWh	44.1	43.5	41.4	41.4	41.2
SO_2 Emissions lb/10^6 Btu*	0.24	0.35	0.52	0.10	0.20
NO_X Emissions lb/10^6 Btu*	0.007	0.014	0.58	0.52	69.6

*Sulfur and nitrogen emissions are given in lb/10^6 Btu of coal fired.

fur compounds, such as mercaptans and thiophenes, on carbon (10) and on zeolites (11-18). Higher temperature sorption processes on various metal loaded sorbents for fuel gas desulfurization have also been studied (19-34, 56-61) but to a lesser degree of understanding. Adsorption processes attaining commercial scale development for fuel gas desulfurization consist of the Calgon process employing metal oxide impregnated carbon (35) and the processes of Haynes (11-13, 15-17, 36, 37) using zeolites.

A summary, collected from the contract and patent literature, of the state of the art of process technology on the hot desulfurization of coal gases relevant to cleanup of low and intermediate BTU gases is presented in Tables 4 and 5. The summary covers the type of sulfur removal agent, applicable temperature range for desulfurization, mode of regeneration, state of development of the process, principal known disadvantages, and literature source. The three general types of desulfurization agents consist of dolomites, molten salts, and metal/metal oxide. Physical configurations consist of fixed bed, fluidized bed and melt.

Table 4. Hot Gas Desulfurization of Coal Syn Gas—Status Summary

Process	Agent	Desulfurization Temp. °C	Desulfurization Attainable H_2S Conc. ppm	Regeneration Temp. °C	Regeneration Agent	State of Development	Disadvantages	Ref.
Dolomite								
Calcined dolomite	CaO-MgO	870-930	600	600-650	CO_2/steam	Bench scale; no current work	Rapid coking; sorbent degradation; reactor corrosion	42
Conoco half calcined dolomite	$CaCO_3$-MgO	900	200-300	700	CO_2/steam	Bench scale; pilot plant; no current work?	Sulfidation kinetics inferior to FeO_x; poor regenerability	32
Westinghouse dolomite	$CaCO_3$-MgO	815-925	<600	800	Air	Bench scale; pilot plant; no current work	Creates excessive non-regenerated ash ($CaSO_4$/MgO)	43
U.S. Steel dolomite	CaO-MgO	1000	<65	Proprietary; produces H_2S		Bench scale Current work?	Produces hydrated unregenerated dolomite ash	44
Molten Metals/Salts								
Battelle molten salt	Li, Na, K, Ca, Carbonate	850-1000	<40	Complex aq. chem. via carbonation with H_2S to Claus		Bench scale?	Complex regen., salt carry over and corrosion	40,41
IGT Meissner	Pb	330-1200	---	>330	Electrolysis	Bench scale expts; current work	Corrosion; Pb volatilization	20

Table 5. Hot Gas Desulfurization of Coal Syn Gas—Status Summary Metal/Metal Oxides

Process	Agent	Desulfurization Temp. °C	Desulfurization Attainable H_2S Conc. ppm	Regeneration Temp. °C	Regeneration Agent	State of Development	Disadvantages	Ref.
Appleby-Frodingham	Iron oxide granules	355-420	150	600-800	Air	Prototype tests? abandoned	Sorbent attrition and pipe corrosion	21,22
MERC	Iron oxide silica	400-750	200-800	950	Air/steam	Bench scale; pilot plant; commercial design	High effluent S; sorbent vulnerable to degradation	23,24,25
Catalysts and Chemicals	ZnO	150-650	---	Not evaluated		Bench scale expts. terminated	Not effective above 650°C	31
IFP	ZnO on support	400-600	10	600-900	Air?	Bench scale current work?	Lack of regenerability	32
METC	Zn ferrite	400-650	10	650	Air/steam	Bench scale current work	Requires good regen. temp. control-sorbent life undetermined	19,50,51
Exxon	La_2O_3-Al_2O_3	150-930	---	150-930	Air/steam	Bench scale expts. terminated	Proprietary	29,30
Foster-Wheeler	Ni	540-790	---	540-790	Air	Bench scale expts. terminated	Not published	28
Johns Hopkins	Cu/Cr/V	350-820	---	---	Air	Bench scale expts. terminated	Failure to sustain sulfur removal capacity	33
Atlantic Refining	Cu/Pb/Zn alumino-silicate	90-540	---	370-540	Air/steam	Bench scale expts. terminated	Low removal efficiency	34
Kennecott	Cu/CuO/$Cu(OH)_2$	480-500	---	820	Air	Terminated	Difficult regeneration	48,27
Esso	Cu/SiO_2	300-980	---	400-740	Cu_2O	Bench scale expts. terminated	Temp. control requires circulation	47
Giner	Cu/ZnO	500-700	< 1	650-700	Air/steam	Bench scale	Demonstrated only as secondary process	60,61

Dolomite Processes

There are several dolomite processes which have been evaluated: the OCR/Air Products and Chemicals/Morgantown Energy Research Center (MERC) (42), Conoco (39), Westinghouse (43) and United States Steel (44). Among these systems, the Conoco process appears to have reached the most advanced stage of development. The other processes do not appear to be under investigation at the present time. In the Conoco process effluent sulfur levels of 200-300 ppm H_2S are achieved at 900°C in the fluidized bed configuration of the half calcined dolomite ($CaCO_3/MgO$). Regeneration of the sulfided sorbent is conducted with carbon dioxide/steam mixtures around 700°C. Disadvantages of the dolomite system include the need for very high temperatures, insufficient sulfur removal, relatively poor sulfidation and regeneration kinetics (inferior to iron oxides), sorbent attrition in the fluidized bed, and accumulation of unregenerated ash ($CaSO_4$).

Molten Salt Processes

The principal molten salt process that has been evaluated for hot fuel gas desulfurization is the molten carbonate eutectic of lithium, sodium and potassium (40-41) with about 17 weight percent $CaCO_3$ added. Effluent sulfur levels of 40 ppm H_2S are attained in the operating range of 850 to 1000°C. Regeneration is achieved via a complex aqueous chemical route involving carbonation and displacement of the sulfur as H_2S. Problems with this process, in addition to the complex regeneration route, are ash accumulation in the melt, carry-over of melt components into the piping, and severe materials corrosion. Battelle Northwest Research Laboratories (40-41) have carried out the recent work on this process, whereas workers at M.W. Kellogg (45) and Rockwell (46) have made early contributions to the technology. An additional melt system examined for coal desulfurization involves molten lead at 327-1000°C (20).

At Battelle, a mixed Li, Na, K and Cu carbonate melt on a solid lithium aluminate support continues the earlier molten carbonate studies (54). Sulfur removal efficiencies up to 99.5% for gasifier fuels with reduction to less than 25 ppm H_2S/COS from 3000-5000 ppm H_2S and 500-1000 ppm COS at 800°C have been obtained. Regeneration with H_2/CO gases displaces the sulfur as H_2S at concentrations up to 50% from the sulfided melt. The use of the supported melt minimizes the volatility and corrosive nature of the earlier molten carbonate melts (55). This system also provides a capability for removal of halogens from the fuel gases that is greater in efficiency than for the sulfur contaminants.

Metal/Metal Oxide Processes

The principal metal/metal oxide hot gas desulfurization sorbents considered for coal gas cleanup have been iron oxides (21-24), copper/copper oxides (27, 33, 34, 47, 48), supported nickel (28, 49), a supported copper/chromium/vanadium sorbent (33), supported copper/-lead/zinc (34), supported La_2O_3 (29, 30), zinc oxide (31) and a zinc ferrite/zinc oxide sorbent (19) (see Table 5). Among these iron oxide, zinc oxide, and copper based systems have been of most interest. Current efforts (56-61) on mixed oxides -specifically copper/zinc oxide- will be discussed later.

Iron Oxide Processes. An iron oxide process, the Appleby-Frodingham (21, 22), demonstrated excessive sorbent degradation and has been abandoned. The MERC iron oxide/silica (fly ash) sorbent (21, 23) is the most advanced of the regenerable fixed bed systems. This process will achieve 200-800 ppm H_2S in the effluent coal gases at temperatures in the range of 400 to 750°C. Regeneration is achieved at 950°C in air or air/steam. This process has been evaluated up through the pilot plant stage and commercial process design.

Zinc Oxide. Unsupported zinc oxide is conventionally used as a low to mid temperature desulfurization but only as a "throw-away" -nonregenerable- system. The supported zinc oxide is under investigation by Giner, Inc. (58-60) and the Institut Francais du Petrole (IFP), (52), where low effluent H_2S concentrations (1 ppm) were shown. However, the development to date has not advanced past bench scale. Evaluations of the IFP zinc oxide at METC (52) show that it was capable of removing H_2S to about 1 ppm but with low sulfur capacity relative to unsupported zinc oxide. Its regenerability was determined to be good, and no loss of surface area was detected after three sulfidation/regeneration cycles. The Giner, Inc. zinc oxide supported on zeolite showed good capcity but poor regenerability (61).

Copper Based Processes. The earlier studies involving copper sorbents (27, 33, 34, 47, 48) are of interest because of their relevance to the current work reported in this paper on copper/zinc oxide. Desulfurization was examined for the copper sorbents in question (Table 5) at temperatures up to 980°C. The most effective regeneration was conducted by solid-solid reaction between Cu_2S and CuO. The principal disadvantages of these copper sorbents were difficult regeneration and failure to sustain desulfurization capacity after regeneration. These studies have been terminated. Results of these studies are examined in some detail below.

About the earliest significant effort to evaluate copper sorbents for hot gas desulfurization was that conducted at Johns Hopkins University (33). The sorbents prepared were mixed oxides of copper with iron, uranium, chronium, tin, manganese, vanadium, molybdenum, antimony annd bismuth as binary and ternary combinations with various binder components such as clay, pumice, alundum and lime. Regeneration of the sulfided sorbents was conducted with air oxidation. High initial sulfur removals in excess of 97% were obtained with some of these sorbents, but removal efficiency declined with desulfurization/regeneration duty cycles. This degradation of desulfurization capacity was attributed to deactivation of the copper by reaction with the binders.

Nachod, et al. (34) used copper, zinc and lead, supported on aluminosilicates to remove sulfur from petroleum distillates and gases containing inorganic (H_2S) and organic sulfur compounds in the temperature range 90 to 550°C. Sulfided sorbents were regenerated with air/steam mixtures at 370-540°C. Unfortunately, the aluminosilicates formulated were very inferior to the hydrothermally stable zeolites currently available. The maximum sulfur removal efficiencies reported were 43%.

Lewis (47) used copper supported on silica to desulfurize H_2S/H_2 gas mixtures over the temperature range 300-1000°C. Ultimate application was intended for coke oven gas, producer gas, water gas and hydrocarbon gases. The desulfurization evaluation tests were conducted with concentrations of H_2S of 1-1.5% in hydrogen with the sorbent in a fluidized bed for stirred gas/solid reactors. Sulfided sorbents were regenerated by oxidation with air or Cu_2O in the temperature range 400-700°C. Metallic copper was considered the preferred sulfur removal agent and efforts were made during regeneration to limit the oxygen concentration so that the regeneration of the sulfided copper produced primarily metallic copper and SO_2. This regeneration mode led to production of high purity SO_2 and a sorbent consisting of 90 wt% Cu and 10 wt% Cu_2S with negligible Cu_2O. Sulfur removal efficiencies of 98.5% (i.e., from 1.5 percent to 225 ppm) were realized in bench scale tests in what was considered the optimum desulfurization temperature range (350-600 °C), with removal efficiencies declining to 80-95% in the higher temperature range of 600-930°C.

Kennecott Copper (27, 48) conducted bench scale desulfurization tests with simulated Lurgi fuel gas containing H_2, CO and 1-1.5% H_2S over the temperature range 480-510°C. Sulfur removal efficiencies in excess of 90% (1.5% to 1500 ppm) were obtained. Sorbents were regenerated at 816°C using solid-solid reaction with CuO. Although the high temperature yielded 99% regeneration, there was considerable sorbent degradation which diminished the efficiency of H_2S removal at the start of the next sulfur removal cycle. Work on this process was terminated and available information is contained primarily from a contract report (48) and in a preprint in a recent meeting (27) where a major problem identified was the hot-solid-transport for a solid-solid regeneration scheme.

Mixed Metal Oxides

Current research on hot regenerable coal gas desulfurization is mainly concerned with mixed metal oxides. There are three different groups, IGT, METC, and Giner, Inc. conducting evaluations on various systems primarily with DOE funding. IGT works with $ZnAl_2O_4$, $ZnCr_2O_4$, MnO, $CoAl_2O_4$, $FeCr_2O_4$, Fe_2SiO_4 and Co_2TiO_4; METC works with a zinc ferrite and Giner, Inc. work is with Cu/ZnO. These studies are reviewed below. It is not claimed that the work reviewed here exhausts the alternatives to coal gas desulfurization but it is considered to represent the prinicpal emphasis of current research.

Mixed desulfurization sorbents have also been under study at IGT (53). The bulk of the sulfur removal occurred over an initial stage with various high capacity mixed oxides and a low effluent sulfur was attained in a second ZnO bed(Fig. 1). An essential feature of this system consisted of a reverse flow oxidation regeneration at 1200-1900^oF of the sulfided sorbent beds with air as a mixture, oxygen (3%)-SO_2 (5%) plus N_2 (92%), to favor recovery of a high percentage of the displaced sulfur as elemental sulfur. Effluent H_2S concentrations of about 10 ppm were attained for several of these sorbents, but

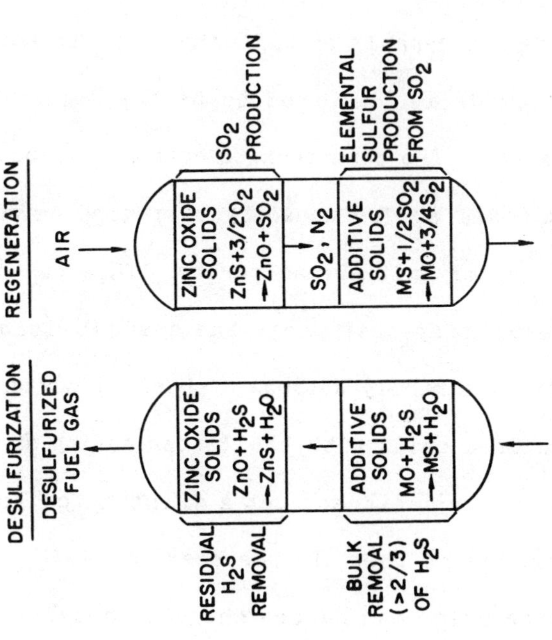

Figure 1. Simplified description of mixed metal oxide approach to fuel gas desulfurization.

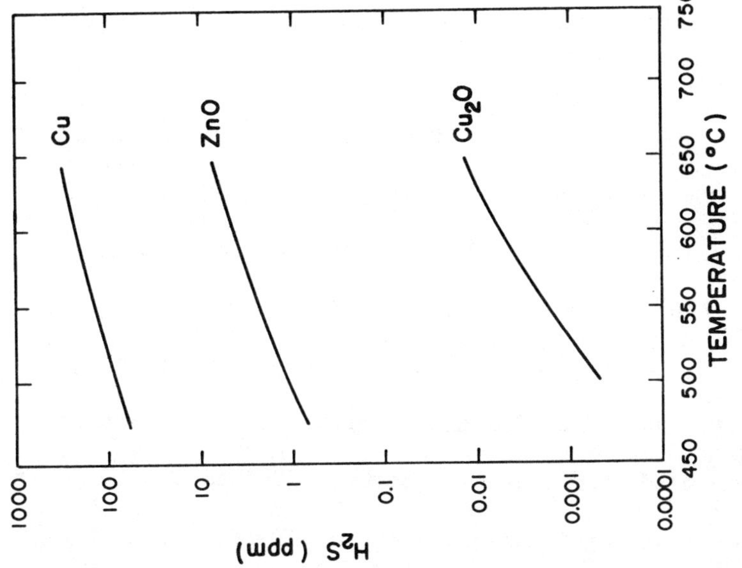

Figure 2. Equilibrium for H_2S with 10% H_2O over Cu, Cu_2O, and ZnO.

regeneration efficiency has so far fell short of expectation.

Investigations at METC which are an extension of earlier work with iron oxides (23, 24, 25) have shown that sulfur levels of 2-10 ppm in the temperature range 550 to 750°C have been achieved for a zinc ferrite sorbent (19, 50, 51,) with simulated Lurgi fixed-bed gasifier off-gas. Extruded zinc ferrite sorbent prepared by a commercial catalyst manufacturer (United Catalyst) is currently undergoing bench scale evaluation at METC (51). Sulfur removal to levels of about 5 ppm H_2S are reported. Since iron oxide is not capable of sulfur removal to this level, the zinc oxide is believed to be the dominant desulfurization agent. The zinc ferrite sorbent has been shown to have improved sulfur capacity over zinc oxide, but with the penalty of exit stream purity. This can be explained by postulating that under the desulfurizing conditions zinc ferrite merely acts as a physical mixture of zinc oxide and iron oxide with the latter acting as a high sulfur capacity constituent but a diluent for low level sulfur removal by ZnO. In the tests at METC, it was shown that zinc oxide can reduce the H_2S concentration to 2 ppm but zinc ferrite at the same conditions (1000^oF with 27.66 percent steam) cannot reduce H_2S levels below 5 ppm which is above the acceptable concentration for the MCFC. Furthermore, regenerability of zinc ferrite at $1000-1200^oF$ is poor (52). Build-up of sulfates, lack of sorbent integrity and regenerability are some of the problems under investigation.

Cu/ZnO Mixed Sorbent studied at Giner, Inc. (56-61), is the only desulfurization sorbent which has demonstrated the capability to reduce sulfur level below 1 ppm. Studies with single and dual composition copper oxide and zinc oxide sorbents with simulated coal gases containing 200 ppm H_2S and 9 to 26% water vapor established that highly dispersed CuO on zeolite, highly dispersed Cu/ZnO on zeolite and a coprecipitated Cu/ZnO provide hydrogen sulfide reductions from 200 ppm to below 1 ppm at $500-650^oC$(Table 6). The desulfurization capacity of a coprecipitated Cu/ZnO sorbent was greater than what would be expected from the desulfurization capacity of either component alone. It is postulated that in the presence of zinc oxide the reduction of CuO to metallic copper is incomplete. As shown in Figure 2, the existence of cuprous oxide, Cu_2O, would make very low

Table 6
Desulfurization Summary

Sorbent	Mole % S Loading	
	Fresh	After (n) Cycles
11% Cu/Zeolite	18	16(4)
18% ZnO/Zeolite	24	7(3)
6% Cu-11% ZnO/Zeolite	15	6(5)
33% Cu-65% ZnO	8	9(7)

Desulfurization conducted at 650°C, prereduced in H_2, 200 ppm H_2S in simulated fuel gas with 6% water partial pressure. Sulfur loadings given as conversion to sulfide (Cu_2S or ZnS) at exit [H_2S] of 10 ppm.

hydrogen sulfide levels possible. This synergistic effect was also evident in the physical blend of Cu/SiO_2 and ZnO as seen in Figure 3. In addition to providing very low hydrogen sulfide levels, the Cu/ZnO exhibited a significant conversion to sulfides at an H_2S concentration below 1 ppm. Although supported copper and zinc oxide showed higher conversion, engineering considerations, such as sulfur capacities on a weight or volume basis indicate a preference for the Cu/ZnO sorbent. A capability to remove COS, CS_2 and CH_3SH was also observed over Cu/ZnO. Efficient oxidation regeneration was obtained with this mixed sorbent at 650-750°C with air and air steam mixtures as well as simulated MCFC exit gases containing 5% O_2, 10% CO_2, 10% H_2O and 75% N_2.

In order to gain insight into the regenerability of the mixed Cu/ZnO sorbent, scanning electron microscopy (SEM) was applied to the prereduced (Figure 4a), sulfided (Figure 4b), and oxidation regenerated (Figure 4c) states. The second micrographs (Figure 4b) show that during sulfidation, a finer grain surface structure has developed with a filling in of the coarser pore spaces. This results apparently from creation of relatively fine crystallites of the segregated copper sulfide and zinc sulfide. The third micrograph (Figure 4c) indicates a return to the coarse open pore structure of the pre-reduced sorbent (Figure 4a) during oxidation of the sulfide.

The conditions of pretreatment used for prereduction might be expected to form a "brass alloy" on the surface of this Cu/ZnO. Sulfidation of this prereduced mixed oxide could be expected to form small crystallites of the new metal sulfide phases at the grain boundaries of the larger "brass crystallites" of the substrate. Exposure to reduction or oxidation conditions would be expected to cause sintering and generate relatively large brass like

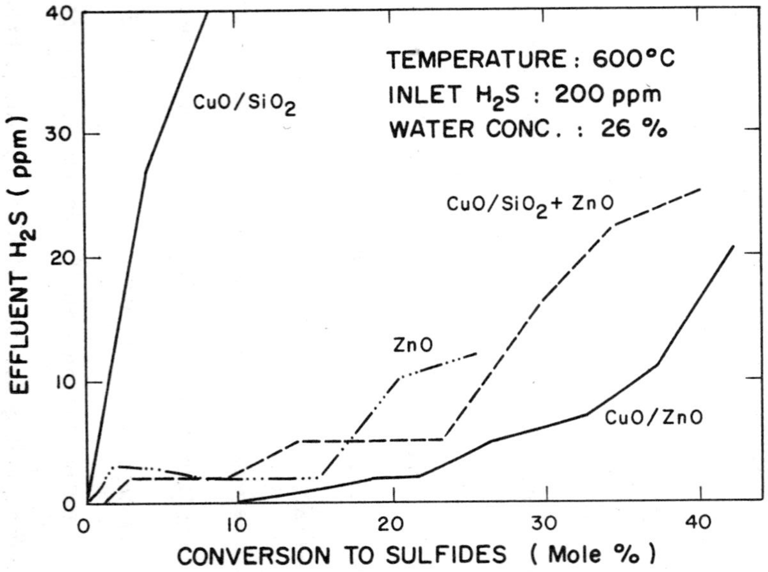

Figure 3. Comparison of desulfurization sorbents.

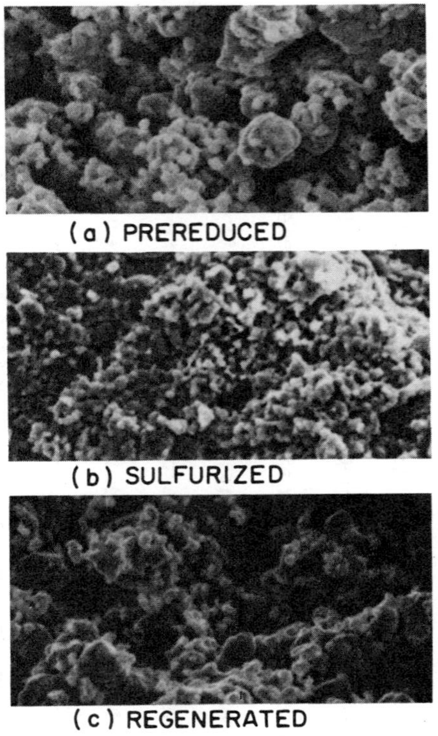

Figure 4. Electron microscopy of CuO/ZnO sorbent.

crystallites, whereas the periodic sulfidation causes a structural reorganization, countering the sintering, and the formation of finer metal sulfide crystallites. This unique structural reorganization mechanism counteracts the loss of surface area due to sintering during regeneration observed for other sorbents and leads to the maintainance of desulfurization capacity after regeneration.

The major limitations of this process are that it is demonstrated only for the secondary sulfur removal (from coal gas containing 200 ppm H_2S) and high temperature (>750°C) regeneration to avoid sulfate formation is required.

FUTURE PROSPECTS

There are substantial incentives to pursue development of coal gas desulfurization, notably hot dry scrubbing. These incentives consist of simplification of process equipment, hence reducing capital if not operating costs, protection of downstream hardware from erosion and corrosion, minimizing secondary waste streams and minimizing ultimate sulfur emission to the atmosphere. The energy conversion systems that will benefit most from successful hot dry sulfur scrubbing processes are combined power cycles with PFBC and the MCFC, and catalytic processes for synthetic fuel production. The impetus to advance development of the most appropriate technology could benefit from a comprehensive assessment of the comparative technical and economic merits of coal precleaning, fuel gas desulfurization following gasification and flue gas scrubbing following combustion.

A wide range of elements have received some consideration for fuel gas desulfurization. Over a period in excess of 50 years representatives from every group of the periodic table have been studied. The elements that have most consistently survived serious consideration are iron, zinc and copper. Other surviving elements are cobalt, chromium and nickel, the alkaline rare earths and the alkali metals principally as oxides but also in the form of carbonates. The current impetus is to use mixed oxide systems notably zinc with metal such as iron and copper. The carbonates currently undergoing evaluation are also mixtures such as dolomites with magnesium or mixed carbonates with Li, Na, K and Ca.

Physical configurations of the desulfurization systems under evaluation continue to be diversified consisting of fixed beds, fluidized beds and melts. All of these systems appear directed toward hydrothermally stable support systems as the most promising for providing a persistent useful life for many cycles of sulfidation and regeneration. Oxidation with air and air steam of sulfided sorbents continue to dominate although studies continue on regenerations with steam/carbon dioxide to recover sulfur as H_2S or with SO_2/O_2 to recover sulfur as elemental sulfur. Recovery of sulfur in elemental form remains an attractive regeneration objective.

REFERENCES

1. Hodge, D.P., Chemical Engineering and News, Vol. 62, No. 22, (1984), p. 5.

2. Markel, K.E., Editor, Third Annual Contaminant Control in Hot Coal Derived Gas Streams, Contractors Meeting Proceedings, DOE/METC84-6, NTIS DE84000216 (1983).

3. Moore, W.E. "Overview of DOE/Fossil Energy's Gas Stream Contaminant Control Program," in K.E. Markel, Ref. 2.

4. United Technologies Corp., "Assessment of Fuel Gas Cleanup Systems," DE81025927, 1980.

5. Marianowski, L.G. "An Update of the Sulfur Tolerance of Molten Carbonate Fuel Cells" in K.E. Markel, Ref.2.

6. Vidt, E.J., "Evaluation of Gasification and Gas Cleanup for MCFC Power Plants," Westinghouse Research, DOE/MC/16220-1551, (1981).

7. Jones, C.H. and J.M Donohue, "Comparative Evaluation of High and Low Temperature Gas Cleaning for Coal Gasification-Combined Cycle Power Systems," Stone & Webster, EPRI AF-416,(1977).

8. Robson, F.L. and W.A. Blecher, "Assessment of Fuel Gas Cleanup Systems," DOE/METC 12050-149, (1981).

9. Lorton, G.A., "Assessment of Sulfur Removal Processes for Advanced Fuel Cell Systems," Final Report EPRI EM-13333, Research Project 1041-5 (1980).

10. Dietz, V.R., "Bibliography of Solid Adsorbents," NBS Circular 566 (1956), Abs. No. 7803; C. Padovani, et al., Riv. Combustibili, Vol. 2, (1948), p. 13.

11. Milton, R.M., "Molecular Sieve Adsorbents," U.S. Patent 2,882,243 (1959).

12. Milton, R.M., "Molecular Sieve Adsorbents," U.S. Patent 2,882,244 (1959).

13. Breck, D.W., U.S. Patent 3,013,982 (1961).

14. Haynes, H.W., Jr., et al., "Regenerating Zeolites Used for Sulfur Removal," J. Oil & Gas, Vol. 78, (1961).

15. Turnock, P.H., et al., "Desulfurization by Selective Adsorption with a Crystalline Zeolitic Molecular Sieve," U.S. Patent 3,620,969 (1971).

16. Turnock, P.H., et al., "Regeneration of Molecular Sieves Having Sulfur Compounds Adsorbed Thereon," U.S. Patent 3,725,299 (1973).

17. Collins, J.J., "Purification of Liquid Hydrocarbons Containing Carbonyl Sulfide," U.S. Patent 3,654,144 (1972).

18. Rosynek, M.P., et al., "Combination Process for Residual Demetalation, Desulfurization and Resulting Coke Gasification," U.S. Patent 3,983,030 (1976).

19. Grindley, T. and G. Steinfeld, DOE/MC16545-1125 (1982).

20. Meissner, H.P., U.S. Patent 3,954,938 (1976).

21. Reeve, L., J. Inst. Fuel, Vol. 319, (1958).

22. Bureau, A.C. and J.J.F. Olden, The Chemical Engineer, (1967).

23. Shultz, F.G. and J.S. Berber, J. Air Pollution Control Assoc., Vol. 20, (1970), p. 93.

24. Oldaker, E.C., et al., "Removal of Hydrogen Sulfide from Hot Low BTU Gas with Iron Oxide-Fly Ash Sorbents," MERC/TPR-75-1 (1975).

25. Westmoreland, P.R. and D. P. Harrison, "Evaluation of Candidate Solids for High-Temperature Desulfurization of low-Btu Gases," Environ. Sci. & Tech, Vol. 10 (1976), p. 659.

26. Joshi, D.K. and E.L. Leuenberger, "Hot Low BTU Producer Gas Desulfurization in Fixed Bed of Iron Oxide-Fly Ash," NTIS No. FE-2033-19 Vols. 1 and 2 (1977).

27. Agarwal, J.C., "Metal Oxide Hot Gas Desulfurization Process," AIChE Conf., Philadelphia, PA (1980).

28. Steiner, P., U.S. Patent 4,039,619 (1977).

29. Wheelock, K.S. and C.L. Aldrige, U.S. Patent 3,974,256 (1976).

30. Wheelock, K.S. and R.S. Say, U.S. Patent 4,002,720 (1977).

31. Gutman, W.R. and R.H. Wright, U.S. Patent 3,441,370 (1969).

32. Onursal, A.B., EPA-700/7-79-169, (1979).

33. Huff, W.J. and L. Logan, AGA Proceedings, Vol. 18, (1965), p. 359.

34. Nachod, F.C., "Desulfurization of Hydrocarbons," U.S. Patent 2,442,982 (1948).

35. Calgon Corp. Bulletin 23-67, 15230 (1978).

36. Haynes, H.W., Jr., Cat. Rev. Sci. Eng., Vol. 17, (1978), p. 273.

37. Haynes, H.W., Jr., U.S. Patent 3,144,307 (1964).

38. TRW Energy Systems, DE/METC-8085-T2 (1979)

39. Curran, G.P., et al., "High Temperature Desulfurization of Low-BTU-Gas," EPA 800/7-77-031, NTIS PB 271008 (1977).

40. Moore, R.H., U.S. Patent 3,919,390 (1975).

41. Moore, R.H., et al., Battelle-Northwest Laboratories, BNWL-SA-6365 (1977).

42. O'Brien, W.G., Jr., et al., NTIS FE-1510-T-2 (1976).

43. Lemezis, S., R&D Report No. 81/Interim Report No. 3, Westinghouse Electric Corp. for ERDA. Contract No. E 49018-1514, July 1974 to June 1975.

44. Feinman, J. and J.E. McGreal, AIChE Conf., Philadelphia, PA, (1980).

45. Cover, A.E., W.C. Schreiner and G.T. Skaperdas, Chem. Eng. Progress, Vol. 69 (1973) p. 3.

46. Botts, W.V., A.L. Kohn and C.A. Trolling, Proceedings 11th Intersociety Energy Conversion Engineering Conf., Vol. 1, (1976), p. 280.

47. Lewis, W.K., U.S. Patent 3,079,223 (1963).

48. Meyer, J.P. and M.S. Edwards, ORNL Rept. TM-6072 (1978).

49. Setzer, H.J. and R.W. Whiting, U.S. Patent 3,485,746 (1969).

50. Grindley, T. and G. Steinfeld, "Development and Testing of Regenerable Hot Coal Gas Desulfurization Sorbents," U.S. Department of Energy, Second Annual Contaminant Control in Hot Coal Derived Gas Streams: Proceedings, Morgantown, WV, NTIS DOE/METC-82-47 (1982).

51. Grindley, T. and G. Steinfeld, "Zinc Ferrite Hydrogen Sulfide Adsorbent," in K.E. Markel, Ref. 2.

52. Steinfeld, G., "Hot Gas Desulfurization," Annual Report, Science Application, Inc., Morgantown, WV, NTIS DE-AC21-81-MC16545 (1982).

53. Anderson, G.L., F.O. Berry, and M. St. Johns, "H_2S Removal Using Mixed Metal Oxides and HCl Removal Using Molten Alkali Carbonates," in K.E. Markel, Ref. 2.

54. Stegen, G.E., "Development of a Solid Absorption Process for Removal of Sulfur from Fuel Gas," DOE/ET11028-9 (1982).

55. Lyke, S.E. and L.J. Sealock, "Development of a Hot Gas Cleanup System for Integrated Coal Gasification/ Molten Carbonate Fuel Cell Power Plants," in K.E. Markel, Ref. 2.

56. Jalan, V. and D. Wu, "Molten Carbonate Fuel Cell Power Plant Desulfurization Systems," NTIS DOE/ET15370, DE83006804 (1979).

57. Jalan, V., "High Temperature Desulfurization of Coal Gases by Regenerative Sorption," 1981 Int'l. Gas Res. Conf.: Proceedings, Los Angeles, CA, 291 (1981).

58. Jalan, V., "Studies Involving High Temperature Desulfurization/Regeneration Reactions of Metal Oxides for the Fuel Cell Program," Giner, Inc., Final Report, ANL-K-83-12, NTIS DE83008714 and DE83008715 (1981).

59. Jalan, V., C. Brooks, C. Georgakis, and M. Desai, "Metal Oxide Absorbents for H_2S Control," U.S. Department of Energy, Second Annual Contractors' Meeting on Contaminant Control in Hot Coal Derived Gas Streams: Proceedings, Morgantown, WV, NTIS DOE/METC-82-47 (1982).

60. Jalan, V., M. Desai, C.Brooks, and R. Waterhouse "Metal/Metal Oxides as High Temperature Desulfurization Sorbents," in K.E. Markel, Ref. 2.

61. Jalan, V., "Studies Involving High Temperature Desulfurization/Regeneration Reactions of Metal Oxides for Fuel Cell Development," Giner, Inc. Final Report, NTIS DOE/MC/16021-1486 (1983).

18

Trace Contaminants Challenge Synfuels Acid Gas Removal

Lun Auyang
Y. C. Liu
Michael Petrik
Larry Siwajek
Helipump Corporation

Removal of trace contaminants, hydrogen sulfide and carbon dioxide from crude gas mixtures is essential to the ultimate success of synfuels production. In the CNG acid gas removal process, the unique properties of liquid carbon dioxide permit efficient absorption of trace contaminants and hydrogen sulfide, and triple-point crystallization of carbon dioxide efficiently separates trace contaminants and hydrogen sulfide from carbon dioxide.

INTRODUCTION

Acid gas removal (AGR) refers to the separation of hydrogen sulfide and carbon dioxide from gas mixtures. The simultaneous removal of other trace sulfurous and nitrogeneous compounds is also usually implied, e.g., carbonyl sulfide, carbon disulfide, mercaptans, hydrogen cyanide. A less emphasized but equally important function of acid gas removal is the separation of carbon dioxide from hydrogen sulfide and other trace contaminants. AGR thus serves to increase the market value of a gas by removing diluents and undesirable compounds, to cleanse a gas of compounds detrimental to downstream processing (catalyst poisons), to separate pollutant-free carbon dioxide for by-product use or atmospheric venting, and to concentrate hydrogen sulfide for sulfur recovery.

The major applications of AGR appear to be: 1) treating natural gas containing sulfurous compounds and carbon dioxide, 2) treating shift gases in substitute natural gas production from coal, 3) treating synthesis gas made by partial oxidation of sulfur-contaminated feedstocks, e.g., gasoline production from coal via synthesis gas (Fischer-Tropsch reaction or through the intermediates methanol or dimethyl ether), 4) acid gas removal in hydrogen production from coal or residual feedstocks for uses such as oil shale refining, and 5) treating synthesis gas made from methane and steam (production of ammonia, hydrogen, and oxo compounds). In short, acid gas removal is a key step in the up-grading of natural gas and synthesis gas made from natural gas, petroleum, or coal, and acid gas removal is an essential step in most envisioned fossil fuel energy conversion processes.

Present gas cleaning processes were developed for the natural gas and petroleum refining industries to treat well-defined gas streams having low acid gas content, a high ratio of hydrogen sulfide to carbon dioxide, and relative freedom from trace contaminants. Present gas cleaning processes require modificiation to treat coal gasifier gas because it is neither well-defined nor of consistent composition. Difficulties include:

- The concentration of hydrogen sulfide in the acid gas stream is too low (usually 25% or less) to support conventional Claus sulfur plants.
- The removal and rejection of trace contaminants is more difficult than removal and rejection of the acid gases hydrogen sulfide and carbon dioxide.
- The sulfur content of vented carbon dioxide is high, typically 100-200 ppm.
- Compared to conventional applications, sharply increased capital and energy costs for cleaning coal gases.

New AGR processes aimed at coal gases combine conventional absorption technology with new solvents and separation techniques to promote sharp separations, improved energy utilization, and reduced capital costs. One such process is the Consolidated Natural Gas (CNG) AGR Process. This paper describes the CNG process and examines its unique features; particularly its ability to absorb sulfur compounds and trace contaminants from crude gas mixtures, and regenerate pure absorbent while concentrating the sulfur compounds and trace contaminants.

AGR TECHNOLOGY FOR SYNFUELS APPLICATION

There are two major types of acid gas removal processes, chemisorption and physical absorption. Typical chemisorption agents are amines and hot potassium carbonate; typical physical absorbents are dimethyl ether of tetraethylene glycol (Selexol) and cold methanol (Rectisol). At low acid gas partial pressure, less than about 200 psia, chemisorption is generally favored; at acid gas partial pressures above 200 psia, physical absorption is generally favored (1). The following discussion assumes a crude coal gasifier gas with pressure and composition favoring physical absorption AGR.

Most gas purification plants consist (see Figure 1) of an absorption tower (absorber) and regenerator (stripper). In the absorber the solvent flows counter-currently to the

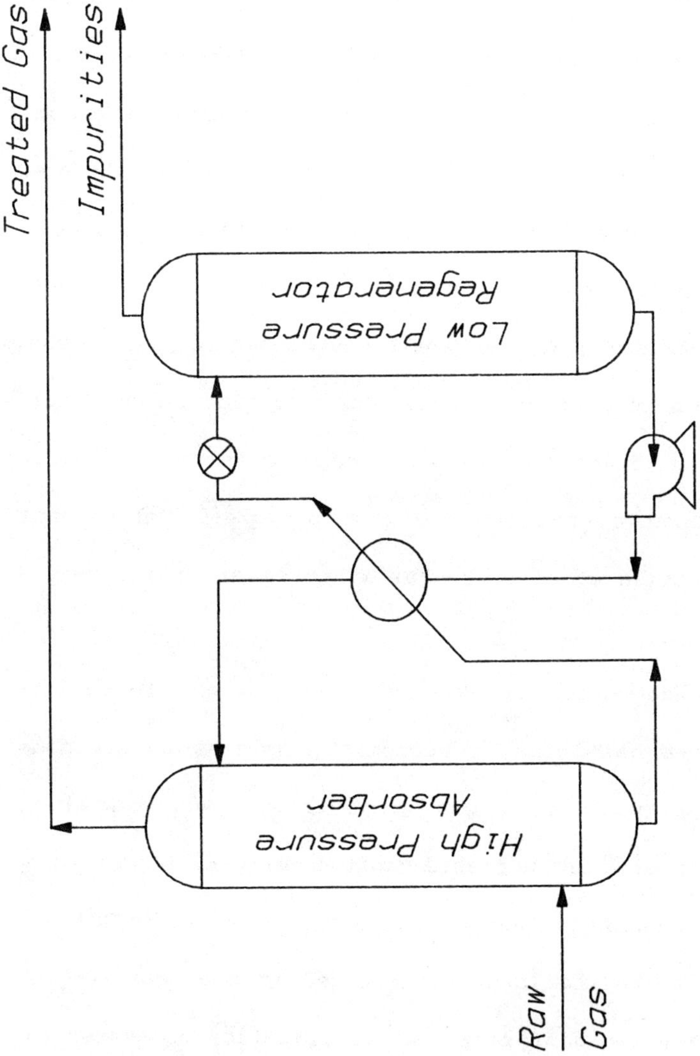

Figure 1. Simple gas purification.

ascending gas, absorbing the removable components. Absorption generally occurs at low temperatures and under elevated pressure, while regeneration is accomplished at high temperature and low pressure. The temperature gradient is utilized by heat exchange between the cool fat solution and the hot regenerated solution, and the pressure gradient energy is recovered with expansion turbines. All absorption-regeneration processes operate continuously.

Existing AGR processes envisioned for use in the production of synthetic fuels from coal face unique challenges because these AGR processes were developed primarily in response to the needs of the petroleum and natural gas industries where crude gas mixtures are relatively well-defined.

Adaptation of existing AGR processes to purify coal derived gas leads to economic penalties to overcome the disparity of feed gas composition. Such penalties are encountered in increased capital cost and increased operating expense. The gas compositions shown in Table 1 (2) illustrate the large differences in coal gas composition between two different coal gasification processes and the large difference between both of these coal gases and one typical refinery sour gas stream. Clearly, purification of

Table 1
Typical Compositions, Dry Gas to Acid Gas Removal, Mol % (2)

Component	Lurgi	K-T	Refinery Sour Gas
Hydrogen	37.6	32.74	-
Oxygen	0.6	-	-
Carbon Monoxide	16.7	57.35	-
Methane	9.1	-	8.4
Ethane	0.7	-	5.2
Propane			4.6
Isobutane			2.5
Normal Butane			7.5
Normal Pentane			3.4
Hexane			1.0
Naphthalene	0.7	-	-
Nitrogen + Argon	0.6	1.16	-
Ammonia	0.2	-	-
Hydrogen Cyanide	2.7	-	-
Carbonyl Sulfide	-	0.114	-
Hydrogen Sulfide	1.1	1.59	62.5
Carbon Dioxide	30.0	7.05	4.9
	100.0	100.00	100.00
CO_2/H_2S	27.3	4.43	0.08
Temp. ^oF	850	2700	122
Press, psia	300	15	22

raw coal gas streams requires a most flexible and adaptable process to meet technical and economic requirements of the various coal gasification processes.

Coal gas differs markedly from the refinery gas of Table 1. For example: 1) the level of CO_2 in the refinery gas is roughly 70% of that for K-T and about 16% of that for Lurgi; 2) the H_2S level in the refinery gas is from 40 to 57 times that for the coal gases; 3) the acid gas ratio, CO_2:H_2S, in the coal gas ranges from about 55 to about 340 times that for the refinery gas; and 4) the coal gas streams contain a large variety of contaminants not present in the refinery gas, some of them highly toxic. Additional "trace" substances can and do appear in some coal gasifier raw gases, examples of which are benzene, toluene, xylene, phenols, thiocyanates, and mercury. These trace contaminants in coal gas have the potential to generate serious gas purification problems through recycle build-up in the liquid absorbent used for gas absorption. Toxic contaminants, such as hydrogen cyanide, carbonyl sulfide, and mercury, are not tolerable in the product gas.

Modification of existing AGR processes has been proposed to permit removal of certain offending trace contaminants. For example, hydrolysis reactors to convert COS to H_2S prior to hydrogen sulfide absorption can effectively eliminate the presence of COS in the sweet gas. The alternative to achieve comparable levels of COS and H_2S in the sweet gas is to increase absorbent flows above those required for hydrogen sulfide absorption (both cold methanol and Selexol solvent absorb H_2S better than COS). These process modifications add to the capital and operating cost of the AGR process, and may solve the trace contaminant problem only with respect to COS.

Acid gas removal using cold methanol has been practiced since at least the 1950's, but only recently have comprehensive studies been done by Rousseau (3) to determine the fate of trace contaminants in a cold methanol AGR process applied to clean coal-derived crude gases. Rousseau's experimental work shows that the unique problems associated with trace contaminant removal from crude coal gasifier gases with cold methanol may as yet be unresolved with present technology. Organic sulfur compounds such as carbon disulfide and thiophene are reported to distribute equally among the product gas, carbon dioxide vent gas, and the acid gas in an AGR system using cold methanol (3). The elemental contaminant mercury is reported to have accumulated in cold methanol to 50 ppb levels after only ten hours of operation (4). The accumulation of simple aromatics (BTX) in cold methanol has also been reported (5). The behavior of Selexol solvent in the presence of contaminants other than COS is not documented, but problems similar to those experienced with cold methanol should be anticipated because of the conventional methods employed for solvent regeneration, i.e., pressure reduction, heating, and inert gas stripping.

CNG PROCESS DESCRIPTION

The CNG acid gas removal process is distinguished from existing AGR processes by three features. The first feature is the use of pure liquid carbon dioxide as absorbent for sulfurous compounds; the second feature is the use of triple-point crystallization to separate pure carbon dioxide from sulfurous compounds; the third feature is the use of a liquid-solid slurry to absorb carbon dioxide below the triple-point temperature of carbon dioxide. Pure liquid carbon dioxide is a very effective absorbent for sulfurous compounds and trace contaminants; triple-point crystallization economically produces pure carbon dioxide and concentrated hydrogen sulfide; for bulk carbon dioxide absorption the slurry absorbent diminishes absorbent flow and controls the carbon dioxide absorber temperature rise without external refrigeration. The sequence of gas treatment is shown in Figure 2, a simplified flow diagram of the CNG acid gas removal process. More detailed descriptions and flowsheets of the CNG process are available (6,7).

Carbon dioxide plays a central role in the CNG process both as a pure component and in mixture with other compounds. The triple-point of carbon dioxide is referred to frequently in the following discussion; it is the unique temperature and pressure at which solid, liquid, and vapor phases of carbon dioxide can exist at equilibrium (-56.6°C, 5.1 atm). The carbon dioxide triple-point is shown in Figure 3, a phase diagram for carbon dioxide.

The CNG process steps which provide the unique capabilities to remove trace contaminants from crude gases and to isolate trace contaminants with the acid gas stream are acid gas absorption with pure liquid carbon dioxide and triple-point crystallization of carbon dioxide. These process steps are illustrated in Figure 4 and discussed below.

Carbon Dioxide Condensation, Hydrogen Sulfide and Trace Contaminant Absorption

Carbon dioxide is condensed by cooling the dehydrated crude gas to about -55°C; the condensate is contaminated

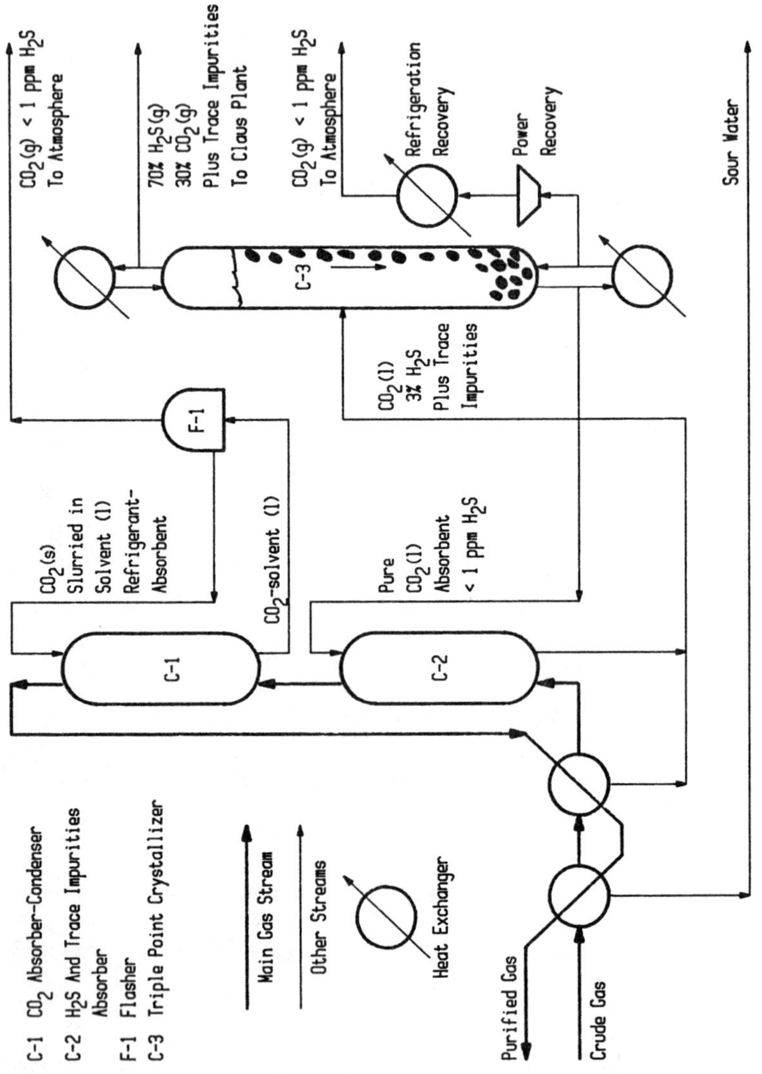

Figure 2. Simplified flow diagram of CNG acid gas removal process.

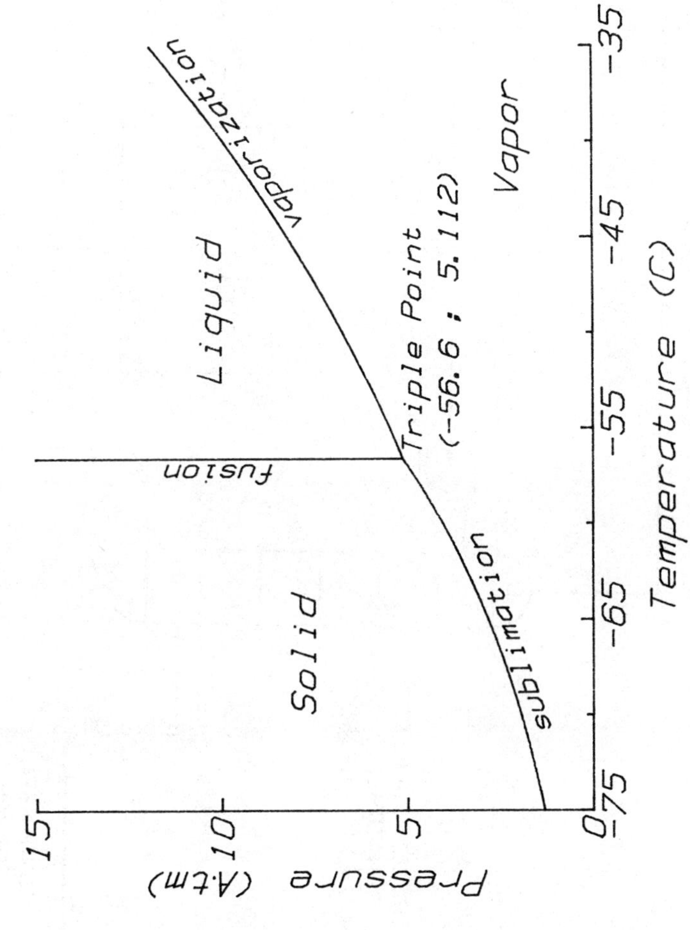

Figure 3. Carbon dioxide phase diagram.

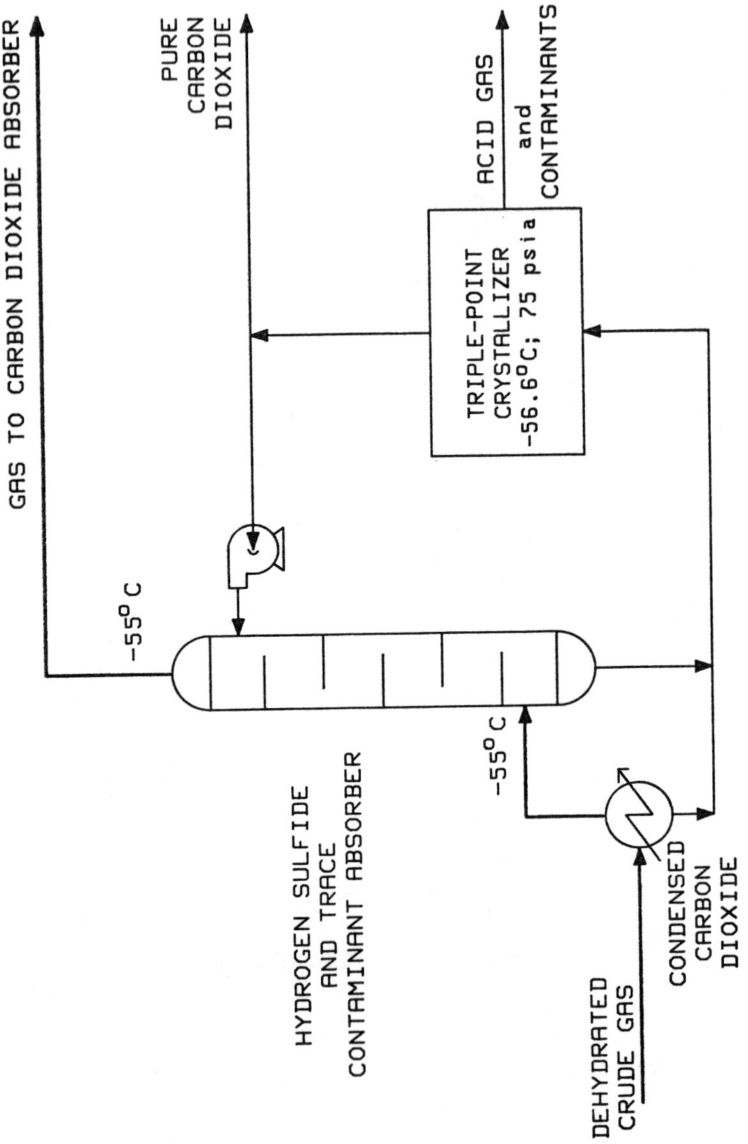

Figure 4. Simplified flow diagram of acid gas absorption and rejection from the CNG process.

with sulfurous compounds. Substantial amounts of carbon dioxide can be condensed if the dew point is relatively warm. For example, a synthesis gas at 1000 psia with 30 mol % carbon dioxide has a dew point of about -30°C, and approximately 65% of the carbon dioxide in the crude gas condenses to a liquid in cooling to -55°C.

The gas at -55°C is scrubbed clean of sulfurous compounds and remaining trace contaminants using pure liquid carbon dioxide. Liquid carbon dioxide has several desirable properties as an absorbing medium for sulfurous compounds. Liquid carbon dioxide absorbs carbonyl sulfide (COS) more effectively than it absorbs hydrogen sulfide as shown by their relative vaporization coefficients, i.e., $K_{COS}<K_{H_2S}$ ($K_i \equiv y_i/x_i$, y_i = vapor mole fraction component i, x_i = liquid mole fraction component i). Thus, an absorber for sulfurous compounds designed to remove hydrogen sulfide using liquid carbon dioxide also removes COS and other less volatile sulfurous compounds from the crude gas. The opposite is true concerning COS for both cold methanol (Rectisol) and the dimethyl ether of polyethylene glycol (Selexol solvent). For comparison, the ratio $K_{H_2S}:K_{COS}$ is plotted in Figure 5 versus liquid phase composition from 100 mol % methanol to 100 mol %

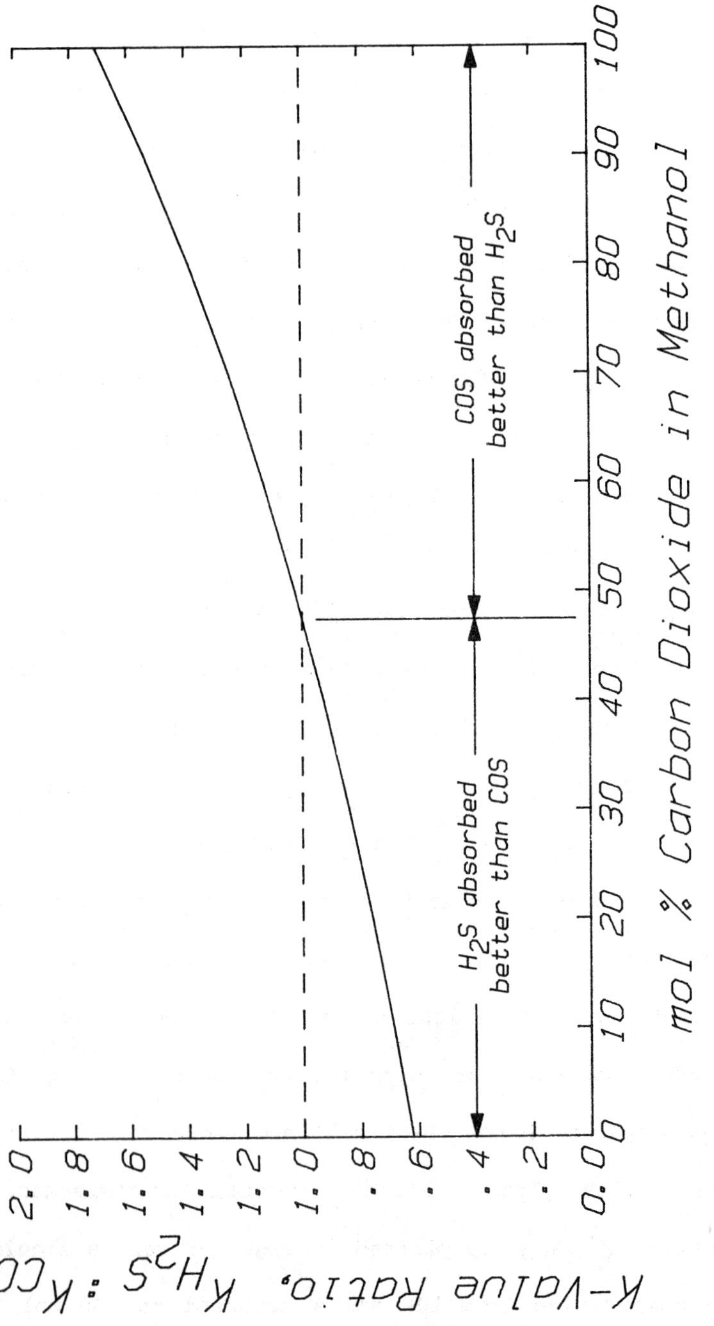

Figure 5. Equilibrium vaporization constant ratios.

carbon dioxide. The ratio $K_{H_2S}:K_{COS}$ for Selexol solvent is even lower than for methanol.

The physical properties of liquid carbon dioxide also enhance its use as an absorbent in general. The low viscosity (about 0.35 cp at -55°C) and high density of liquid carbon dioxide promote high stage efficiency. For comparison, the viscosity of methanol is 7 to 8 times greater over the temperature range 0 to -55°C (Figure 6), and Selexol solvent's viscosity is 5 to 10 centipoise around room temperature. Liquid carbon dioxide is denser (specific gravity 1.17 at -55°C) than most absorbents which, together with its low viscosity, enables high liquid and gas flow rates in absorption and stripping towers. For comparison, methanol has a specific gravity of about 0.85 (-40°C), and Selexol solvent about 1.03 at room temperature. Carbon dioxide has a relatively low molecular weight of 44 compared to Selexol solvent's molecular weight of over 200 (8). Low molecular weight favors high gas solubility per unit volume of solvent.

Liquid carbon dioxide absorbent, with sulfurous compounds, other trace cominants, and perhaps some co-absorbed light hydrocarbons such as methane and ethane, is combined with the contaminated liquid carbon dioxide

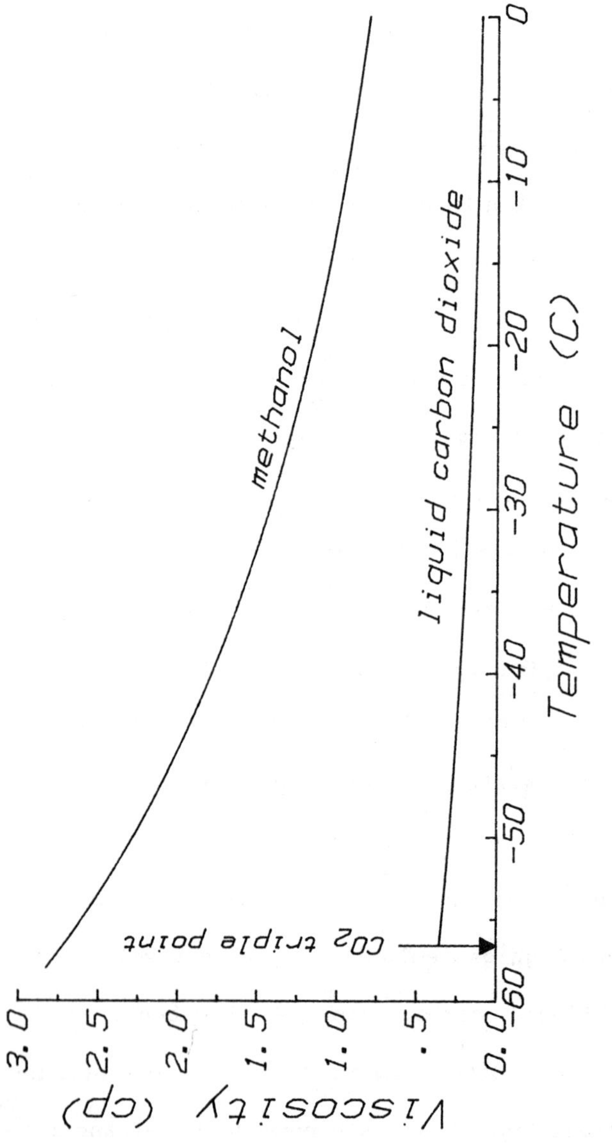

Figure 6. Viscosity comparison.

condensed in precooling to -55°C. The combined carbon dioxide stream, typically 3 to 5 mol % hydrogen sulfide, is stripped of light hydrocarbons if necessary, and sent to the triple-point crystallizer where pure carbon dioxide absorbent is regenerated. The treated gas contains less than 1 ppm H_2S, and leaves the sulfur absorber at essentially -55°C with carbon dioxide the only significant impurity yet to be removed.

Carbon Dioxide Regeneration by Triple-Point Crystallization

Crystallization of carbon dioxide produces pure carbon dioxide (less than 1 ppm total sulfur compounds) and concentrated hydrogen sulfide (up to 75 mol %). The great advantage of crystallization is that pure solid carbon dioxide crystals form from mother liquor containing sulfurous and various other trace compounds. In contrast, separation of pure carbon dioxide from hydrogen sulfide by distillation is difficult because of the very small relative volatility between carbon dioxide and hydrogen sulfide at low hydrogen sulfide concentrations (9). Solid carbon dioxide is not known to form solid solutions with any of the trace compounds likely to be absorbed by liquid carbon dioxide in the hydrogen sulfide absorber. Consequently, trace contaminants are rejected with the hydrogen sulfide rich acid gas stream.

Triple-point crystallization is performed in a continuous separation cascade, similar to continuous distillation. A typical triple-point crystallization cascade is illustrated in Figure 7. The cascade operates at pressures (and temperatures) near the triple-point of carbon dioxide (-56.6°C, 75.1 psia) which permit gaseous, liquid, and solid carbon dioxide to coexist. Pressure differences throughout the cascade cause carbon dioxide to alternately freeze and melt, thereby providing the means of separation. Solid carbon dioxide is formed by flashing and is melted by direct contact with compressed carbon dioxide vapor. Crystal formation occurs at pressures slightly below the triple-point; crystal melting occurs at pressures slightly above the triple-point (only a few psi in each case). Both freezing and melting occur by direct contact, eliminating the need for intermediary heat exchange surface. Compression of flashed carbon dioxide vapor provides the pressure driving force for melting. Extremely pure carbon dioxide can be achieved by only a few stages of recrystallization.

Triple-point crystallization profits from the high triple-point pressure (75.1 psia) of carbon dioxide. The

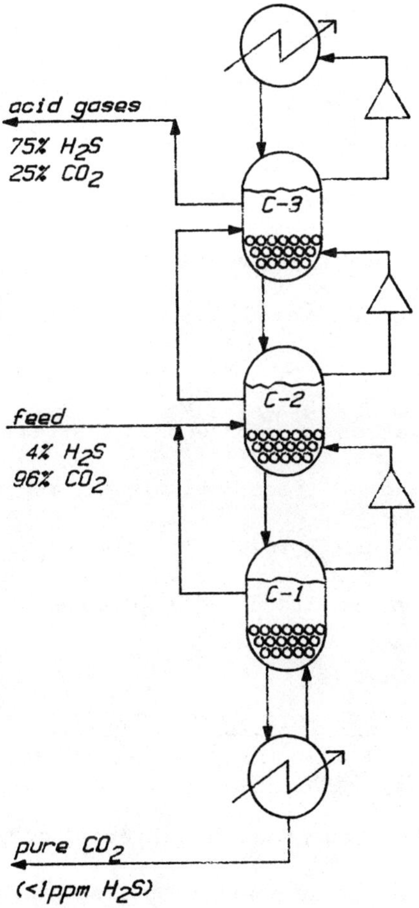

Figure 7. Triple-point CO_2 crystallizer.

relatively high density of carbon dioxide vapor at the triple-point permits use of conventional compression machinery. The above ambient triple-point pressure prevents air leaks to the crystallizer which otherwise would degrade the efficiency of direct contact heat transfer.

Pure carbon dioxide produced by triple-point crystallization is split into two streams, one is absorbent recycled to the hydrogen sulfide absorber, the other is sent back through the process for refrigeration and power recovery, and is delivered as a product stream or is vented to the atmosphere. The other product of triple-point crystallization is a concentrated stream of hydrogen sulfide containing the trace contaminants removed from the crude gas. Triple-point crystallization of carbon dioxide in the CNG process is discussed in greater detail by Liu (10).

TRACE CONTAMINANT ANALYSIS IN THE CNG PROCESS

Trace Contaminant Identification

Coal is a heterogeneous raw material comprised mainly of amorphous carbon, but containing also hydrogen, oxygen, nitrogen, sulfur, and perhaps up to fifty other elements in trace amounts. Many high molecular weight aliphatic and aromatic hydrocarbon compounds, as well as compounds of sulfur and nitrogen, are present in coal. The gasification of coal volatilizes many of the compounds and elements present in the raw coal and causes them to react with one another and with the primary gas phase components hydrogen, carbon monoxide and water. The trace compounds are primarily reduced compounds of carbon, sulfur and nitrogen, a result of the strongly reducing hydrogen atmosphere in the gasifier. A list of compounds most often cited in the literature (11,12,13,14,15,16,17,18,19) as trace contaminants in coal gasifier gas is presented in Table 2.

Trace Contaminant Analysis - Assumptions

The assumptions made to predict the path of trace contaminants in the CNG process are presented here. The assumptions concerning equilibrium distribution of contaminants between vapor and liquid phases are necessary because contaminant equilibrium data is virtually non-existent at low temperature, high pressure, and low concentration.

Contaminant Partial Pressure. Dalton's law of partial pressures, $y_c P_{tot} = P_c$, is used to calculate the change in contaminant concentration with temperature as the crude gas stream drops in temperature from +25°C down to -55°C through the CNG process. Thus, a contaminant is assumed to

Table 2
Trace Contaminants in Coal Gasifier
Exit Gas

Hg, Mercury	i-C_5H_{12}, iso-Pentane	C_6H_6, Benzene
HCN, Hydrogen Cyanide	t-C_5H_{12}, tert-Pentane	C_7H_8, Toluene
NH_3, Ammonia	C_6H_{14}, Hexanes	o-C_8H_{10}, ortho Xylene
CH_3NH_2, Methyl amine	C_2H_4, Ethylene	m-C_8H_{10}, meta Xylene
$C_2H_5NH_2$, Ethyl amine	C_3H_6, Propylene	p-C_8H_{10}, para Xylene
CH_3SH, Methyl mercaptan	1-C_4H_8, 1-Butene	C_6H_5OH, Phenol
C_2H_5SH, Ethyl mercaptan	cis-2-C_4H_8, cis-2-Butene	CH_3OH, Methanol
C_4H_4S, Thiophene	trans-2-C_4H_8, trans-2-Butene	C_2H_5OH, Ethanol
C_2H_6, Ethane	1-2 C_4H_6, 1-2 Butadiene	n-C_3H_7OH, n-Propanol
C_3H_8, Propane	1-3 C_4H_6, 1-3 Butadiene	i-C_3H_7OH, i-Propanol
n-C_4H_{10}, n-Butane	C_2H_2, Acetylene	C_2H_6O, Dimethyl ether
i-C_4H_{10}, iso-Butane	C_3H_4, Propyne	$C_4H_{10}O$, Diethyl ether
n-C_5H_{12}, n-Pentane		

begin condensing at temperature T_d where the contaminant partial pressure P_c equals the pure contaminant vapor pressure $P^o{}_c(T_d)$. The initial condensation temperature T_d is the raw gas stream's dew point with respect to the contaminant; no contaminant can be removed from the gas phase by condensation until its dew point is reached. As the temperature falls below T_d, progressively more contaminant condenses.

Contaminant Equilibrium Vaporization Constants. Equilibrium vaporization constants K_c ($K_c = y_c/x_c$, y_c = vapor phase mole fraction of c, x_c = liquid phase mole fraction of c) for each contaminant c are determined by comparing the contaminant vapor pressure to the vapor pressure of hydrogen sulfide. We define K_c by the following equation:

$$K_c(T) = (P^o{}_c/P^o{}_{H_2S}) * K_{H_2S} \qquad (4\text{-}1)$$

where $P^o{}_c$ is the contaminant vapor pressure and $P^o{}_{H_2S}$ is the vapor pressure of hydrogen sulfide. Trace contaminant volatilities, in general, are closer to the volatility of hydrogen sulfide than the volatility of other significant crude gas components. Hence, for lack of actual data, the vapor-liquid phase nonideal behavior of trace contaminants is represented by the nonideal behavior of hydrogen sulfide as embodied in the term $K_{H_2S}(T)$. Trace contaminant solubilities in the gas phase calculated via equation (4-1) are greater than would be predicted if ideal gas behavior were assumed, i.e., vaporization constants K_c obtained from equation (4-1) predict more difficult absorption than would ideal gas behavior.

Equilibrium vaporization constants K_c are assumed to vary with system pressure as follows:

$$K_c(P) = K_{c,1000} * (P/1000) \tag{4-2}$$

where $K_{c,1000}$ = trace contaminant equilibrium constant at 1000 psia, and P = system pressure, psia. Trace contaminants are assumed to behave ideally with respect to changes in total system pressure. Thus, the vapor phase and liquid phase activity coefficients are assumed invariant with system pressure. These assumptions are forced because of the sparse equilibrium data at typical system temperatures and pressures, and because of the extremely low concentrations of most contaminants.

Absorption Calculations. Simplified equilibrium stage calculations were performed using the absorption/stripping factor method discussed in Treybal (20) to determine contaminant distributions in the CNG process absorbers and strippers.

The basic assumption of the absorption factor method for calculating ideal stages is that both the operating and equilibrium lines are straight in the concentration region of interest, in this case the extremely dilute trace contaminant concentrations encountered in the AGR system.

Trace Contaminant Removal Prior to AGR

Most trace elements and high molecular weight compounds not volatilized in the gasifier are removed with the ash or slag. Lighter elemental contaminants, particulate matter, and water-soluble contaminants such as ammonia, hydrogen cyanide, and phenols, are effectively removed by a cyclone, venturi scrubbing, and/or a water quench. Subsequent dehydration of the raw gas in the CNG process with molecular sieves also serves to remove elemental contaminants such as mercury present in the sour gas. Mercury contamination of the sieve, however, is inconsequential over the sieve life (21).

The most probable contaminants in the pretreated raw gas stream entering an acid gas removal process are shown in Table 3. This list of contaminants summarizes the most

Table 3
Trace Contaminants Most Likely to be Present in Raw Coal Gasifier Gas Entering an Acid Gas Removal Process

Contaminant	Maximum Concentration	Freezing Pt °C	Vapor Pressure @ 25°C
Hg	.1 ppm	-38.9°	1850 E - 6 mm Hg
HCN	2 ppm	-13.2	760 mm Hg
NH_3	0.6 mole %	-77.7	10 atm
CH_3SH	2 ppm	-123	2 atm
C_4H_4S	2 ppm	-38.2	80 mm Hg
C_2H_6	1.6 mole %	-183.2	40 atm
C_3H_8	0.2 mole %	-187.1	10 atm
C_2H_4	0.5 mole %	-169	60 atm
C_2H_2	0.2 mole %	-81.5	50 atm
C_6H_6	0.01 mole %	5.5	100 mm Hg
C_6H_5OH	100 ppm	41	1 mm Hg @ 40.1°

frequently cited contaminants and maximum concentration levels, independent of any particular coal gasifier or coal type.

Trace Contaminant Removal in the CNG Process

The distribution of trace contaminants in product streams leaving the CNG process is shown in Figure 8. As shown, the significant troublesome contaminants leave with the sour water condensate, and with the concentrated acid gas stream.

Sulfurous Compounds, Aromatics, and C_3+ Hydrocarbons: Sulfurous compounds, aromatics, and C_3+ hydrocarbons are sparingly soluble in water, and pass essentially untouched through the CNG process dehydration system and into the hydrogen sulfide absorber. These contaminants are scrubbed from the raw gas by pure liquid carbon dioxide absorbent to very low levels of the order 1 ppm or less. Liquid carbon dioxide has the unique property among known acid gas solvents that it absorbs COS more easily than H_2S (see Figure 5). Hence, a flow of pure liquid carbon dioxide sufficient to scrub H_2S from the raw gas to the 1 ppm level also removes COS and other less volatile trace contaminants such as thiophene, carbon disulfide and other organic sulfur compunds to

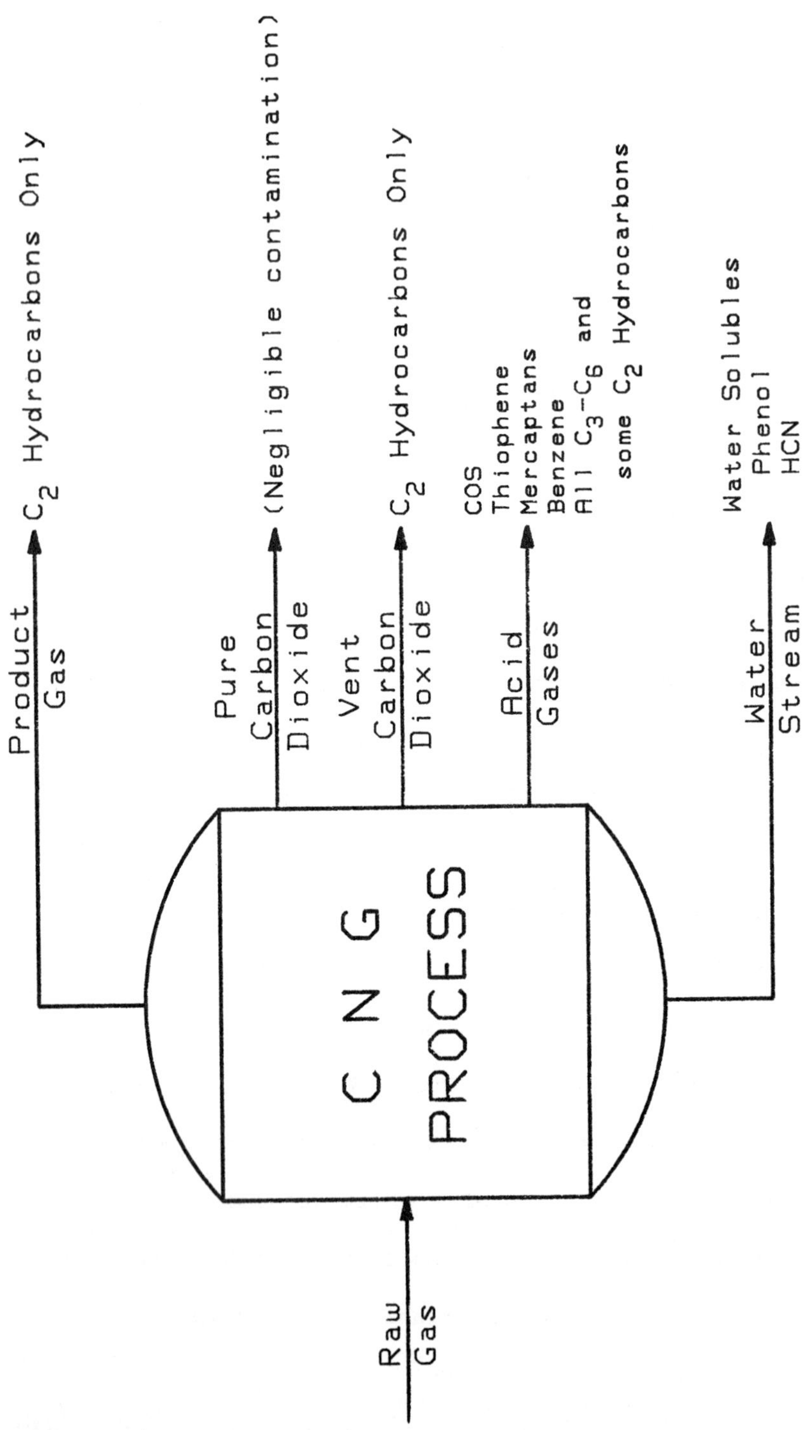

Figure 8. Distribution of trace contaminants in CNG process product streams.

the 1 ppm level or less.

Pure liquid carbon dioxide for recycle to the hydrogen sulfide absorber is regenerated by triple-point crystallization. Liquid carbon dioxide absorbent containing the hydrogen sulfide and trace contaminants is crystallized to yield very pure crystals of carbon dioxide.

Experimental work has demonstrated that dramatic separation of carbon dioxide from trace contaminants is possible by triple-point crystallization. Concentration changes of the order 500 to 5000 have been achieved in one stage of crystallization in a pilot unit triple-point crystallizer. Representative sulfurous compounds and trace contaminants examined include hydrogen sulfide, carbonyl sulfide, methyl mercaptan, ethane, and ethylene. No evidence has been found to suggest that carbon dioxide forms solid solutions with any of the remaining contaminants. Consequently, these contaminants, together with a relatively small amount of water soluble contaminants co-absorbed by the liquid carbon dioxide in the hydrogen sulfide absorber, exit the CNG process with the hydrogen sulfide (acid gas) product stream.

C_2 Hydrocarbons: Ethane, ethylene, and acetylene are the only contaminants expected to break through the hydrogen sulfide absorber and proceed with the sulfur-free crude gas to the carbon dioxide removal portion of the CNG process. Absorption calculations show that approximately 1-5% of the C_2 hydrocarbons are absorbed by liquid carbon dioxide in the hydrogen sulfide absorber. These C_2 hydrocarbons are rejected in the carbon dioxide triple-point crystallizer, and exit the CNG process with the acid gas product stream. An additional 30-55% of the C_2 hydrocarbons initially present in the crude gas are co-absorbed in the carbon dioxide removal step of the CNG process. These hydrocarbons are ultimately rejected to the atmosphere with vented carbon dioxide. The residual 40-70% of C_2 hydrocarbons remain with the treated product gas from the CNG process. Problems such as methanation catalyst contamination or product degradation due to the presence of C_2 hydrocarbons in the product gas are believed minimal (22).

Benzene: The trace contaminant benzene is disussed separately for several reasons: 1) benzene has a relatively high melting (freezing) point, +5.5°C, 2) in the CNG process, the dehydrated raw gas is sensibly cooled several tens of degrees Celsius below the benzene freezing point before liquid carbon dioxide condenses, inviting the possibility of benzene precipitation, and 3) in the CNG triple-point crystallizer, temperatures down to about -80°C exist and the solubility of benzene in liquid mixtures of carbon dioxide and hydrogen sulfide may be exceeded. Hence, the simplified calculation procedures described earlier were abandoned in favor of a more rigorous method for determining the fate of benzene in the CNG process.

The solubility of benzene in raw gasifier gas as a function of temperature and operating pressure was calculated using the techniques described by Prausnitz (23,24,25). The solubility of benzene in 1000 psia crude gasifier gas with composition given in Table 4 is shown in Figure 9. The carbon dioxide dew point of the gasifier gas is about -27°C. At -27°C, the solubility of solid benzene in the gasifier gas is about 0.03 mol %. The average benzene concentration in coal gasifier gas is about 0.01 mol % (14,18,19) and hence, we do not anticipate benzene precipitation in the precooling heat exchangers before liquid carbon dioxide condensation commences at -27°C.

Table 4
Raw Gas Composition (Contaminant-Free) Entering the CNG Acid Gas Removal Process

Component	Mole %
H_2	40.67
CO_2	32.41
CO	13.56
CH_4	12.18
H_2S	01.11
H_2O	00.07
COS	00.01
TOTAL	100.00

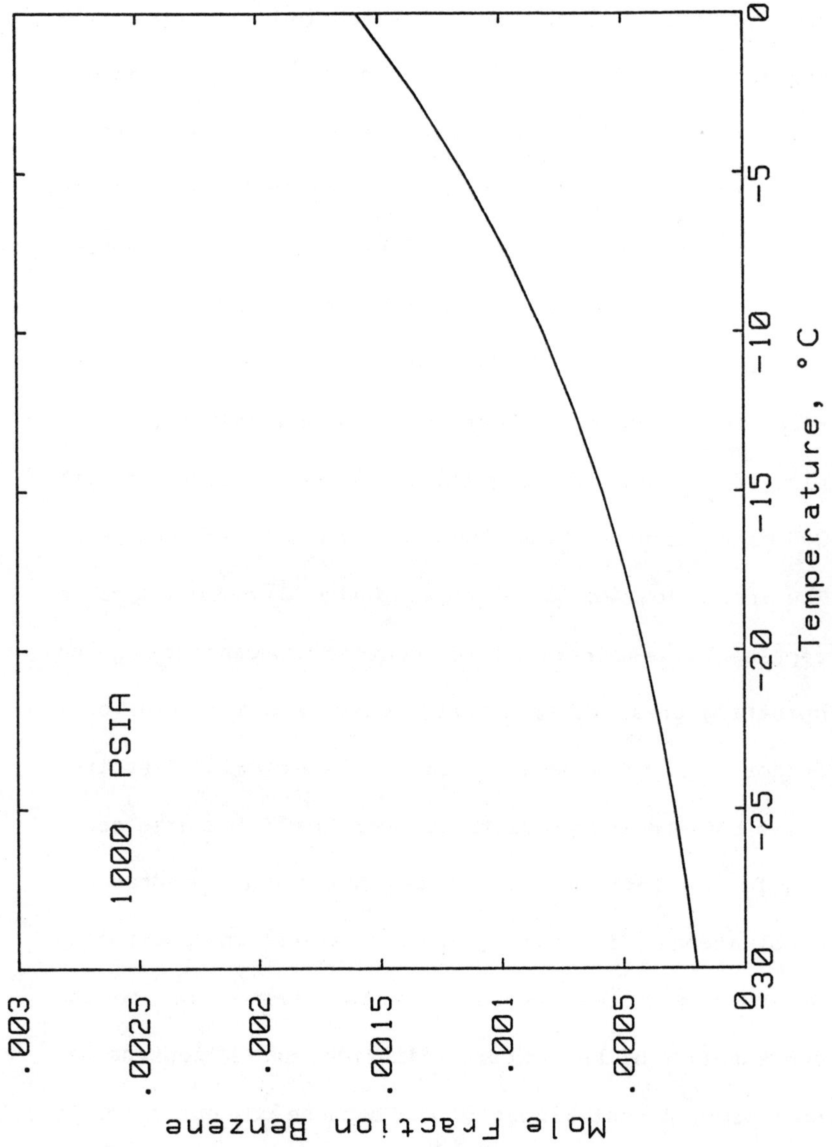

Figure 9. Solubility of benzene in crude gas.

Figure 10 is useful for predicting if various combinations of benzene concentration and operating pressure are conducive to the formation of solid benzene prior to carbon dioxide condensation. Figure 10 shows the solubility of benzene in gasifier gas as a function of operating pressure with temperature as a parameter; benzene solubility is a strong function of temperature, but above 300 psia, is relatively invariant with pressure. The carbon dioxide dew point of the gasifier gas is plotted as a dashed line; the dew point is about -55°C at 300 psia, about -30°C at 800 psia, and -27°C at 1000 psia. The area below and to the right of the CO_2 dew point curve represents combinations of benzene concentration and operating pressure such that liquid carbon dioxide condenses prior to benzene. Liquid carbon dioxide dissolves solid benzene and prevents accumulation. The area above and to the left of the CO_2 dew point curve represents combinations of benzene concentration and operating pressure such that benzene precipitates prior to the condensation of liquid carbon dioxide. Hence, benzene may be a problem contaminant if the benzene concentration is high, and the operating pressure is low.

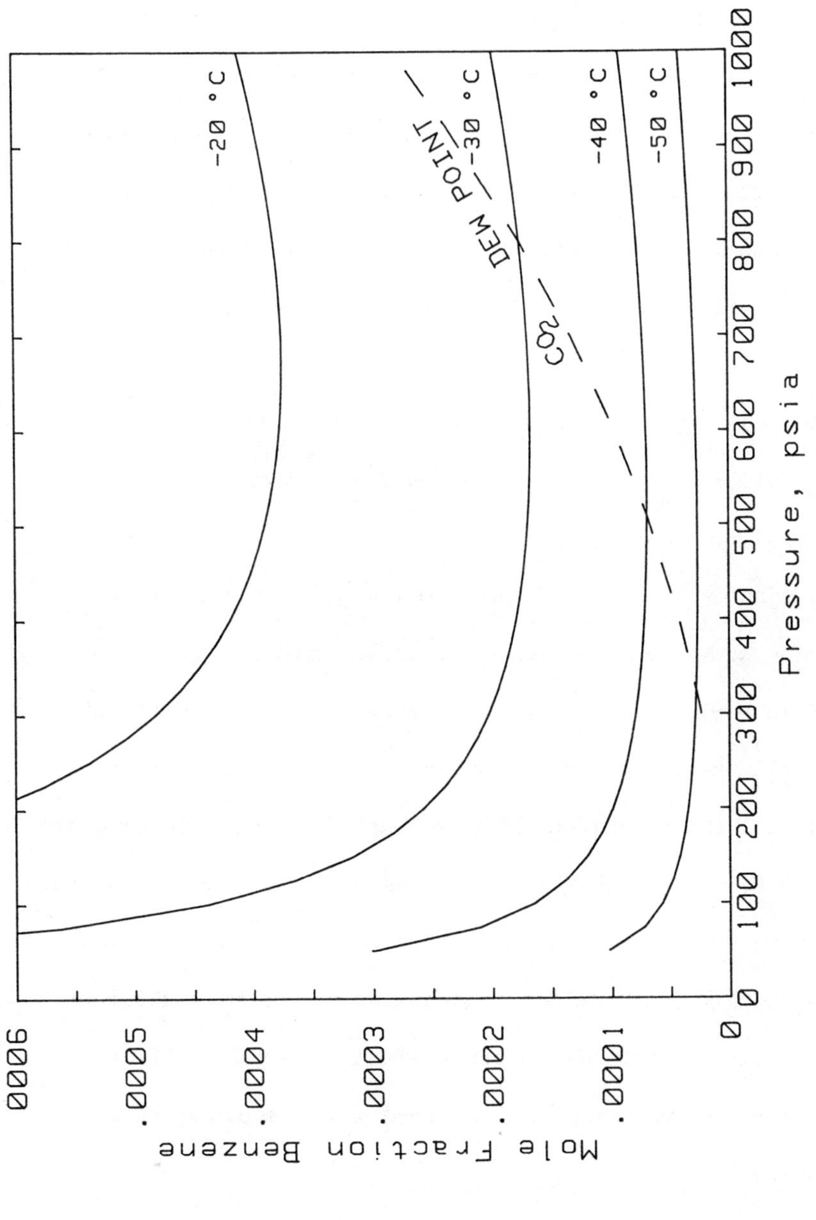

Figure 10. Solubility of benzene in crude gas.

Although carbon dioxide condenses from the gasifier gas as the temperature is reduced below -27°C, the possibility of benzene precipitation still exists. However, our concern shifts from the solubility of benzene in gasifier gas to the solubility of benzene in liquid carbon dioxide -- will the co-condensing benzene dissolve completely in the liquid carbon dioxide, or will solubility limits be exceeded? The equation of ideal solubility from Prausnitz (23) was used to calculate the liquidus line for the benzene-carbon dioxide system. The liquidus line for the benzene-carbon dioxide system is shown in Figure 11. The lower temperature limit which can exist in the carbon dioxide condenser is -55°C; Figure 9 shows that solid benzene does not form at -55°C until the liquid phase concentration of benzene is about 30 mol %. The probability of reaching 30 mol % benzene in the liquid carbon dioxide condensate is extremely remote; hence, we conclude that benzene does not solidify in the liquid carbon dioxide condensate, or in the subsequent sulfurous compound absorber where hydrogen sulfide and trace contaminants are absorbed with liquid carbon dioxide.

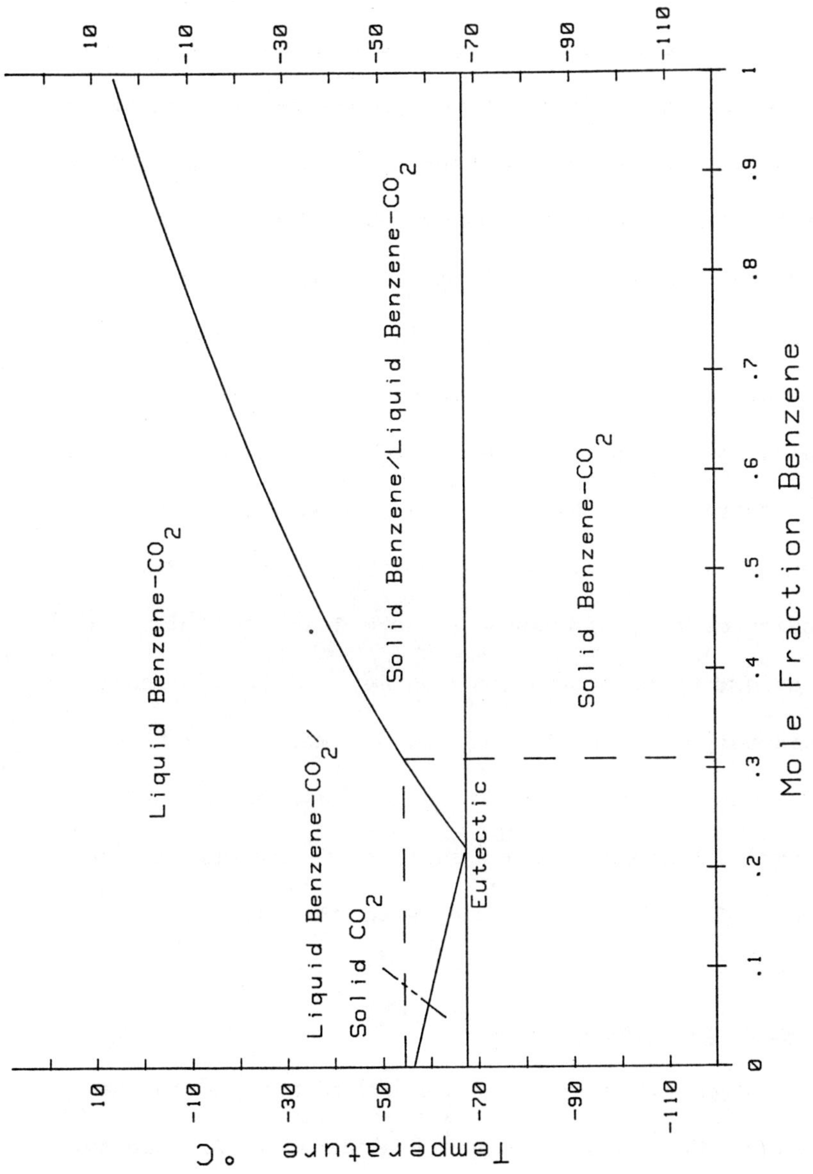

Figure 11. Liquidus line for benzene-CO_2.

Two conditions exist in the carbon dioxide triple-point crystallizer which warrant investigation for possible benzene solidification: 1) the temperature of the acid gas stream reaches about -76°C, and 2) the concentration of benzene increases to about 5 mol % as hydrogen sulfide is concentrated in the (liquid) acid gas stream. The acid gas stream is about 75 mol % hydrogen sulfide; hence, the solubility of benzene in hydrogen sulfide is calculated using the equation of ideal solubility from Prausnitz (23). The ideal liquidus line for the system benzene-hydrogen sulfide is shown in Figure 12. Figure 12 indicates no solid phase is formed at -76°C until the liquid phase concentration reaches about 18 mol % benzene (probably greater than 18 mol % because liquid carbon dioxide is also present in the acid gas stream). Consequently, benzene does not solidify in the crystallizer, and leaves the CNG process with the liquid acid gas stream.

SUMMARY AND CONCLUSIONS

A general overview of the path of trace contaminants in the CNG process is given in Figure 8. All contaminants, with the exception of C_2 hydrocarbons which remain with the treated gas, are essentially completely removed

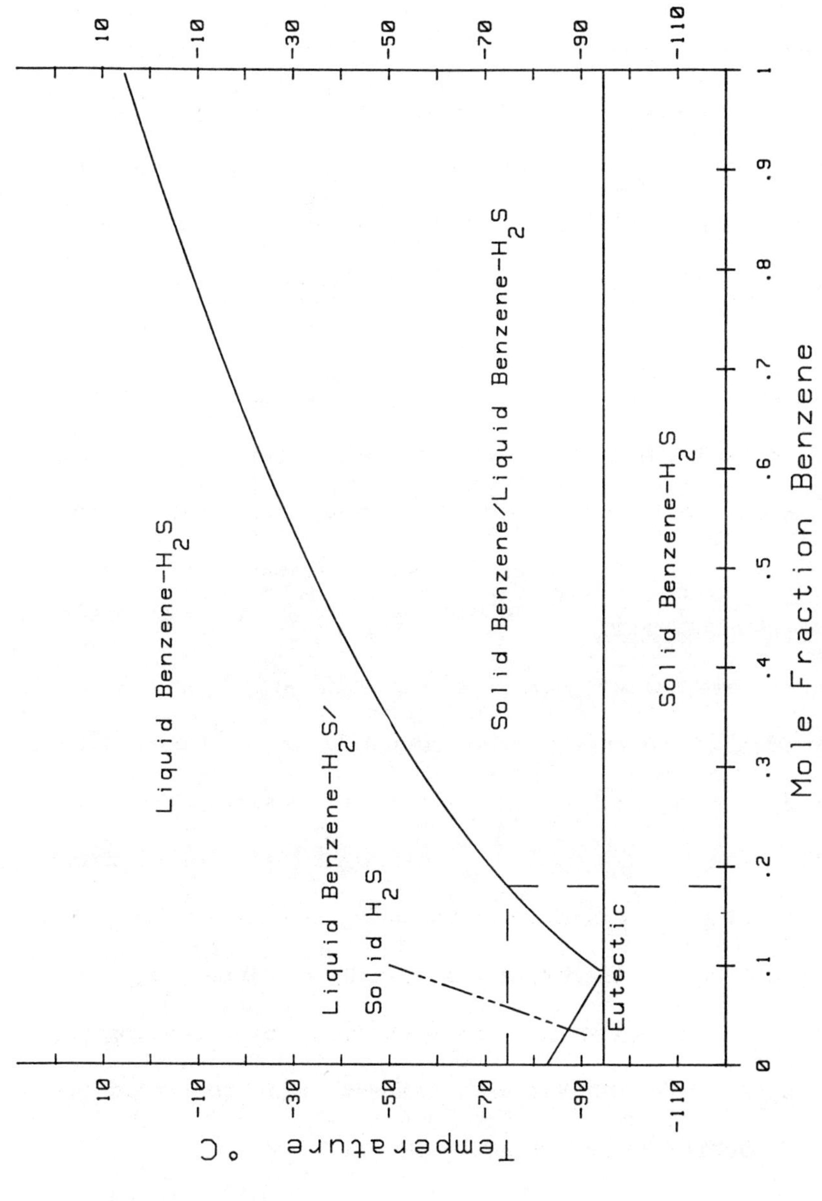

Figure 12. Liquidus line for benzene-H_2S.

from the product gas and rejected from the CNG process with sour water waste streams and the acid gas stream. Pure liquid carbon dioxide absorbent effectively scrubs the gasifier gas of hydrogen sulfide, carbonyl sulfide, and trace contaminants. Triple-point crystalliztion of carbon dioxide regenerates pure liquid carbon dioxide absorbent and rejects contaminants with the acid gas stream. Except for the remote possibility of benzene as discussed above, contaminants are not recycled with the carbon dioxide absorbent, and no accumulation of contaminants in the CNG process is envisioned.

ACKNOWLEDGEMENTS

Development work on the CNG acid gas removal process, and triple-point crystallization of carbon dioxide in particular, has been sponsored since 1976 by the CNG Research Company, a subsidiary of the Consolidated Natural Gas Company. For the past three years, the U.S. Department of Energy, Morgantown Energy Technology Center, has co-sponsored the work in a joint development program with CNG Research Company under contracts DE-AC21-80MC14399 and DE-AC21-83MC20230.

REFERENCES

1. Christensen, K.G., and W.J. Stupin, "Merits of Acid Gas Removal Processes", Hydrocarbon Processing, Feb. 1978, p. 125.

2. Ghassemi, M., Strehler, D., Crawford, K., and Quinlivan, S., "Applicability of Petroleum Refinery Control Technologies to Coal Conversion", EPA Report No. EPA-600/7-78-190, NTIS No. PB-288 630, October, 1978.

3. Ferrell, J.K., et al., "Coal Gasification/Gas Cleanup Test Facility: Volume III. Environmental Assessment of Operation with New Mexico Subbituminous Coal and Chilled Methanol", prepared for U.S. Environmental Protection Agency, Cooperative Agreement No. CR804811.

4. Rousseau, R.W., "Physical Absorption of CO_2 and Sulfur Gases from Coal Gasification -- Pilot Plant Results", paper presented at the Second Annual GRI Gas Separations Workshop, October 21-22, 1982, National Bureau of Standards, Boulder, Colorado.

5. Rousseau, R.W., J.K. Ferrell and R.M. Kelly, "Packed Absorber and Stripper Operation on Gases Produced from Coal", presented at PACHEC '83, May 1983, Seoul, Korea.

6. Adler, R.J., et al., "The CNG Process, A New, Low Temperature, Energy Efficiency Acid Gas Removal Process", presented at the Symposium on Industrial Gas Separations, American Chemical Society, Washington, D.C., June 14-15, 1982.

7. Adler, R.J., Brosilow, C.B., Brown, W.R., and Gardner, N.C., "Gas Separation Process", U.S. Patent 4,270,937, June 2, 1981.

8. Sweny, J.W., "The Selexol Solvent Process in Fuel Gas Treating", paper presented at 81st National AIChE Meeting, Kansas City, Missouri, April 11-14, 1976.

9. Sobocinski, D.P., and Kurata, F., "Heterogeneous Phase Equilibria of the Hydrogen Sulfide - Carbon Dioxide System", AIChE J., Vol. 5, No. 4, p. 545, 1959.

10. Liu, Y.C., et al., "Triple-Point Crystallization Separates and Concentrates Acid Gases", paper 52A, presented at AIChE Spring National Meeting, Houston, Texas, March 30, 1983.

11. Jonardi, R.J., et al., "Environmental Assessment of the Hygas Process", DOE Report Contract #EX-76-C-01-2433, June, 1979.

12. "Coal Derived Fuel Gases for Molten Carbonate Fuel Cells", DOE Report Contract #EY-TI-C-21-8085, November, 1979.

13. Koppenaal, D.W., "Trace Elements Study in Coal Gasification Process Streams", Ph.D. dissertation, University of Missouri-Columbia, September, 1978.

14. "Environmental Assessment Fuel Gas Demonstration Plant Program", DOE Report Contract #EW-78-C-02-5065, February, 1979.

15. Felder, R.M., et al., "Evolution and Removal of Pollutants from the Gasification of a Subbituminous Coal in a Fluidized Bed Reactor", ACS Div. of Fuel Chem., Vol. 27(1), Las Vegas, Nevada, March 28-April 2, 1982.

16. Clean Fuels from Coal: Symposium Papers, presented September 10-14, 1973, sponsored by Institute of Gas Technology, Chicago, 1973.

17. Kronosky, R.M., et al., "Gasification of Iowa Coal in the SYNTHANE PDU Gasifier", ACS Div. of Fuel Chem., 173 National Meeting, March 21-25, 1977.

18. Forney, A.J., et al., "Analysis of Tars, Chars, Gases, and Water found in Effluents from the SYNTHANE Process", U.S. Bur. Mines, TPR 76, January, 1974.

19. Somerville, M.H., et al., "A Comparative Study of Effluents and Their Control from Four Dry Ash Lurgi Gasification Plants", DOE Report Contract #EX-76-X-02-4035, July, 1978.

20. Treyball, R.E., Mass Transfer Operations, McGraw Hill, New York, 1980.

21. personal communication, Linde Division of Union Carbide, Cleveland; see also Linde Molecular Sieve Brochure F-1979 D 88-0414, Linde Division of Union Carbide.

22. Eisenlohr, K.H., F.W. Mueller, and M. Dry, "Effect of Certain Reaction Parameters on Methanation of Coal Gas to SNG," Methanation of Synthesis Gas, Advances in Chemistry Series 146, Len Seglin, ed., 1975.

23. Prausnitz, J.M., Molecular Thermodynamics of Fluid Phase Equilibria, Prentice-Hall, Englewood Cliffs, N.J., 1969.

24. Prausnitz, J.M. and Chueh, P.L., Computer Calculations for High Pressure Vapor-Liquid Equilibria, Prentice Hall, Englewood Cliffs, N.J., 1968.

25. Prausnitz, J.M., et al., Computer Calculations for Multicomponent Vapor-Liquid Equilbria, Prentice Hall, Englewood Cliffs, N.J., 1967.

19

The Ryan/Holmes Technology—An Economical Route for CO_2 and Liquids Recovery

James M. Ryan
Manager, Research and Development
Gas Processing Division, Koch Process Systems, Inc.

Fred W. Schaffert
Vice-president, Business Development
Gas Processing Division, Koch Process Systems, Inc.

INTRODUCTION

The Ryan/Holmes Technology represents a unique, distillative approach to the recovery of CO_2 and hydrocarbon liquids from natural gas streams. The technology is a very flexible and ecomomic solution to treating gas streams with very high CO_2 concentrations typically encountered with EOR projects utilizing CO_2 injection. Recent experience has shown that appreciable cost savings can be realized in many EOR applications with a particular Ryan/Holmes four column distillation sequence in contrast to the three column arrangement discussed in previous articles.

PROCESS PRINCIPLES

The Ryan/Holmes Technology is a family of distillation separations using NGL-based additive introduced to the top or near the top of the distillation column. The additive, which is recycled from the bottom of a column in the distillation sequence, permits the desired separations to occur. Typically, the recycled additive is a butanes-plus mixture. All the distillations use conventional processing equipment.

The proprietary separation features involved are summarized in Table 1. The demethanizer column separates methane as an overhead product from CO_2, C_2, H_2S, C_3 and heavier hydrocarbons. Recycled additive is introduced to the condenser or to the top of the column in order to avoid freezing in the column. Only a small amount of additive is needed to avoid freezing. The relative volatility of C_1 to CO_2 is also improved by the presence of additive. The overhead C_1 fraction can be designed for 1 or 2% CO_2, as representative of pipeline grade gas, or can be designed for 50 ppm CO_2 or lower, suitable for feed to a liquefaction unit.

Table 1
Function of Additive

DEMETHANIZER—SEPARATE C_1 FROM CO_2, C_2, H_2S, C_{3+}

- AVOID FREEZING
- RAISE TEMPERATURE
- INCREASE RELATIVE VOLATILITY

ETHANE RECOVERY COLUMN—SEPARATE CO_2 FROM C_2, H_2S, C_{3+}

- CO_2/C_2 BREAK AZEOTROPE
- CO_2/H_2S ENHANCE RELATIVE VOLATILITY

As more additive is circulated, the overhead of the demethanizer column becomes warmer as does the point of closest approach to freezing. Increased flow of additive can result in a demethanizer column where the coldest temperature is -35°F and the condenser cooling requirement can be accomplished by standard propane refrigeration systems. It should be noted that H_2S present in the feed is readily eliminated from the overhead C_1 product to levels below 1 ppm.

Ethane recovery columns are operated to produce a CO_2 overhead product and C_2, H_2S, C_3 and heavier hydrocarbons as a bottom product. The use of additive in this column breaks the azeotrope which forms between CO_2 and C_2. CO_2 and H_2S do not form an azeotrope, but on the CO_2 rich end the relative volatility is very poor. The additive functions to increase the relative volatility between CO_2 and H_2S. Additive is introduced about ten trays down from the top of the column so as to provide a knock-back section at the top of the column which prevents the C_4+ additive from being carried into the CO_2 overhead product. The separation of CO_2 and C_2 or H_2S can be accomplished to whatever extent desired.

These basic separations have been tested in a pilot plant program conducted with ARCO Oil and Gas at their Covington, Oklahoma gas plant and reported to the 1982 GPA Annual Convention (1). Detailed studies of vapor liquid equilibrium were conducted in phase equilibrium cells to obtain relevant data for the entire range of concentrations involved. The CO_2 Pilot Plant is now located at Koch's Corpus Christi Refinery where a program of further separation and column internal studies is in progress.

EOR GAS TREATING REQUIREMENTS

Enhanced oil recovery (EOR) projects using CO_2 as a miscible injectant are being used extensively to recover additional oil from existing reservoirs. Typically, these projects involve injecting CO_2 into the formation on a schedule in which CO_2 injection and water injection are alternated in a so called "WAG" operation. Oil production is stimulated by the presence of miscible CO_2. Concurrent with oil production rate increases, CO_2 is also produced because it is present in the oil and CO_2 "breakthrough" occurs. The separator gas requires processing in order to recover the valuable CO_2 for recycle as well as to eliminate CO_2 from the residue gas product and from recovered natural gas liquids. In many cases H_2S is also present in the formation. Gas processing is required to eliminate H_2S from the residue gas and from the CO_2 product.

A wide range of gas flows and compositions must be processed over the life of an EOR project. Figure 1 is a representative projection of the gas volumes to be processed over a period of 15 years. Usually, formations for which CO_2 injection are considered have a declining pattern of hydrocarbon gas volume production rate. After injecting CO_2, the hydrocarbon rate will climb, peak out and then resume a declining pattern. Along with the hydrocarbon gas, CO_2 is produced giving a total gas production rate which may increase quite considerably. In the example shown here, the total rate rises from about 15 MMSCFD at the beginning of the project to 100 MMSCFD nine or ten years later. The hydrocarbon gas volume increases by almost a factor of two and peaks out earlier than the CO_2 production. Of the hydrocarbon production, methane is typically less than 60% of the total and over the years declines as a fraction of the hydrocarbon gas. The NGL content of the hydrocarbon fraction increases over the years and may rise from values of 10-15 gal/MCF to

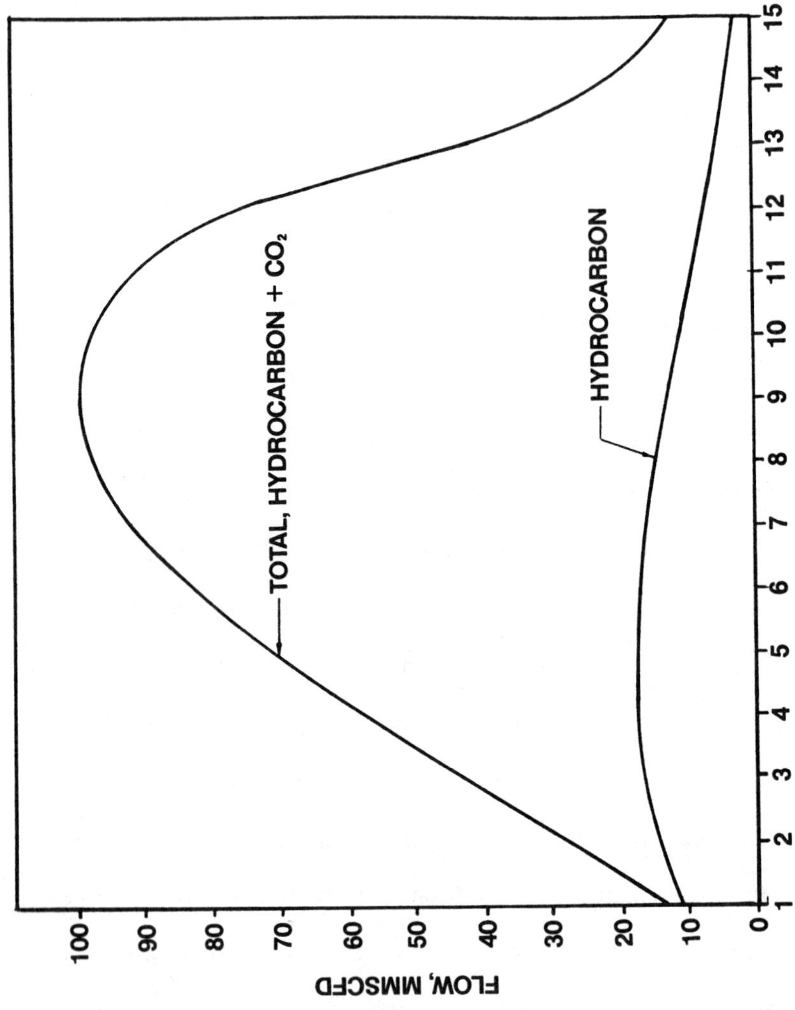

Figure 1. Projected associated gas volumes.

values well over 20 gal/MCF. CO_2 content in the total gas volume will usually rise quickly to above 30 or 40% and often times will spend many years at 70 to 90%. The gas processing plant built to handle a gas of such varying volume and composition must have a great deal of inherent flexibility.

Typical key specifications for the three products of an EOR gas processing facility are shown in Table 2. The methane product can be sold or used as fuel. Typically, in the early years there is an excess of methane required over the plant fuel needs which can be sold if it conforms to conventional sales gas specifications. Carbon dioxide content of the residue gas is therefore usually limited to about 2% and H_2S is removed to the pipeline standard of 1/4 grain/100 SCF.

The key specifications for the CO_2 product typically include a limitation of about 4% methane and lighter constituents in order to maintain miscibility of the injectant at pressures well below the reservoir fracture limit. H_2S may be limited to 100 ppm for safety reasons. A pressure of somewhat over 2000 psig is commonly needed for reinjecting the CO_2.

Table 2
Typical Key Specifications

SALES GAS—FUEL	
CO_2	2%
H_2S	¼ GRAIN/100 SCF
550 PSIG	PIPELINE OR FUEL SYSTEM
CO_2 PRODUCT	
C_1	4%
H_2S	100 PPM
2200 PSIG	REINJECTION
NGL PRODUCT (SOUR)	
CO_2/H_2S	2/1
C_2 RECOVERY	90 - 60%

NGL produced from a Ryan/Holmes facility will contain any H_2S present in the feed. In this regard, the requirement for the NGL product is that the CO_2 to H_2S ratio be less than 2 to 1. This makes the acid gas compatible with Claus unit operation without further selective separation. A conventional amine unit is able to remove the last of the acid constituents which can then be fed to the Claus unit.

Ethane recovery is typically specified between 60 and 90%. The limiting design condition may provide for a 60% recovery. When gas volume throughputs or the concentration of CO_2 is less, recoveries will rise to 90% or more. The recovery of propane and heavier hydrocarbons is essentially 100%.

RYAN/HOLMES THREE COLUMN SCHEME

The three column Ryan/Holmes scheme has been presented in the literature previously (1, 2) and consists of a demethanizer column, an ethane recovery column and an additive recovery column in sequence as shown in Figure 2. The demethanizer column and ethane recovery column are supplied with recycled additive to make possible the specified separations.

The demethanizer column separates the feed into a overhead fraction which is methane and lighter components such as nitrogen. The bottom product of the demethanizer is CO_2, C_2 and heavier hydrocarbons including C_3, C_4+ and additive. Any H_2S in the feed will be produced in the bottoms fraction.

In the ethane recovery column, CO_2 is produced as an overhead fraction. C_2, H_2S, C_3, C_4+ and additive are produced from the bottom of the column and flow to the additive recovery column. The role of the additive recovery column is to remove lighter NGL's to prevent them from being recycled. Butanes-plus is produced from the bottom of the column providing the net heavier NGL make and recovered additive to be recycled to the first two columns. H_2S, if present in the plant feed, will be

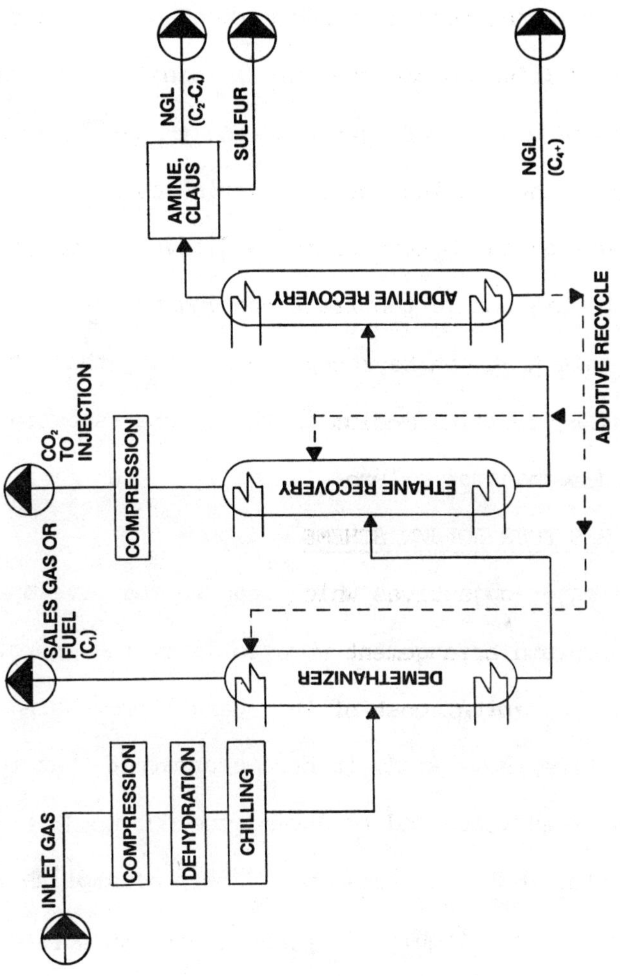

Figure 2. Three-column Ryan/Holmes process.

produced in the overhead of the additive recovery column. The sour NGL's of the overhead are processed by conventional amine treatment to produce sweet NGL product.

The additive column pressure is usually high enough to avoid the use of refrigeration and to permit overhead condensation by cooling water or aerial coolers. The temperature of the bottom of the additive column is typically the highest in the fractionating process. To some extent, this heat can be recovered by using the bottom stream as a reboiling medium in the ethane recovery column and the demethanizer column.

RYAN/HOLMES FOUR COLUMN SCHEME

The major objectives which lead to the development of the four column arrangement were to lower the energy consumption and capital cost of the overall processing scheme. After some study it became apparent that reducing the flow of additive and producing the CO_2 product as a liquid column bottoms stream would help accomplish these objectives. If a liquid CO_2 product bottoms stream were to be produced, it would have to be from a column into which no additive was fed. Figure 3 shows the resulting four column flow schematic.

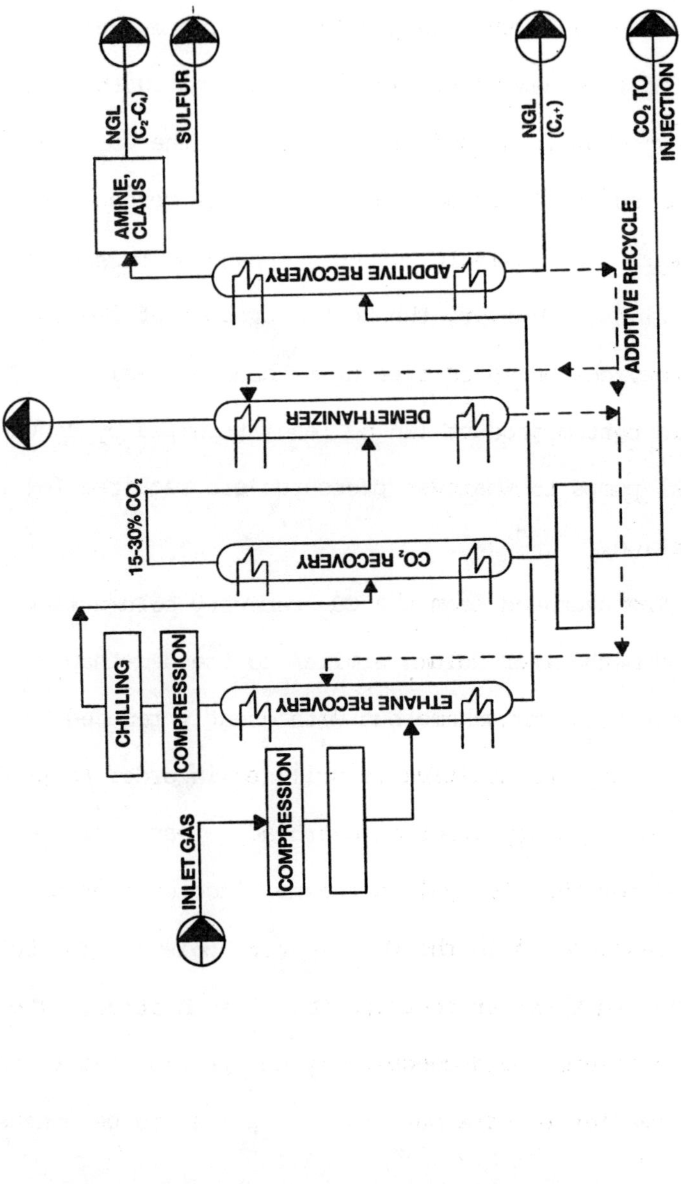

Figure 3. Four-column Ryan/Holmes process.

The first column entered is the ethane recovery column in which the split between CO_2 overhead and a bottom product of C_2, H_2S, C_3+ and additive is effected. The overhead is then compressed, chilled by refrigerant and fed to a new column which is designated the CO_2 recovery column. No additive is used in the CO_2 recovery column so the overhead cannot be a pure methane fraction but will contain 15 to 30% CO_2, depending on specific conditions. However, the bottom product of the CO_2 recovery column is as free of methane as desired. The liquid bottom product can be readily pumped by conventional pumps to whatever pressures are required for reinjection.

The overhead from the CO_2 recovery column flows next to a demethanizer column similar to the demethanizer of the three column scheme but with greatly reduced CO_2 content. Recycled additive is utilized in order to produce a sales gas quality fraction overhead. Less additive is required for the four column scheme since much of the CO_2 has been removed in the CO_2 recovery column. The bottoms of the demethanizer contains CO_2 plus additive. The bottom stream could conceivably be fractionated to recover additive for recycle but that has proven to be unnecessary. The bottoms can be used to fulfill some of the additive requirement in the ethane recovery column, the remainder of that requirement being made up with fresh recycle additive. By this means, the additive which is used in the demethanizer is used again.

The bottoms from the ethane recovery column flow to the additive recovery column in which a light NGL-heavy NGL split is made. Typically, C_2 through some of the C_4 is taken overhead including any H_2S, if present in the inlet gas. The remaining heavy NGL bottoms comprises additive recycle and NGL product make.

COMPARISON OF THREE VS. FOUR COLUMN APPROACH

A general comparison between the three column and the four column processes is shown in Table 3. The first column seen by feed in the three column process is run at a temperature of -40 to -120°F. In the four column process, the coldest conditions in the first column are -10 to +40°F. Because the process tends to fractionate water away from the cold process temperature zones, the dehydration processing required for the four column scheme is not as severe. Dehydration for the three column process should be to a -100°F dew point specification or even lower. In the case of the four column process, the dehy-

Table 3
General Comparison, 3-Column vs. 4-Column Process

	3 COLUMN	4 COLUMN
FIRST COLUMN, °F	−40 TO −120	−10 TO 40
DEHYDRATION, °F	−120	−50
DEHYDRATION PROCESS	SUPER GLYCOL OR MOL SIEVE	STANDARD GLYCOL OR SUPER GLYCOL
CO_2 PRESSURIZING	COMPRESSION	PUMPING
CO_2 PROCESSING SEQUENCE	INTERNAL PRESSURE REDUCTION	EFFICIENT PRESSURE SEQUENCE
ADDITIVE	MORE—NO "DOUBLE USE"	LESS—USED "DOUBLE"
REBOILER TEMPERATURES	HIGHER	LOWER
POWER, HP	HIGHER	LOWER

dration objective can be a dew point of only -50°F.

In order to obtain the required dehydration for a three column process, the glycol will have to be super stripped, such as by gas stripping or alternatively by an azeotropic entrainer. Alternatively, molecular sieves may be used at a somewhat greater expense. With a four column process, a less severe method can be used.

The sequence of CO_2 processing represents a major difference between three column and four column processing. CO_2 is often 70-90% of the inlet gas in an EOR application. In the three column method, the CO_2 in the feed is compressed to about 600 psig, chilled and liquefied before entering the demethanizer and leaves the demethanizer in the liquid phase. The pressure is then dropped to about 300-400 psig in the ethane recovery column where the CO_2 is produced as an overhead vapor product. Pressurization of CO_2 in the three column process is accomplished by compressing the overhead vapor product from the ethane recovery column. The four column scheme avoids liquefying and revaporizing the CO_2 and also steps the CO_2 up in pressure increments and so avoids the pressure energy losses inherent in the three column approach. In the four column process, CO_2 pressurization is accomplished by pumping the liquid bottoms product of the CO_2 recovery column, thereby saving compression costs.

To some degree, additive is used "doubly" in the four column method. Reboiler temperatures in the four column process are lower than the three column process which reflects the lower quantity of recycle additive needed for the four column case.

The conclusion derived from several specific designs for EOR applications shows that the total horsepower (process compression, refrigeration compression and pumping) for the four column method is about 75% of the total for the three column method. Heating medium energy input is about 65% of the three column process and additive circulation is 60% of that required for the three column process. All of these factors contribute to a more efficient and lower cost plant design.

For streams with lower CO_2 concentrations three column processing might be favored. However, in the cases of processing associated gas from a CO_2-based EOR project, the four column method is the clear winner.

ADVANTAGES OF RYAN/HOLMES TECHNOLOGY

The advantages of the basic Ryan/Holmes approach for an EOR application are summarized in Table 4. Comparison of the Ryan/Holmes approach to other processing schemes serves to highlight the distinct advantages of the distillative approach for most EOR applications.

Most other processing schemes involve first absorbing the CO_2 from the gas stream with a solvent having CO_2 affinity. Typically, the CO_2 must be regenerated from the solvent at low pressure and the CO_2 recompressed to the high reinjection pressure required for recycle to the flood. This recompression step entails a considerable amount of additional investment and operating costs. Following CO_2 removal, the natural gas stream must be dried and treated in a conventional liquids extraction plant in order to recover the NGL product.

Membrane schemes also suffer from the same disadvantage since the CO_2 becomes the low pressure permeate stream and must be recompressed. Also, most other processing alternatives are unable to separate selectively any H_2S present in the feed gas.

Table 4
Advantages of Ryan/Holmes Technology

FLEXIBLE, RELIABLE DISTILLATION

FEED DEHYDRATION BY TEG

NON-CORROSIVE, NON-FOAMING

NO CHEMICALS

CO_2 PRODUCED DRY, HIGH PRESSURE

HIGH NGL'S RECOVERY

H_2S CONCENTRATED IN LOW VOLUME NGL STREAM

SELECTIVE H_2S REMOVAL

CO_2/H_2S RATIO CONTROLLED FOR CLAUS FEED

ECONOMICAL

An article by Shell Oil Company presented at the 1983 GPA Annual Convention (3) compared various process routes for recycling CO_2 from an EOR project. The basic four column Ryan/Holmes approach is shown to have the lowest cost. As a result of their evaluation work, Shell opted to license the Ryan/Holmes technology for the Wasson Denver Unit in West Texas. However, Shell did elect to use their own selective H_2S absorption process prior to feeding the gas to the Ryan/Holmes portion of the plant.

COMMERCIAL STATUS

Presently, Koch Process Systems has licensed four applications of the Ryan/Holmes Technology. These include the Seminole Unit operated by Amerada Hess, the Willard Unit, Wasson Field operated by ARCO Oil and Gas (4), the GMK South Field operated by Mobil Producing Texas and New Mexico, Inc., and the Wasson Denver Unit operated by Shell Oil Company. Several other projects are actively considering application of the Ryan/Holmes Technology.

REFERENCES

1. Holmes, A.S., J.M. Ryan, B.C. Price and R.E. Styring, "Pilot Tests Prove Ryan/Holmes Cryogenic Acid Gas/ Hydrocarbon Separations," Proc. Gas Proc. Assn., (1982), p. 75.

2. Holmes, A.S., J.M. Ryan, B.C. Price and R.E. Styring, "Process Improves Acid Gas Separation," Hydrocarbon Processing, Vol. 61, No. 5, (1982), p. 131.

3. Flynn, A.J., "Wasson Denver Unit - CO_2 Treatment," Proc. Gas Proc. Assn., (1983), p. 142.

4. Price, B.C., and F.L. Gregg, "CO_2/EOR: From Source to Resource," Proc. Gas Proc. Assn., (1983) p. 131, (Oil and Gas Journal, Vol. 81, No. 34, (1983) p. 116).

20

Natural Gas Processing with PRISM® Separators

D. J. Stookey
Development Manager, Natural Gas Market
Separations Business, Monsanto Company

T. E. Graham
Manager, Market and Systems Development
Separations Business, Monsanto Company

W. M. Pope
Senior Process Engineer
Separations Business, Monsanto Company

INTRODUCTION

The growing interest in carbon dioxide miscible flooding as an enhanced oil recovery (EOR) technique has sparked renewed interest in improved methods for separation of carbon dioxide from hydrocarbon gases. Particular emphasis has been placed on developing low energy separation methods. Simplified flow sheets have also been of keen interest as a means for enabling reduced operator attention and for easy turn-up and turn-down capability at remote sites. Gas separating membranes address these issues and offer distinct advantages over

Based on material © 1984 by the American Institute of Chemical Engineers (AIChE) contained in *Environmental Progress,* October 1984, by permission of AIChE.

conventional acid gas removal techniques. Thus it is not surprising that membrane systems are gaining acceptance for carbon dioxide recovery in EOR projects (1, 2, 3). Similarly, membranes offer the natural gas processing industry an alternative to absorption processes for gas treating facilities.

The ability of polymeric membranes to separate gases was initially applied on an industrial scale to the recovery of hydrogen. Monsanto began applying its hollow fiber membranes in its own commercial plants in early 1977. Since their market introduction under the tradename PRISM separators in 1979, Monsanto has sold 55 membrane-based gas separation systems worldwide. Hydrogen recovery systems are operating successfully in refineries, ammonia plants, and methanol plants. Monsanto has also been extending the application of PRISM separators to other services. Extensive field testing has been completed using small scale systems to demonstrate the applicability of membranes for removal of carbon dioxide and hydrogen sulfide from hydrocarbon gases. Results of these studies have lead to the successful design, installation, and operation of three commercial scale units for carbon dioxide removal from EOR casinghead gas, from landfill gas, and from high pressure natural gas. Fabrication and installation of a second unit for each of these applications is now underway.

General Concepts

A general understanding of the theory behind membrane gas separation will aid understanding of the advantages, limitations, and applicability of this technology to natural gas processing. Polymeric membranes separate gases by selective permeation of gas species in contact with the membrane. The gases dissolve in the polymer and are transported across the membrane barrier under an imposed partial pressure gradient. The pressure gradient is accomplished by feeding high pressure gas to the outside of the hollow fiber membrane while the permeate side is operated at substantially lower pressure. Typically the ratio of feed to permeate pressures is in excess of three. The relative rates at which gas species will permeate across the membrane are also dependent upon the polymer selected for the membrane. In the PRISM separators currently marketed, water vapor, carbon dioxide, and hydrogen sulfide are "fast" species in comparison to methane, nitrogen, and other hydrocarbon gases. Thus their utility in acid gas removal from natural gases.

A number of articles have appeared in the published literature describing in detail the principles of gas separation using membranes (4, 5). The reader is referred to these references for details of the mechanisms involved. Monsanto's PRISM separators utilize bundles of hollow fibers enclosed in pressure vessels. Their geometry is similar to a shell and tube heat exchanger. Feed enters on one end of the shell side. Product gas exits the other end of the shell. Acid gas components which permeate to the bores of the hollow fibers pass through an internal tubesheet and are discharged through the closure flange on the pressure vessel. Several separators may be connected in series and parallel arrangements to meet specific requirements. Monsanto markets these systems as shop assembled, skid-mounted units which can easily be installed and started-up at remote field locations.

Natural Gas Processing Application

Membranes are particularly suited to applications where a high partial pressure differential can be applied to the faster permeating species. This condition is readily satisfied in carbon dioxide removal from EOR casinghead gas and landfill gas. These gases, typically contain from 50% to 80% carbon dioxide. In such cases a high partial pressure differential can be achieved even at modest (200-300 psig) feed pressures when the permeate is collected at atmospheric pressure. The high partial pressure differential results in high acid gas fluxes through the membrane and thereby reduces the size of the membrane system required to process a specific amount of feed.

The hollow fibers in Monsanto's PRISM® separators are capable of withstanding very high differential pressures. Systems in hydrogen service routinely operate with 1600 psi differential between shell and bore sides of the hollow fibers. In acid gas service, differentials as high as 1000 psi have been applied. Many natural gas streams are available at pressures in the 600 to 1000 psig range. Thus in natural gas service good partial pressure differentials can be achieved with as little as 10% acid gas in the feed. Feeds containing much lower concentrations can be processed via membranes. Indeed, one of our commercial units processes a natural gas stream containing only 3% carbon dioxide to a pipeline specification of 1.5% maximum carbon dioxide. The point to be made is that at lower acid gas concentrations the membrane system size increases due to the lower partial pressure differial driving force but can still leave membranes economical.

All gas species have a finite permeation rate and will transport through the membrane. The selective removal of only the acid gas components is not feasible with currently available membranes. Some methane and hydrocarbon gases permeate along with the acid gases. Figure 1 shows the typical relationship between the composition of the permeate and the

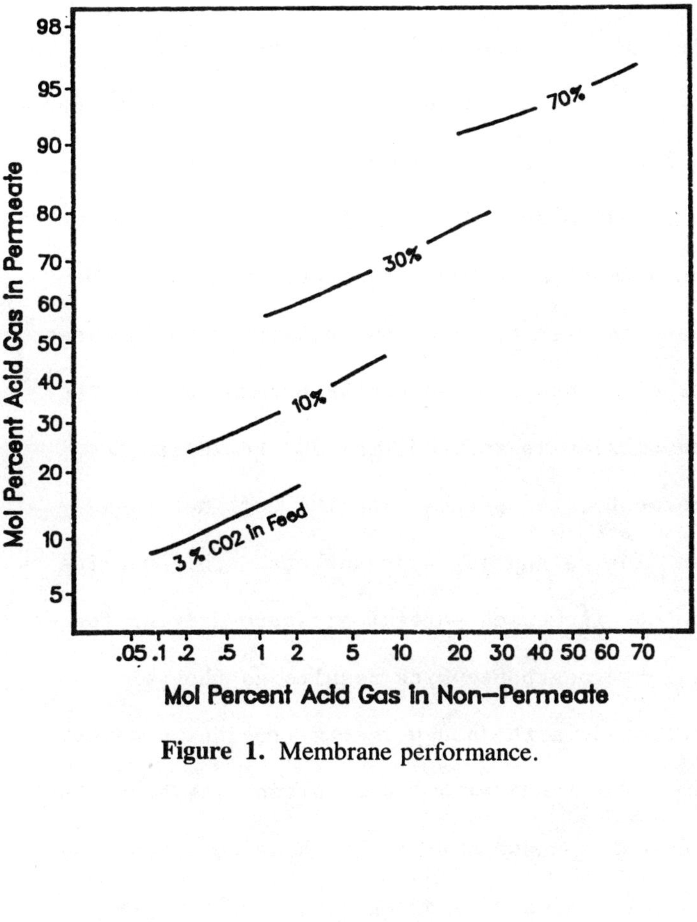

Figure 1. Membrane performance.

non-permeate streams produced by one polymeric membrane when various feeds are processed. The various curves represent the operating points for single stage processing of the respective feeds. The uppermost curve shows that the highest permeate purity which can be produced from a feed containing 70% carbon dioxide and 30% hydrocarbons would be about 95% carbon dioxide and 5% hydrocarbon. Note that the permeate is richest in carbon dioxide when the least amount of carbon dioxide has been removed. As the carbon dioxide is reduced from 70% to 20% in the non-permeate, the permeate purity declines to 90% carbon dioxide and 10% hydrocarbons. This decline is due to the increased partial pressure driving force for the hydrocarbons which results as they are concentrated in the non-permeate (residue product) stream. For feeds with lower carbon dioxide content, the permeate produced will also have correspondingly lower carbon dioxide and correspondingly higher hydrocarbon content.

In Figure 1 and in succeeding examples we have referred only to carbon dioxide as the acid gas component present. Since hydrogen sulfide permeates at the same rate and selectivity in the PRISM separator, the compositions reported are applicable to carbon dioxide alone, hydrogen sulfide alone, or the total of these two components.

Hydrocarbon loss to the permeate as shown in Figure 1 may not be serious. In some cases, fuel must be supplied anyway to incinerate the acid gases. Slippage of hydrocarbons to the membrane permeate in such case does not represent a loss. Hydrocarbon losses can also be minimized by a multistage membrane arrangment. Figure 2 describes a two stage system. In this arrangement, the first stage is designed to produce the desired natural gas product. Permeate from the first stage is compressed and fed to a second membrane unit. Non-permeate from the second stage is then recycled to the first stage. All acid gases removed by the system are concentrated in the second stage and leave in its permeate.

Figure 1 can be used to illustrate the utility of a two stage arrangement. Suppose we are to process

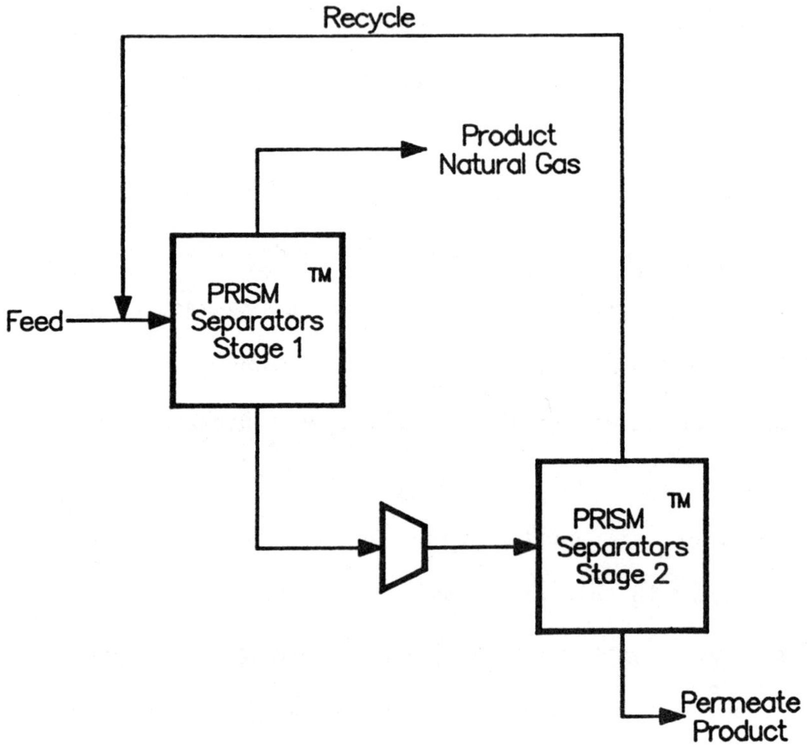

Figure 2. Two-stage flow diagram.

gas containing 10% carbon dioxide to a natural gas product containing 1% carbon dioxide. The 10% feed curve in Figure 1 shows that a permeate containing 30% carbon dioxide would be produced in the first stage when the 1% carbon dioxide level is reached in the natural gas product. Using these compositions to perform a component balance reveals 31% of the feed will permeate, resulting in a 24% hydrocarbon loss for a single stage. This proving unacceptable suggests a second stage is indeed required. A 30% carbon dioxide feed can be split to a 70% carbon dioxide permeate and a 10% carbon dioxide non-permeate to be recycled. A similar component balance around the second stage shows that 33% of its feed permeates and that hydrocarbon loss is reduced to only 4.4% while 91% of the carbon dioxide has been removed.

Broadly speaking, Figure 1 shows that membranes are not today a panacea to all acid gas removal needs. Feed conditions and processing and product requirements vary broadly within the natural gas industry. The present task is to identify applications and processing schemes where membranes offer practical solutions. While current membranes may not offer the total solution, often opportunities do exist for their advantageous utilizations. Applications for acid gas removal in natural gas processing generally fall within the following categories:

1. Bulk removal of acid gases.
2. Production of pipeline quality natural gas.
3. Upgrading the heating value of low BTU fuel.

In the bulk removal applications, membranes are used to remove the bulk or major portion of the acid gases from the stream. Final processing to pipeline quality product is accomplished by conventional technology as is illustrated in Figure 3. For new facilities the synergism of combined membrane and absorption or adsorption technologies can result in significant capital and operating cost savings over a stand-alone absorption unit. The combination also offers greater flexibility in processing streams with fluctuating flow rates and compositions since the membrane's

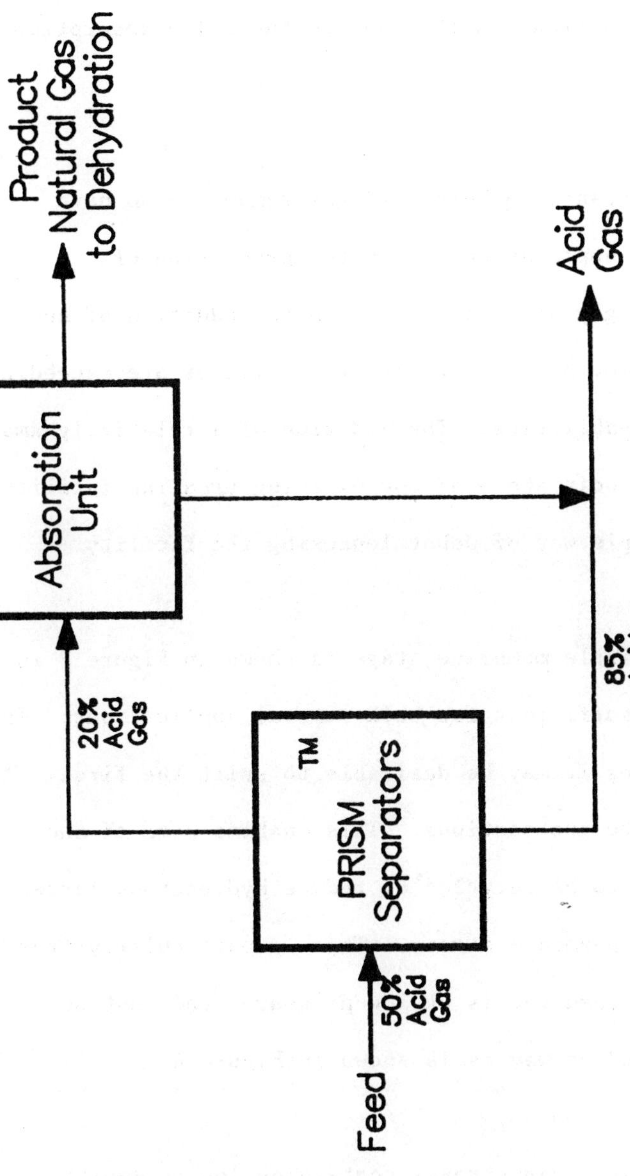

Figure 3. Bulk acid gas removal.

excellent turn-up and turn-down capability can smooth out fluctuations in the feed to the amine absorption unit.

Membranes in bulk acid gas removal also have retrofit application. With the production of increasingly sour gases or with the addition of new wells, many acid gas treating facilities are exceeding their capabilities. The addition of a relatively small membrane unit ahead of the existing treating facility is a simple way of debottlenecking the facility.

A single membrane stage as shown in Figure 3 is usually sufficient for bulk removal applications. In some cases it may be desirable to split the first stage into two sections. This enables some of the permeate to be recycled to reduce hydrocarbon losses with the permeate stream. This is particularly true when the feed gas is at low pressures and must be compressed anyway as is shown in Figure 4.

Natural gas streams containing low to medium levels of carbon dioxide can be processed to pipeline

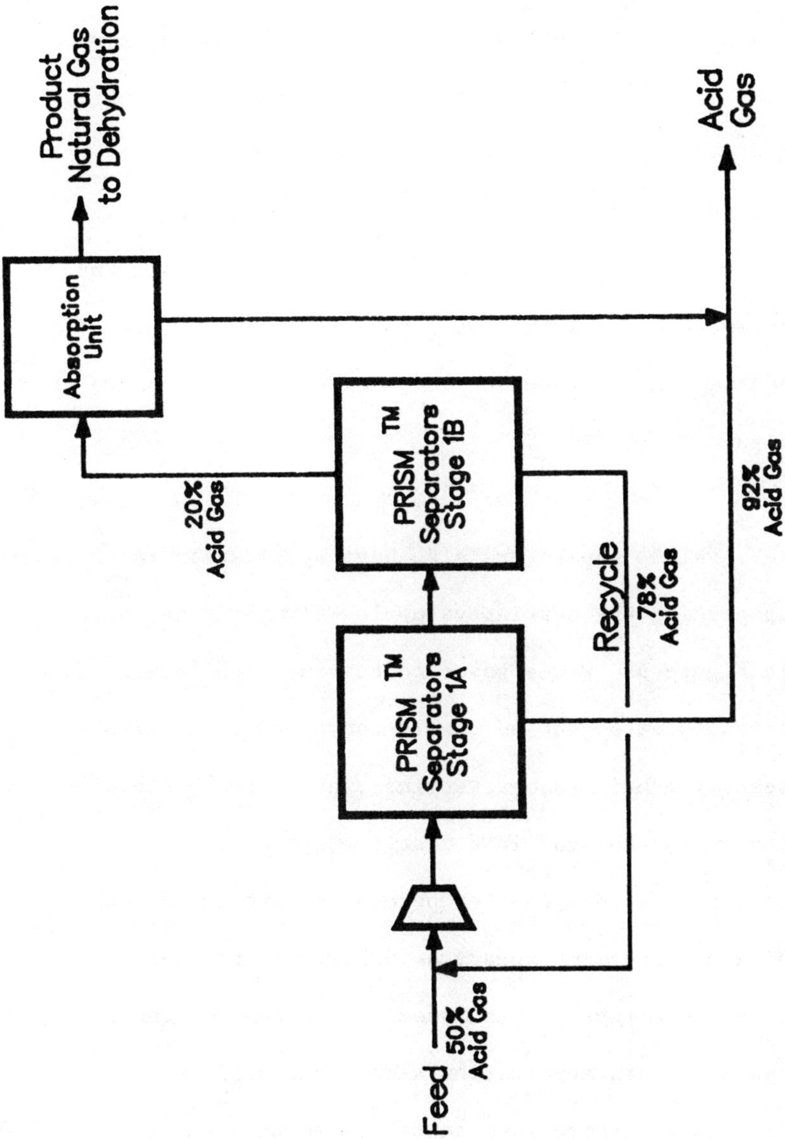

Figure 4. Bulk acid gas removal with recycle.

quality natural gas in a single stage as shown in Figure 5. In the example shown, the carbon dioxide level is reduced from 5% to 1.5% in a single stage. The permeate in this case contains 21 % carbon dioxide. As such it is a good medium-BTU fuel which can supply on-site fuel requirements. If the permeate exceeds the local fuel requirements, the excess can be processed in a second stage as shown in Figure 2.

PRISM® separators also have application in up-grading the heating value low-BTU gases as shown in Figure 6. Waste gases containing high levels of carbon dioxide can be easily converted into suitable feeds for boilers, gas engines, and fired heaters by processing the gas in a single stage of PRISM separators. Since water permeates with the carbon dioxide, the membranes also dehydrate the fuel, thus reducing corrosion and freezing problems in the fuel system. This application also finds utility in producing on-site fuel needs at remote gathering units.

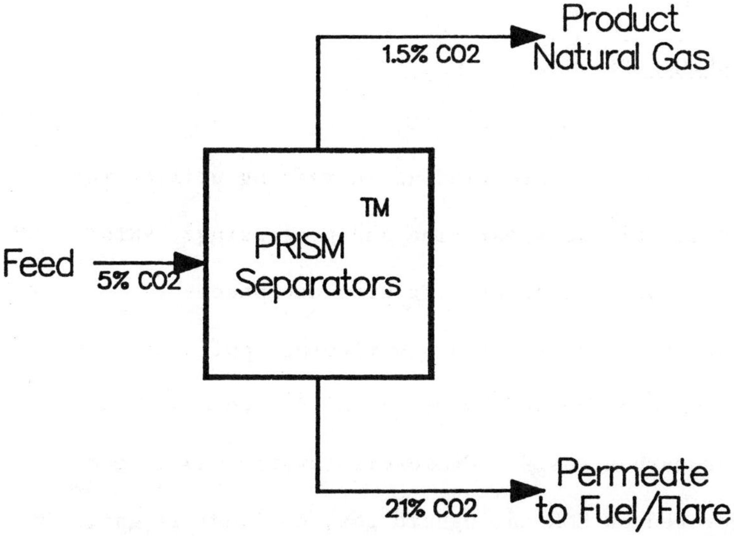

Figure 5. Producing pipeline Spec natural gas.

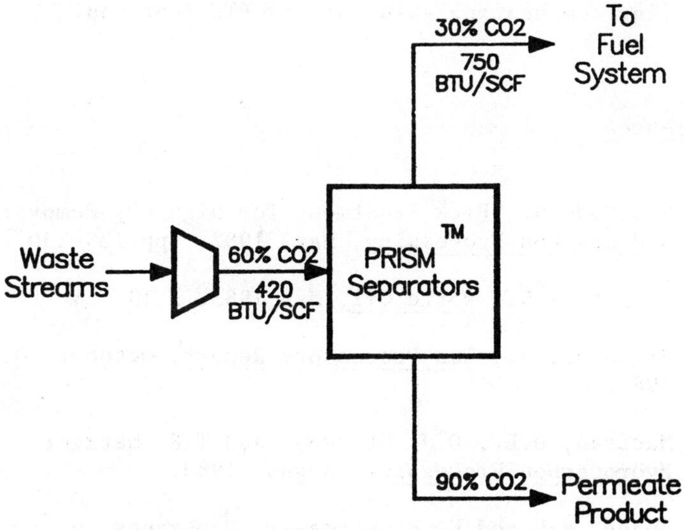

Figure 6. Upgrading heating value of fuel.

Conclusions

Membranes are finding increasing utility in industrial gas separation and processing. PRISM separators, after gaining wide acceptance in hydrogen recovery, are now finding application in carbon dioxide and hydrogen sulfide removal from hydrocarbon gases. Commercial systems have been operated on EOR casinghead gas, on landfill gas, and on high pressure natural gas. Systems can be provided for bulk removal of acid gases, for production of pipeline quality natural gas, and for upgrading the heating value of low BTU fuel gas.

References

1. C.S.Goddin, "Pick Treatment for High CO_2 Removal", Hydrocarbon Processing, May, 1982. pp 125-130.

2. Wright, T.R., World Oil, May 1983, p 19.

3. Brewster, J., Gas Processors Report, October 10, 1983.

4. MacLean, D.L., D.J. Stookey, and T.R. Metzger, Hydrocarbon Processing, August 1983.

5. Hwang, S-K and K. Kammermeyer, Membranes in Separations, John Wiley, New york, 1975.

21

The Operation of the DELSEP Membranes in the Oil Patch

Fred G. Russell
Vice-president, Technology
ENSTAR Engineering Company

SUMMARY

After six years of field testing of the DELSEP membranes utilizing feed streams of CO_2 and H_2S and hydrocarbons, this membrane system has proved its performance capabilities and is offered to the gas processing industry on a fully warranted basis. The most recent testing on the latest membrane material has been used to confirm performance characteristics including flux rates, separation factors (alphas), cooling effect and partial pressure limitations for CO_2 and H_2S components.

Operating characteristics and design considerations for Enhanced Oil Recovery (EOR) processes as well as gas processing plant applications are reviewed.

FIELD TESTING

The DELSEP membrane process has been extensively tested at field sites in Canada and the United States. The testing program was begun in 1977 and is continuing at this time. These tests have been conducted to observe the performance characteristics of the DELSEP membrane process when the following important process parameters have been varied to cover the large range of expected applications:

- feed pressure
- permeate pressure
- CO_2 content
- H_2S content
- feed temperature
- feed flow rate per element

To date over five hundred test points have been analyzed to determine the optimum combination of the above noted process parameters. Figure 1 gives a range of operating conditions tested to-date. A review of this table will indicate the extensive test program that was conducted

Four Inch Diameter Commercial Element

Feeds Rates Per Element	1,000 - 11,860 SCFH
Inlet Pressure	270 - 924 psia
Inlet Temperature	68 - 135° F
Permeate Pressure	15 - 150 psia
Inlet Gas CO_2 Content	0.11 - 92.9 mol %
Inlet Gas H_2S Content	100 ppmv -32.5 mol %
Inlet Gas CO_2 Partial Pressure	0 - 585* psia
Inlet Gas H_2S Partial Pressure	0 - 216* psia
Water Content	Dry - Saturated*

*Not recommended for continuous operation.

SI Conversion: kPa = 6.895 x psi; m^3/hr = SCFH x 0.02832
°C = (°F-32)/1.8

Figure 1. Range of tests data.

to assure adequate knowledge of the DELSEP membrane performance to be expected in actual field applications.

CURRENT STATUS

As a result of the field test program conducted by ENSTAR and a laboratory test program conducted by Envirogenics, the DELSEP membrane system is now offered on a commercial basis with complete processing warranties as usually offered in gas processing facilities. Field testing as well as laboratory work is continuing to broaden the operating experience with the membrane. The established operating parameters of the DELSEP membrane system are shown in Figure 2.

APPLICATIONS

Gas Sweetening

Membranes offer an attractive alternative to some gas sweetening applications. Higher CO_2 feed concentrations (above 15 mol percent) which require only moderate CO_2 removal (down to 3 mol percent in the residue gas) and are located near facilities where low BTU/c.f. fuel gas can be utilized are ideal applications for gas sweetening using the DELSEP membrane system. A recent application for this type of system is summarized below:

Inlet Pressures	200 - 2,000 psia
Inlet Temperatures	32° - 140° F
Inlet CO_2 Content	Up to 95%
Inlet Water Content	Dry to 80% R.H.
Permeate Pressures	15 - 1000 psia
Maximum Differential Pressure	1100 psi
Residue CO_2 Content	Down to 1%
Element Sizes Available	2", 4", 8" Diameter
Feed Flow Rate	Varies with Element Size
Maximum CO_2 Partial Pressure	500 psia
Maximum H_2S Partial Pressure	100 psia

SI Conversion: kPa = psi x 6.895; °C = (F-32)/1.8
m = inches x 0.0254

Figure 2. DELSEP operating parameters.

	mol Percent		
	Feed	Residue	Permeate
N_2	0.65	0.74	0.40
CO_2	16.75	4.98	50.31
C_1	76.70	86.90	47.62
C_2+	5.90	7.38	1.67
Volume	7.7 MM/D	5.7 MM/D	2.0 MM/D
Pressure	571 psig	561 psig	50 psig
Temperature	103° F	90° F	90° F

SI Conversion: kPa = psi x 6.895; m^3/D = Scf/d x 0.02832

°C = (°F-32)/1.8

In this particular application a large fuel supply was needed for various lease facilities and the 2.0 MM/D (56,640 m^3/D) low BTU/c.f. permeate stream was mixed with higher BTU gas to provide the required energy.

Gas sweetening of isolated, one well gas streams may also be attractive because of the extreme simplicity of operations involved in a membrane process.

EOR Applications

Applications of CO_2 removal from produced gas in Enhanced Oil Recovery (EOR) operations where CO_2 has been injected into the reservoir present several challenges for the membrane system. To better understand the overall system, its possible problems, and the recommended solutions, the following example application is presented.

Example EOR Problem. A three phase separator operates at 50 psig (345 kPa) and 100° F (37.8° C) to separate produced gas, oil and water. The projected produced gas (at year three after CO_2 breakthrough has occurred in the producing wells) will have an estimated gas composition as given in Figure 3. A residue gas with 15% CO_2 content is acceptable since this gas will be used primarily for lease fuel. The separated CO_2 should be approximately 95% pure but under no circumstances should have more than five percent methane content. The CO_2 is required at 2500 psig (17,236 kPa) for transportation back to wells for reinjection into the reservoir.

A preliminary design of the membrane yields a residue gas with nearly all of the propane and heavier components from the feed gas still present since these components have negligible permeation rates through the membrane. A dew point calculation on this residue gas indicates that liquids will condense from this residue gas if the temperature falls below 200° F (93.3° C) when a membrane processing pressure of 575 psia (3,964 kPa) has been established.

Component	Mol %
Nitroget	0.34
Carbon Dixide	85.00
Hydrogen Sulfide	0.53
Methane	7.08
Ethane	2.71
Propane	2.26
Isobutane	0.37
N. Butane	0.90
Isopentane	0.30
N. Pentane	0.24
Hexanes Plus	0.27
Pressure	50 psig
Temperature	100° F
Volume	10,000 Mscfd

SI Conversion: kPa = psi x 6.895; °C = (°F-32)/1.8
m3/D = SCFD x 0.02832

Figure 3. Produced gas EOR project.

Since the membrane must operate below 140° F (60° C), either heavy components must be removed from the gas or the operating pressure (and corresponding dew point temperature) must be lowered. The membrane processing pressure was chosen to be as high as possible, but the CO2 partial pressure must not exceed 500 psia (3,447 kPa).

If the operating pressure is lowered to reduce the dew-point temperature to an acceptable level (i.e., below the membrane operating upper temperature limit of 140° F [60° C]) then the available driving force to remove the CO_2 from the inlet gas is reduced and a larger membrane area will be necessary to accomplish the desired CO_2 removal.

If heavy hydrocarbons are removed utilizing additional facilities including coolers, reboilers, condensers, fractionators, reflux pumps and a refrigeration system then a high feed gas pressure can still be used and the larger membrane area requirement will not be encountered. The recovered liquids from the fractionator system are a saleable product and this revenue can be used to offset the additional facilities costs.

After reviewing the above two alternates for lowering of the gas dew-point temperature, it was determined the best approach was to recover the heavy hydrocarbons. Figure 4 is a process schematic for a fractionation system required to recover a propane plus liquid stream from the produced gas given in Figure 3. It was determined that 45% recovery of the propane component and the limiting of the CO_2 content in the recovered liquids to 1000 ppmv would yield the proper membrane feed gas, and at the same time maintain a rather simple liquid recovery facility to yield a desirable hydrocarbon liquid product. Figure 5 gives a summary of equipment sizes which the liquid recovery facilities required for the EOR example problem.

All of the design work discussed up to this time for the example problem has ignored any temperature effects that may be caused by CO_2 removal utilizing a membrane system. Field testing has indicated a cooling effect on both the residue gas and the permeate gas caused by the removal of CO_2 from the feed gas. Figure 6 is a graphical picture of this cooling effect on the residue gas for a 4" element. No units are given on this curve since the data is considered proprietary at this time. As a point of reference, it has been noted[1] that a 40° F (22.2° C)

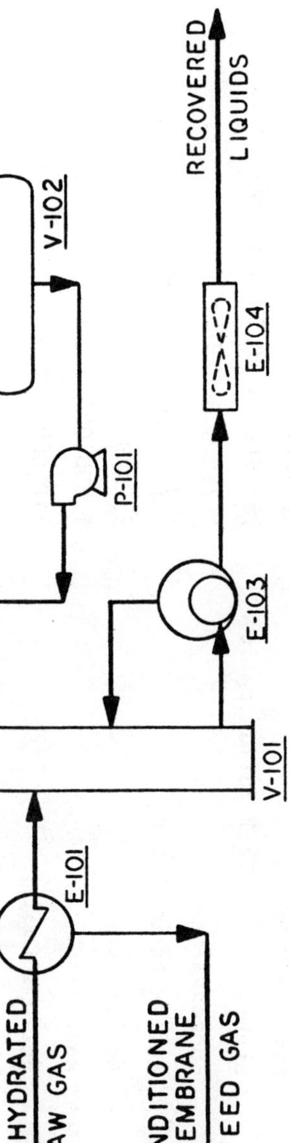

Figure 4. Process schematic liquids recovery unit.

	Equipment	Sizing Data*
E-101	Feed Exchange	1,400,000 BTU/hr
V-101	Stabilizer (0.30:1 reflux ratio)	36" dia. x 10 Theoretical Trays
E-102	Condenser	1,750,000 BTU/hr
V-102	Reflux Accumulator	9,700 M/D & 29 gpm
P-101	Reflux Pump	29 gpm
E-103	Reboiler	1,600,000 BTU/hr
E-104	Bottoms Cooler	110,000 BTU/hr

* Includes no safety factors. Data directly from process calculations.

SI Conversion: Watts = 3.413 BTU/hr; m = inches x 0.0254
L/s = gpm x 0.0631

Figure 5. Hydrocarbon dew point control facilities.

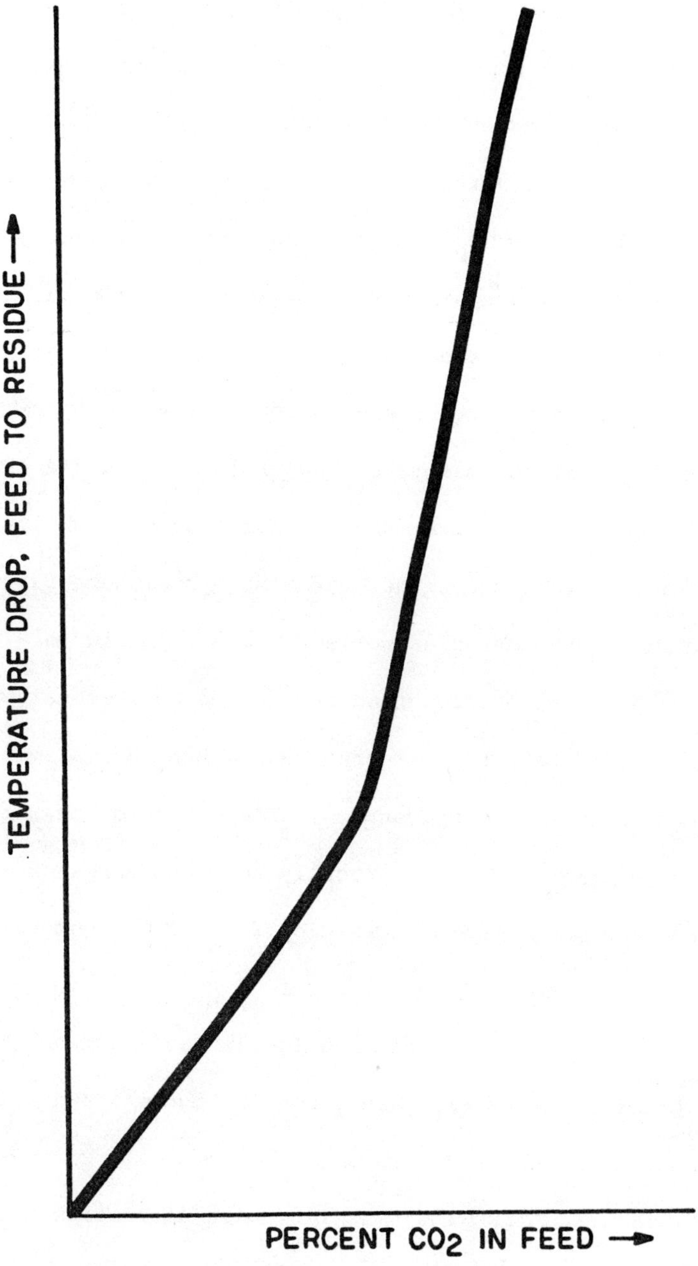

Figure 6. Cooling effect for CO_2 removal, constant feed rate, 4-in. element.

temperature change between the feed temperature and residue/permeate stream temperatures was observed across one element when the feed concentration of CO_2 exceeded 60 percent.

In order to compensate for the cooling effect caused by the removal of CO_2 from the gas stream, reheating schemes at various points in the flow will be required. The design of the reheating system must consider an upper temperature level set by the maximum allowable operating temperature of the membrane (140° F [60° C]) and a lower temperature level set by the dew-point temperature of the membrane residue gas. The operating system temperature profile for the example problem is shown in Figure 7. Two reheating steps are required during the CO_2 removal process to maintain operating temperatures within the maximum and minimum allowable temperatures. The source of heat for the reheating steps will probably be hot compressor discharge gas developed as a result of inlet gas compression or CO_2 injection gas compression.

The entire gas processing facilities required for the example EOR problem are shown in Figure 8 and a mol balance is given in Figure 9.

In considering the cost for these facilities the question arises as to the optimum feed and permeate pressures

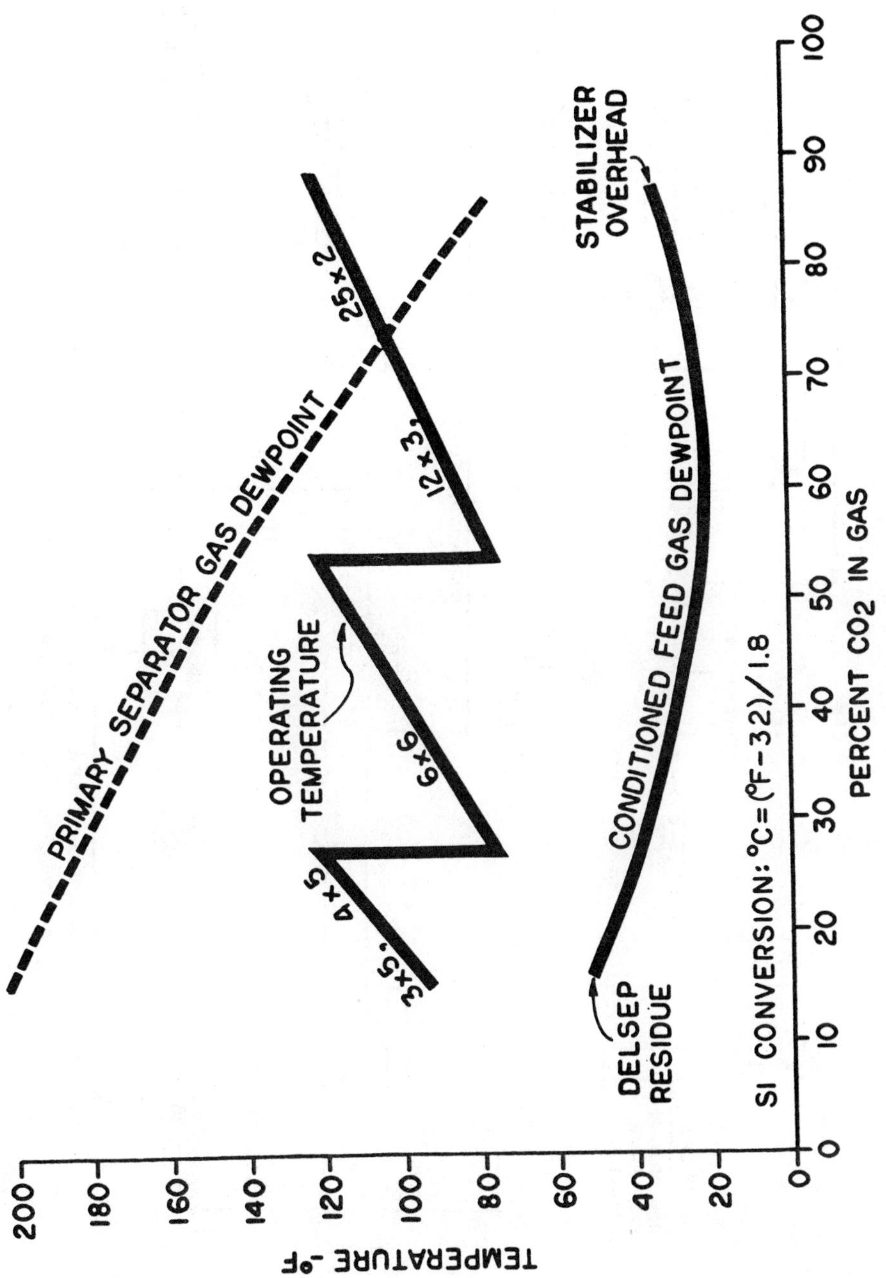

Figure 7. System temperature profiles.

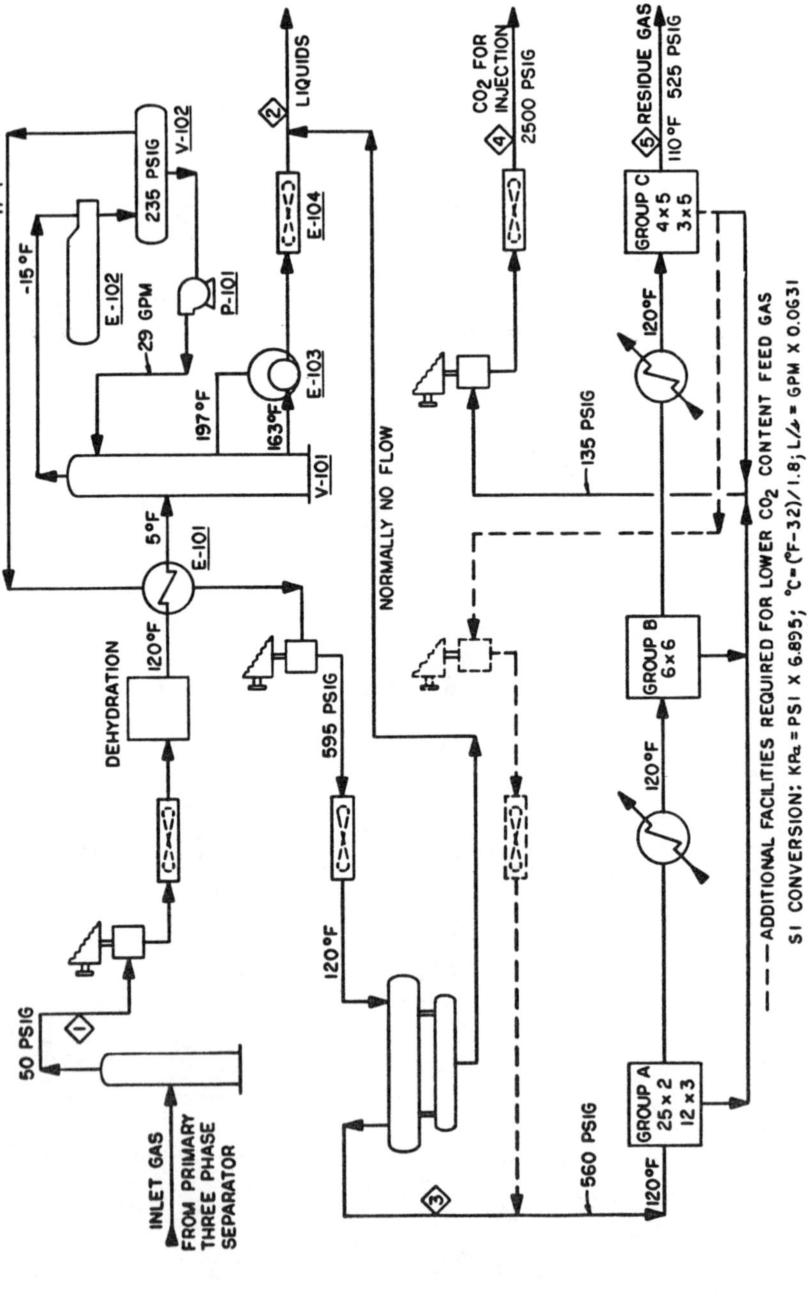

Figure 8. CO_2 recovery facility.

	Mols per Hour				
Stream No.	1	2	3	4	5
Component	Inlet Gas	Recovered Liquids	Feed to Membranes	CO_2 Product	Residue Gas
N_2	3.7	-	3.7	1.9	1.8
CO_2	933.3	-	933.3	920.4	12.9
H_2S	5.8	-	5.8	5.8	-
C_1	77.7	-	77.7	39.3	38.4
C_2	29.8	-	29.8	6.7	23.1
C_3	24.8	11.2	13.6	1.3	12.3
iC_4	4.1	3.5	.6	-	.6
nC_4	9.9	9.3	.6	-	.6
iC_5	3.3	3.3	-	-	-
nC_5	2.6	2.6	-	-	-
C_6+	3.0	3.0	-	-	-
Total, m/hr	1098.0	32.9	1065.1	975.4	89.7
MSCFD	10,000	-	9,700	8883	817
GPM	-	6.8	-	-	-

SI Conversion: m^3/D = Scf/DX 0.02832; L/s = GPM x 0.0631

Figure 9. Example of EOR problem.

for the membrane system when considering not only membrane cost but also feed compressor and CO_2 injection compressor costs. ENSTAR made an optimization study of these cost factors and the results are shown in Figure 10. For the example EOR problem, the maximum inlet pressure as set by the maximum allowable partial pressure of CO_2 in the feed gas and a permeate pressure of 150 psia (1,034 kPa) should be used. This type of optimization study, considering both capital costs and operating costs, should be performed on each membrane system to establish the proper operating parameters.

FUTURE DEVELOPMENT

Field and laboratory tests will continue to improve performance prediction techniques. Also membrane development work will continue to expand operating parameters of the DELSEP system. The use of membranes for the removal of CO_2 and H_2S from light hydrocarbon gases needs to be considered as one of the several ways to remove these unwanted ingredients. The best removal method will vary from application to application depending on such parameters as product purity requirements, quality required and value of fuel gas, operating temperatures and pressures, and the utilization of existing facilities.

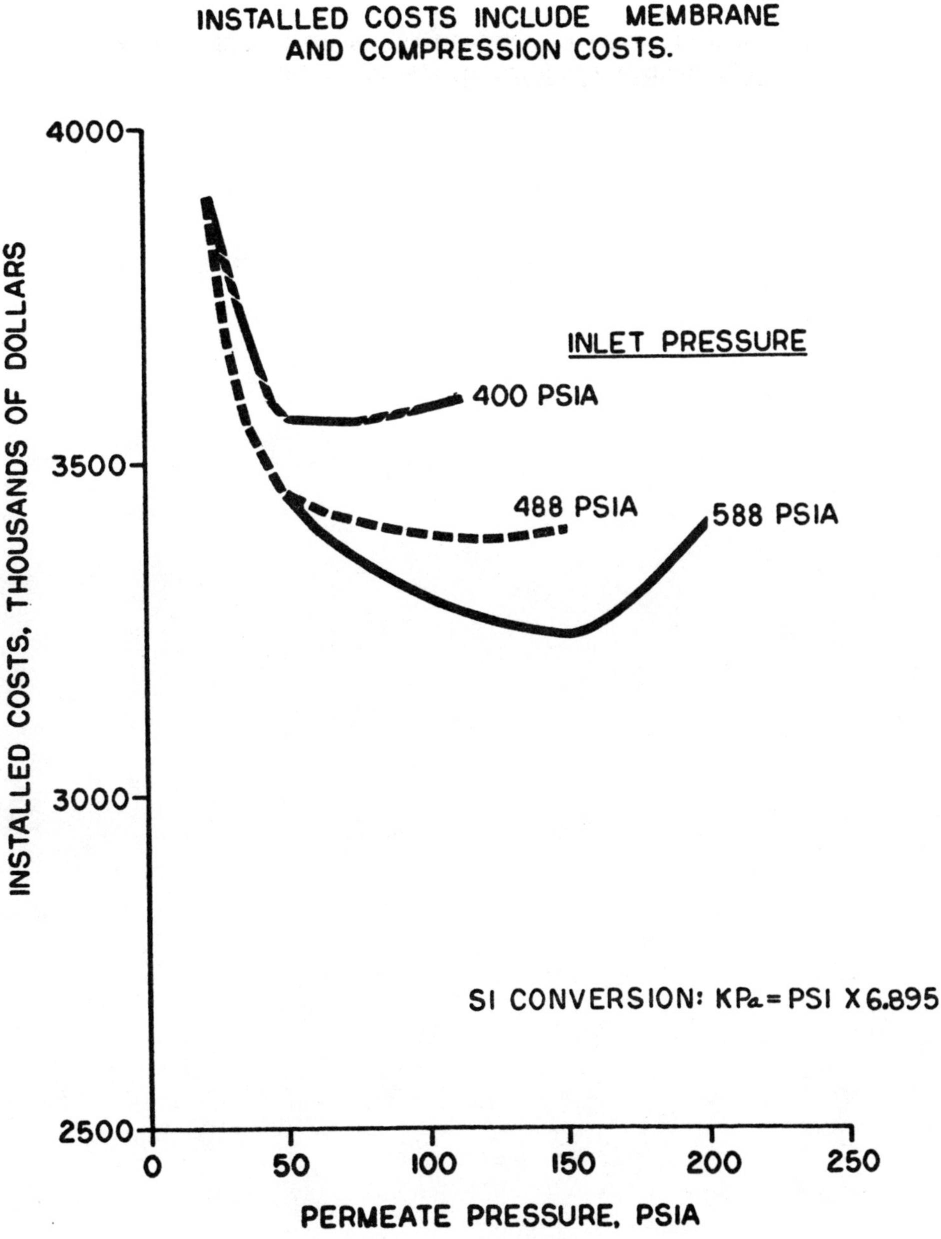

Figure 10. Optimum feed/permeate pressure.

REFERENCES

(1) Russell, Fred G., "Field Tests of DELSEP Parameters", Hydrocarbon Processing, (August, 1983)

22

Catalytic Conversion of COS for Gas Clean-Up

Niels R. Udengaard
Manager, Process Engineering
Haldor Topsoe, Inc.

Valdis Berzins
Business Development Executive
Haldor Topsoe, Inc.

INTRODUCTION

Carbonyl sulfide, COS, will always be present in the process gas leaving the gasifier in plants based on partial oxidation of sulfur containing oil or coal.

In layouts where hydrogen sulfide, H_2S, is removed upstream of the shift section, practically all COS is converted to H_2S across the shift catalyst and removed in subsequent purification units.

If H_2S is not removed upstream of the shift section, COS will normally be at chemical equilibrium at outlet of the shift catalyst bed. However, if the exit temperature from the shift section is high, or part of the gas by-passes the shift reactor (in case of production of synthesis gas for methanol or SNG, Fig. 1), the concentration of COS may be unacceptably high for the downstream purification units. In such cases, a catalytic reaction step is required downstream the shift section for further conversion of COS.

COS CONVERSION

COS can be converted to H_2S according to the following two reactions:

$$COS + H_2 = CO + H_2S \quad 1)$$

$$COS + H_2O = CO_2 + H_2S \quad 2)$$

The sum of the two reactions is the shift reaction.

Reactions 1) and 2) are both catalyzed by hydrogenation catalysts such as cobalt molybdate shift catalysts. Reaction 2) is catalyzed by alumina type catalysts which are not active for shift reaction. The TOPSOE sulphur tolerant shift catalyst, named SSK, is an excellent COS conversion catalyst. Since the COS equilibrium is established with a close approach at the exit of the last shift reactor in shift units based on this catalyst, no extra COS conversion is possible in cases where the exit temperature from the shift unit is low. Therefore, no

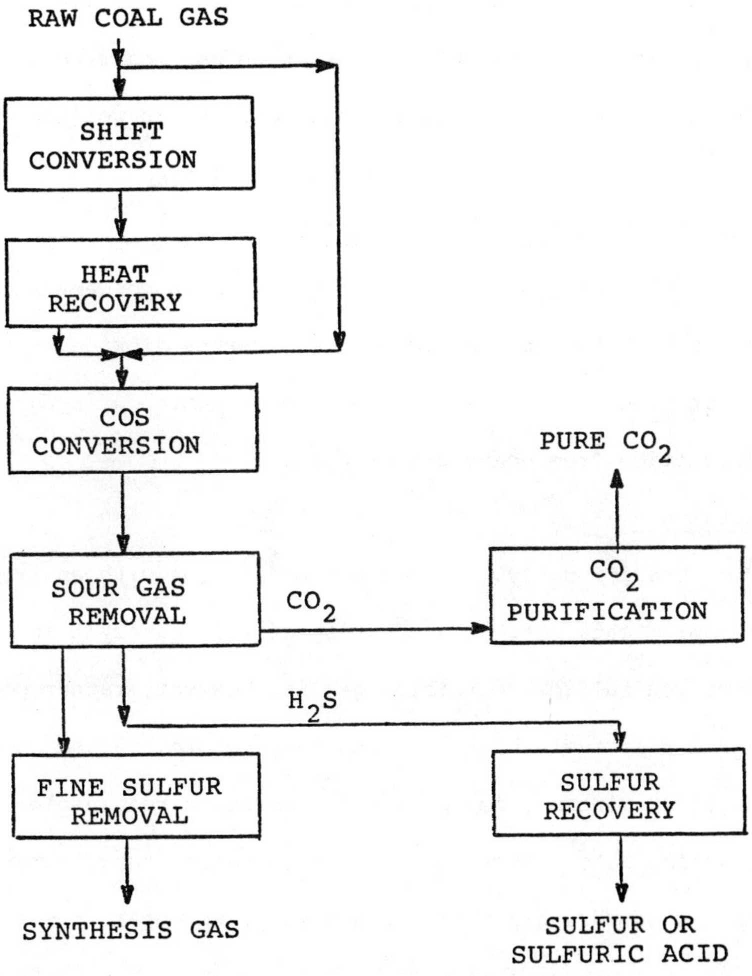

Figure 1. Production of synthesis gas for methanol or SNG.

COS conversion stage is included in processes for production of hydrogen or ammonia synthesis gas. Fig. 2 is an example of such operating conditions from an industrial plant where ammonia is produced from heavy oil (Texaco gasification). The COS concentration was measured to be 4 ppm in the dry exit gas. This corresponds well to the calculated equilibrium concentration of 5 ppm.

SULFUR-TOLERANT SHIFT PROCESS AND CATALYSTS

Efficient conversion of the very high concentration of carbon monoxide to hydrogen and carbon dioxide by the shift process is essential for the economics of syngas production from heavy oil or coal.

Catalysts

Traditionally, the conversion is done with an iron-chromium catalyst. The activity of this catalyst in hydrogen sulfide containing gas is, however, rather poor, and a relatively high reaction temperature, in the order of 350°C minimum, is required to obtain a reasonable reaction rate. Because of the high temperature, a rather high steam/dry gas ratio is necessary to obtain a sufficiently low equilibrium concentration of carbon monoxide.

Recently, TOPSOE introduced a new sulfur-tolerant shift process. The process is based on the use of a

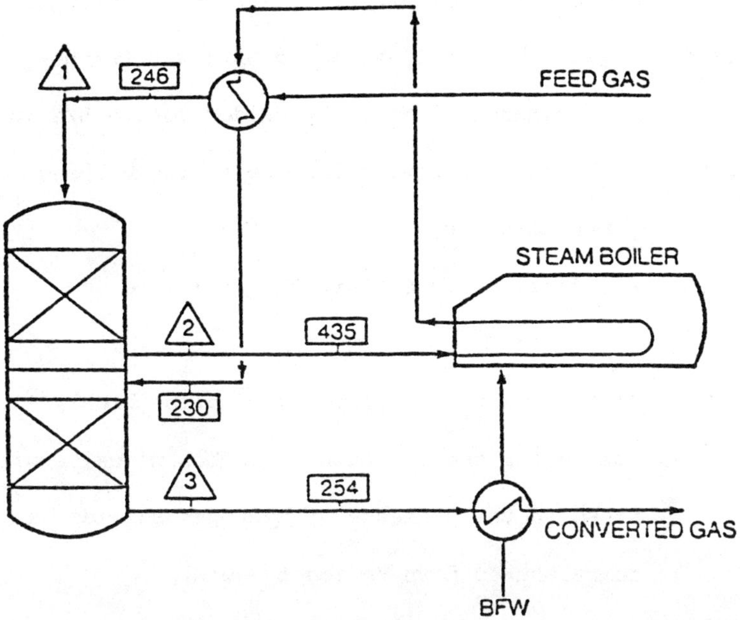

		1	2	3
H_2	VOL % (DRY)	45.12	60.73	62.18
CO	VOL % (DRY)	46.05	4.48	0.62
CO_2	VOL % (DRY)	6.58	33.16	35.64
CH_4	VOL % (DRY)	0.10	0.09	0.08
INERTS	VOL % (DRY)	1.65	1.18	1.14
H_2S	VOL % (DRY)	0.50	0.36	0.34
STEAM/DRY GAS RATIO		1.08	0.49	0.43

Figure 2. Sulfur-tolerant shaft unit in industrial ammonia plant.

highly active sulfur-tolerant shift catalyst, which was invented at the Exxon Research & Development Laboratories in Baton Rouge. The catalyst, named SSK, was further developed for commercialization by Haldor Topsøe A/S in Denmark, which has a worldwide license to produce and market this new catalyst.

The main features of SSK are the following:

- It has the unique property of being highly active for the reaction of carbon monoxide with steam in the presence of hydrogen sulfide, at temperatures down to about 200°C.
- It maintains its activity over a very wide range of operating conditions including temperatures up to 475°C.
- No specific catalyst poisons are known for SSK. It should especially be mentioned that it is insensitive to the presence of even relatively large amounts of chlorine in the process gas.
- A minimum concentration of hydrogen sulfide in the process gas is required to maintain the activity of SSK. The required level depends on the operating conditions, particularly the

temperature. In some applications, it may be as low as a few ppm hydrogen sulfide in the dry gas. There is no upper limit to the acceptable sulfur content.

Process

The best way to take advantage of the special features of the new catalyst SSK will be somewhat different for different coal gasification schemes. The main considerations would, however, be more or less the same in each case.

As an example, we have revamped the sulfur-tolerant shift section of an ammonia plant based on partial oxidation of heavy oil. The shift section consisted of three bed CO conversion (two reactors) with water quench addition between the beds. With a conventional HTS catalyst, a CO leakage of 3 to 4% was obtained.

The new layout for the shift section is shown on Fig. 2. Due to the high activity of SSK catalyst and the high conversion per bed, there is no need for the second HTS reactor.

The SSK catalyst has now been in operation for more than four years and has shown excellent performance. In spite of the total catalyst volume having been reduced from $15.8m^3$ to $9.6m^3$ (and the number of beds reduced from 3 to 2), we have been able to lower the CO leakage from 3.5% (dry) to 0.6% (dry) after installation of SSK catalyst. Furthermore, high pressure steam has been obtained by means of a waste heat boiler which has replaced the quench vessel between the first and second bed. Operating data from the shift section have been included as seen in Fig. 2. It is seen that across the first bed, a CO conversion from 45% to 4.6% (dry) is obtained at the high dry gas space velocity of 5000 Nm^3/m^3 cat./h.

The gas composition exit first and second bed are both close to equilibrium.

In production of synthesis gas for methanol or SNG or in gas purification in power plants, no shift conversion is desired at the COS conversion stage. In such cases, an alumina type catalyst is recommended.

Fig. 3 shows, for a number of cases, expected operating conditions in the COS conversion stage. The last example in the table illustrates the conditions in the COS conversion stage in the TVA Ammonia from Coal Project. The installation of a COS conversion stage in this ammonia project is due to special design considerations.

CASE	1	2	3	4
P, ATM A	31.0	24.2	24.2	31.6
T, °C	224	191	235	204
DRY GAS COMP.				
H_2 VOL. %	47.9	44.6	35.8	53.2
CO VOL. %	16.0	22.3	44.6	2.9
CO_2 VOL. %	32.8	29.8	18.0	42.2
INERTS VOL. %	3.3	2.6	0.6	0.7
H_2S VOL. %	0.1	0.7	1.0	1.0
COS IN/OUT, PPM	70/3	400/8	500/6	51/10
H_2O/DRY GAS	0.27	0.26	0.68	0.41

CASE 1: SYNTHESIS GAS FOR SNG

CASE 2: SYNTHESIS GAS FOR MeOH

CASE 3: GAS FOR COMBINED CYCLE POWER PLANT

CASE 4: TVA AMMONIA FROM COAL PROJECT

Figure 3. Typical operating conditions in COS conversion units coal based plants.

Performance of the COS hydrolysis catalyst at operating conditions close to those shown in the examples in Fig. 3 has been confirmed by extensive pilot plant runs, partly in our own laboratories, and partly in experiments carried out by others.

OPERATING CONDITIONS

The highest conversion of COS to H_2S is obtainable at lowest possible temperature. The minimum operating temperature is, as for all other catalytic units, restricted by the dew point of the process gas. The temperature at the reactor inlet must be safely above the dew point of the process gas in order to avoid the risk of water condensation in the pores of the catalyst particles.

Low operating temperatures also mean that the reaction rate is lower, thus requiring larger catalyst volumes. This is normally only a problem in large plants, where maximum reactor sizes must be taken into consideration.

The equilibrium reaction is also influenced by the concentration of the reactants and the products. High concentration of H_2O will favor the conversion of COS. Pressure has no direct influence on the conversion, but indirectly defines the dew point of the gas. For the same water concentration, a higher pressure requires a higher operating temperature in order to maintain same approach above the dew point.

CATALYST

TOPSOE has developed a catalyst, named CKA, for the COS hydrolysis reaction. The CKA catalyst consists mainly of activated alumina. It is available in $3x10^{-3}$ or $6x10^{-3}$m extrudates. The bulk density is 705 Kg/m^3.

The CKA catalyst as delivered requires no activation before use. It must be heated in nitrogen to a temperature of approximately 38 °C above the dew point of the process gas before the process gas is introduced. The CKA catalyst has a stable and high activity for the hydrolysis of COS, and it can be used in the temperature range from 120 to 400 °C. Normally, the operating temperature is in the range from 200 to 245 °C.

COS is converted to a concentration close to chemical equilibrium which, in most cases, corresponds to 1 - 10 ppm. The corresponding space velocity is between 1000 and 10000 v/v/hr.

The catalyst is, as are all other catalysts of this type, sensitive to poisoning by oxygen due to formation of oxygenated sulfur compounds on the catalyst surface. It is, therefore, important to make certain that oxygen is removed before feeding the gas to the COS-conversion unit.

The CKA catalyst is extremely selective for the COS hydrolysis reaction, and no side reactions are expected at normal operating conditions.

Most components other than COS will pass unreacted through the CKA catalyst bed; however, should the process gas contain components such as carbon disulfide, CS_2, careful consideration of the final layout must be observed in order to determine the effect of the following possible reaction:

$$CS_2 + CO_2 \longrightarrow 2\ COS$$

EXPERIENCE

The CKA catalyst was developed several years ago. It has been tested thoroughly in our test facilities at numerous different operating conditions including runs at high and low steam to dry gas ratios (Fig. 4).

The CKA catalyst is today in operation in the TVA Coal to Ammonia Plant at Muscle Shoals, where it has

CKA TEMP., DEG. C	206	182	180-210	182
PRESSURE, ATM. ABS.	30.0	30.0	30.0	30.0
CO, % DRY GAS	3	40	40-50	20-40
CO_2, % DRY GAS	42	20	15	15-23
CH_4, % DRY GAS	0.1	0.2	0.5	0.15
(C_2-C_3 TRACES)				
H_2S, % DRY GAS	0.9	1.0	1.0	0.9
H_2 BALANCE				
STEAM/DRY GAS	0.39	0.16	0.1	<0.1
DEW POINT, ΔT, DEG. C	32	36	50-80	50-100
S.V. WET N1/1/H	40000	30000	35000	3000
COS PPM DRY FEED	23	300-400	200-300	100-200
COS PPM DRY EXIT	6	15-30	APP. 50	18-50
SHIFT REACTION	NONE	NONE	NONE	NONE

Figure 4. Typical data sets from CKA test runs.

successfully demonstrated the expected performance (Fig. 5).

	Design	Sample 1	Sample 2	Sample 3
Dry SV, Nm^3/m^3, H	6050	6510	6040	5920
S/DG	0.41	0.42	0.30	0.37
Inlet, temp., °C	205	214	192	192
Inlet pressure, ATM A	31.3	31.2	31.4	31.1
% H_2	52.4	52.4	55.9	56.7
% CO	2.9	2.83	3.05	2.98
% CO_2	41.6	42.6	39.6	38.8
% H_2S	0.9	0.75	0.75	0.73
ppm COS in/out * dry	51/10	22/7	9.3/7.0	10/6.7
CH_4+Ar+N_2	balance	-	-	-

*Measured on Gas Chromatograph.

Figure 5. Performance of COS-hydrolysis catalyst type CKA.

23

H_2S Removal by the Stretford Process—Further Development by the British Gas Corporation

Brian M. Wilson
Engineering Services Department, North West Gas
Manchester, England

Roger D. Newell
Research and Development Division, British Gas Corporation
London Research Station, England

Abstract

The British Gas "Stretford" H_2S removal process is successfully applied in over 100 plants worldwide. This paper describes recent further development work by British Gas in the fields of mass transfer, high pressure operation and effluent treatment. New laboratory, pilot, and full-scale plant work is presented on the investigation of microbiological phenomena associated with the process.

1. INTRODUCTION

The British Gas Corporation's Stretford Process is a liquid phase H_2S removal process which was developed in the late 1950s for the treatment of coal derived town gas.

Since that time it has been applied to a wide variety of sour gases in a number of different industries.

The particular advantages of the Process are that it is H_2S selective and capable of reducing H_2S to very low levels at all pressures in an integrated system producing elemental sulphur; secondary off-gas treatment steps are not needed, and the operating temperature is relatively low (c.$40^{\circ}C$).

A list of applications is given in Table 1. At the time of writing, a total of 105 plants are either operating, under construction or being designed. An additional 50 plants have operated but are now demolished, generally as a result of conversion to natural gas in the UK. The largest plant constructed to date removed 70 tonnes/day of H_2S from a gas flow of 225,000 Nm^3/h.

The most recent applications include treatment of gas on an off-shore production platform, treatment of Flexi-coker gas and treatment of sour off-gas from geothermal power plants. The process has been operated at near to atmospheric pressure but British Gas has recently completed operation of a high pressure natural gas pilot plant which has demonstrated its potential for high pressure applications.

Table 1
Schedule of Licensed Plants Operating or Under Construction

PRESSURE LESS THAN 1 BAR	
Claus Tail Gas	37
Coke Oven Gases	23
Regeneration Off-Gas	
Rectisol	1
Amine	2
Selexol	1
Benfield	1
Low Btu Gases	
Coal	4
Petroleum Coke	1
Miscellaneous	
Geothermal Vent Gas	9
Coal Liquefaction (Misc. Vent Gas)	1
Reformer Gas	4
Natural Gas	2
Refinery Gas	2
Oilfield Gas	1
Ore Roasting	1
PRESSURE 1 - 6 BAR	
Natural Gas	3
S.N.G. (Desulphiding)	3
Refinery Gas	2
Reformer Gas	1
PRESSURE 6 - 15 BAR	
Natural Gas	1
S.N.G. (Desulphiding)	1
PRESSURE 15 - 20 BAR	
Coal Gasification	1
PRESSURE GREATER THAN 20 BAR	
Natural Gas	2

The Stretford Process is offered by several engineering contractors under licence. British Gas offers operational design and technical support to its licencees and customers, and has continued a programme of development to satisfy the changing demands of new applications and more stringent environmental requirements.

The purpose of this paper is firstly to outline the recent development work that has been carried out in the areas of high pressure operation, mass transfer and effluent treatment, and secondly to present the results of microbiological investigations that have formed part of the continued support work to Process operators.

2. OUTLINE OF THE STRETFORD PROCESS

The chemistry of the Process has been fully described in several previous papers (1-4).

An outline flow diagram is given in Figure 1.

H_2S is absorbed in the alkaline Stretford liquor. This consists primarily of an aqueous solution of sodium carbonate/bicarbonate, sodium vanadate and the sodium salt of anthraquinone disulphonic acid (ADA). Also formed in solution are the by-product salts sodium sulphate and thiosulphate, and thiocyanate if HCN is present in the gas.

The H_2S is converted initially to HS^- ion which reacts

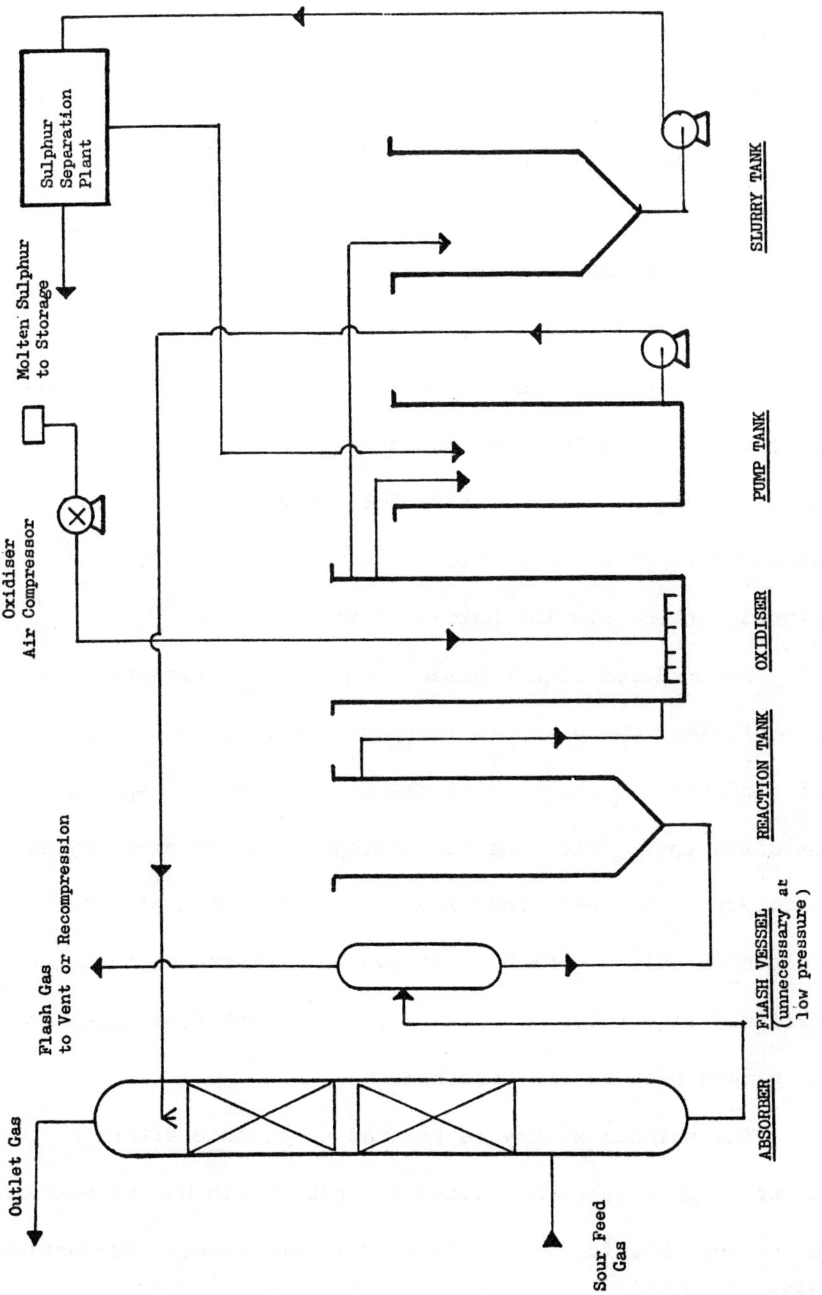

Figure 1. General arrangement of Stretford process.

with the oxidised vanadium in the liquor to form elemental sulphur.

$$HS^- + 2V^{5+} + OH^- \longrightarrow H_2O + 2V^{4+} + S^o$$

A considerable proportion of the HS^- is rapidly oxidised so that virtually no H_2S equilibrium vapour pressure is present above the loaded solution. This means that very low outlet H_2S concentrations can be achieved. Several minutes are necessary for the reduction of the vanadium to go to completion, and a reaction tank provides this solution hold-up time.

The reduced liquor passes from the reaction tank to the oxidiser where air is introduced through an arrangement of diffusers at the base of the tank. The V^{4+} species is oxidised to V^{5+} with the ADA acting as an oxygen-carrying catalyst. The aeration also serves to "froth off" the suspended sulphur which overflows to a slurry storage tank. Oxidised liquor overflows to a pumping tank from which it is pumped back to the absorber.

The sulphur slurry is treated to produce either a moist sulphur cake, or molten sulphur, which has been found to be suitable for sulphuric acid manufacture. Production of moist sulphur cake is achieved by vacuum filtration or centrifugation, normally in a solid bowl decanter. Molten sulphur is produced by batch melting of this cake under pressure, and decantation of the sulphur from the aqueous phase. Alternatively, it can be produced continuously by re-slurrying the moist cake to a pumpable form and passing this through a heat exchanger and separation vessel under pressure. In the case of continuous melting, the filtration stage is often not used (to save on capital cost and operational labour), and the sulphur slurry is melted directly as described above, the penalty being of course increased heating costs.

3. STRETFORD PROCESS DEVELOPMENT

3.1 High Pressure Operation

Operation of a high pressure pilot plant at Partington, near Manchester, has recently been completed. This plant was capable of removing up to 10 kg of H_2S in 3,500 Nm3/h of natural gas at pressures up to 40 bar. Operation of this plant has demonstrated the potential of the Stretford Process for treatment of high pressure gas, and has provided mass transfer data to allow more accurate sizing of absorbers to be made.

The experimental part of the plant was the absorption

system which consisted of two absorbers of 200 mm and 380 mm diameter, each packed with 12m of splash plate packing. The towers were arranged in reversible series and could be operated counter-currently or co-currently. An outline of this system is shown in Figure 2.

The plant was arranged to take its supply from the high pressure natural gas supply to the site. H_2S was manufactured on site by the high temperature hydrolysis of CS_2 and was added to the feed gas prior to the absorption system. The cleaned gas was returned to the low pressure distribution system. Facilities were available for the addition of CO_2 to the gas, and recycling of the outlet gas if required.

The remainder of the plant consisted of conventional vessels for low pressure liquor regeneration and sulphur handling (Figure 3).

The plant was commissioned in the summer of 1981 and operated until spring 1983. During this time approximately 13 tonnes of H_2S were treated. H_2S outlet concentrations were generally less than 1 ppm (v/v) and at times approached the limit of detection.

Mass transfer data were gathered over a range of CO_2 partial pressures of 0.05 to 1.5 atm and over a wide range

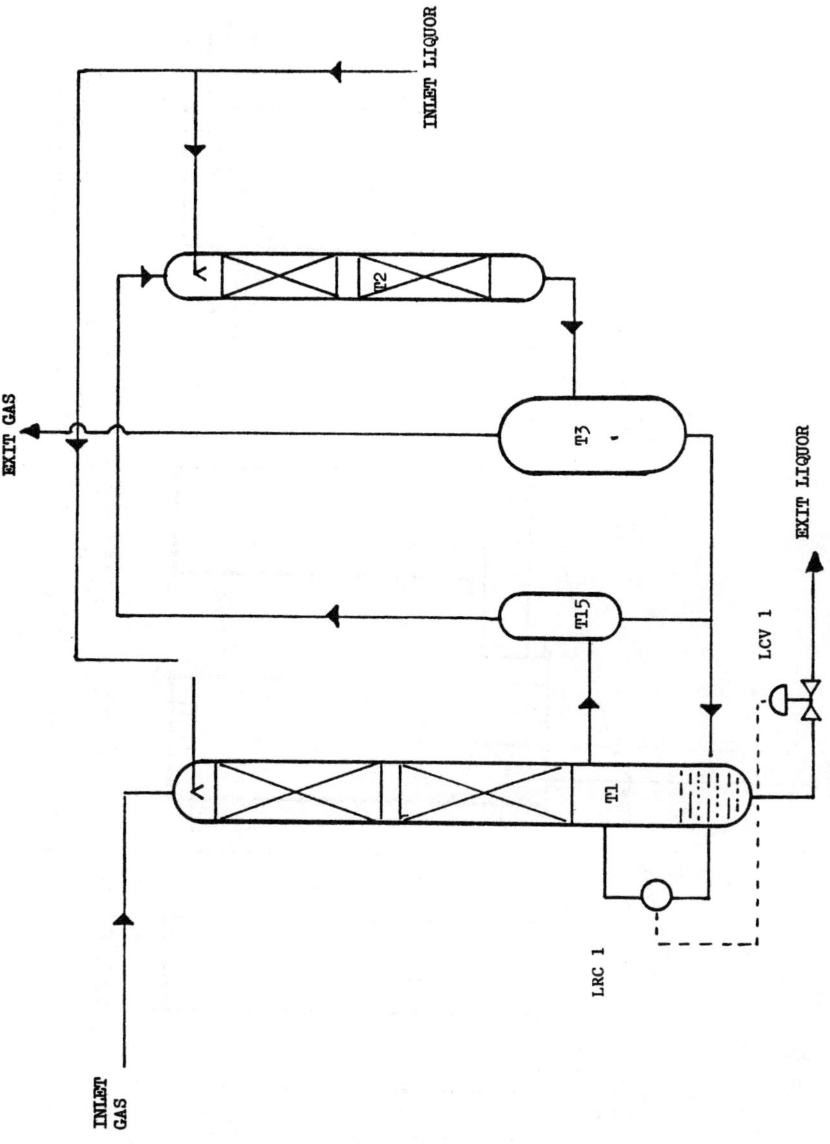

Figure 2. Partington pilot plant—general arrangement of absorption system for co-current operation.

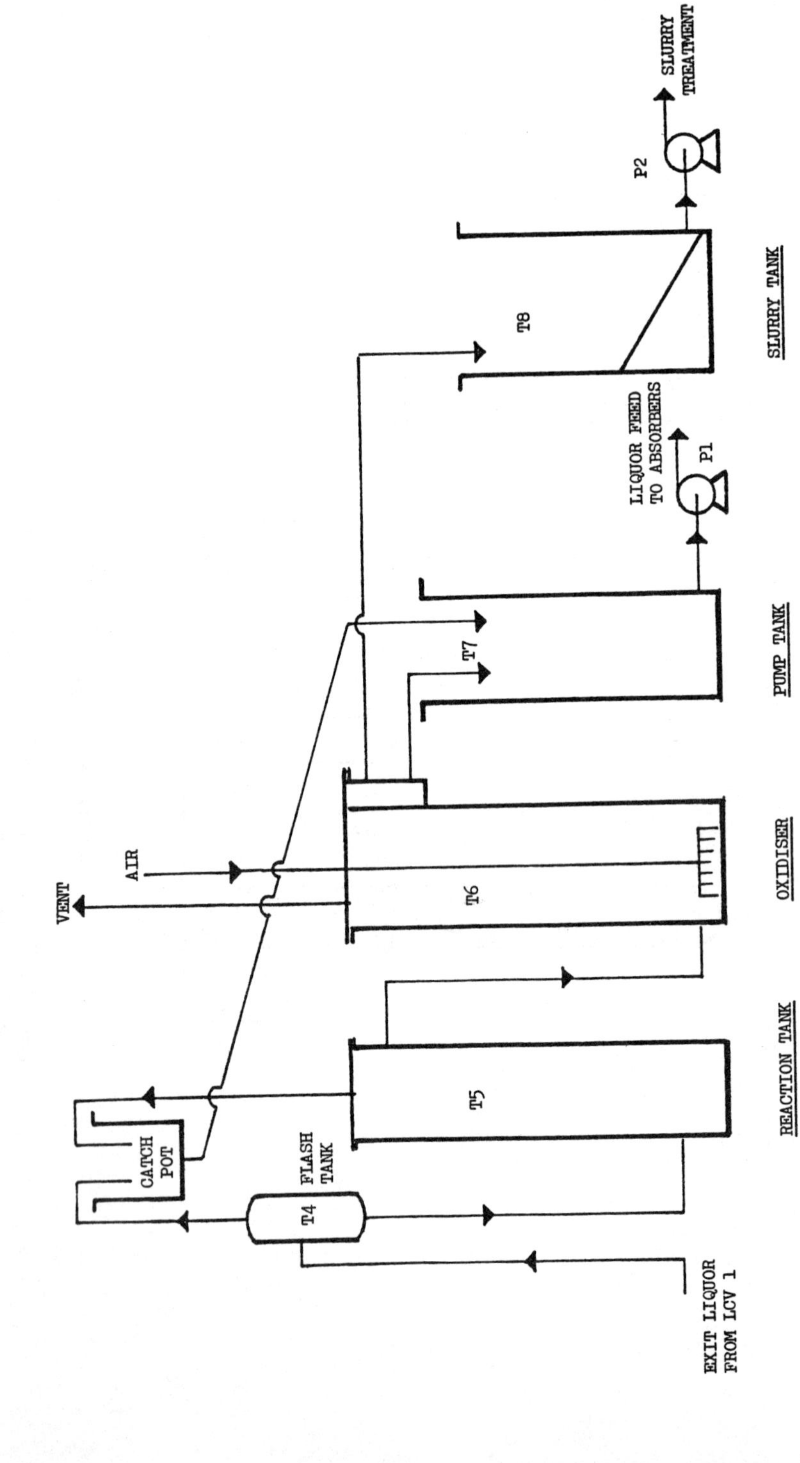

Figure 3. Partington pilot plant—general arrangement of low-pressure section.

of gas and liquor velocities and pressures. It was found that the effect of total pressure on mass transfer was much less than predicted from theoretical considerations, and that the effect of CO_2 partial pressure was much less than had been found at atmospheric operating pressure. This latter effect has been previously reported (5) for work carried out on a small scale Stretford Process plant operating at high pressure with a wooden grid-type packing but at much lower flow rates than used at Partington.

Since these effects were much less than had been expected, British Gas is now able to design absorbers of a practical size for high pressure duties. In Table 2, the size of absorber that would have been predicted previously is compared with the absorber design which can be made using the results from the Partington plant. The case considered is for treatment of natural gas at 70 bar pressure with a CO_2 partial pressure of 1.5 bar, removing H_2S from 500 ppm to 1 ppm. Now only one tower is needed, whereas two towers in series may have been necessary to achieve the packed height calculated from the earlier data.

Trials were made on co-current operation, and the results obtained were very encouraging. Outlet concentrations of 1 ppm (v/v) and less were achieved consistently

Table 2
Comparison of Absorber Sizing Methods

SIZING BASIS	COUNTER CURRENT THEORETICAL DATA	COUNTER CURRENT USING H.P. PILOT PLANT DATA	CO-CURRENT USING H.P. PILOT PLANT DATA
ABSORBER DIAMETER m	4.25	4.25	1.8
PACKED HEIGHT m	65	12	17
OVERALL ABSORBER HEIGHT m	88	21	26
PACKED VOLUME m^3	940	174	44

The above absorber sizings have been prepared based on a typical natural gas treatment duty, and illustrate the improvements in absorber size brought about by operation of the high pressure pilot plant. The potential benefits of co-current operation are also illustrated.

over fully loaded solutions. Some further development of the technique is necessary before it can be generally applied, but as can be seen from the size comparison in Table 2, there is potential for significantly decreasing constructional costs of the absorber in high pressure applications.

The particular benefits of co-current operation are that flooding is not a constraint in determining absorber diameter, and that as the Stretford Process operates with very little or no equilibrium H_2S vapour pressure, very low levels can be achieved in a single stage. It was also found that the absorber level control was less sensitive to minor process changes when operating co-currently, which would make general operation of a plant much easier.

3.2 Mass Transfer

The majority of Stretford plants have operated at low pressure, and until the mid 1970s most operated with gases of low CO_2 content. As the range of Process applications expanded, it became apparent that there was a need to determine accurately the effect of CO_2 partial pressure on H_2S mass transfer coefficient.

The primary effect of CO_2 partial pressure is to depress the pH of the solution according to the equilibrium:

$$Na_2CO_3 + H_2O + CO_2 \rightleftharpoons 2NaHCO_3$$

as the pH falls, the mass transfer coefficient is reduced. Pilot plant work was instigated to quantify this effect.

Figure 4 shows a flow diagram of the pilot plant used for these investigations. The work was carried out at atmospheric pressure using a 610 mm diameter absorber packed with 3m of splash plate packing.

The results obtained confirmed the expected depression of mass transfer coefficient with increasing CO_2 partial pressure, but also showed further interesting effects. It was found that the effects on mass transfer coeffficient of gas velocity and irrigation rate were in themselves functions of CO_2 partial pressure. These effects are shown in Figure 5. It is considered that at low CO_2 partial pressure the mass transfer is predominantly gas phase controlled, and as the CO_2 partial pressure rises, the liquid phase assumes greater influence until at 1 atm CO_2 partial pressure the mass transfer is virtually wholly liquid phase controlled. This is due to the fact that at higher CO_2 partial pressures, the H_2S/HS^- reaction in the liquid phase is slowed down by the lower pH. This reaches

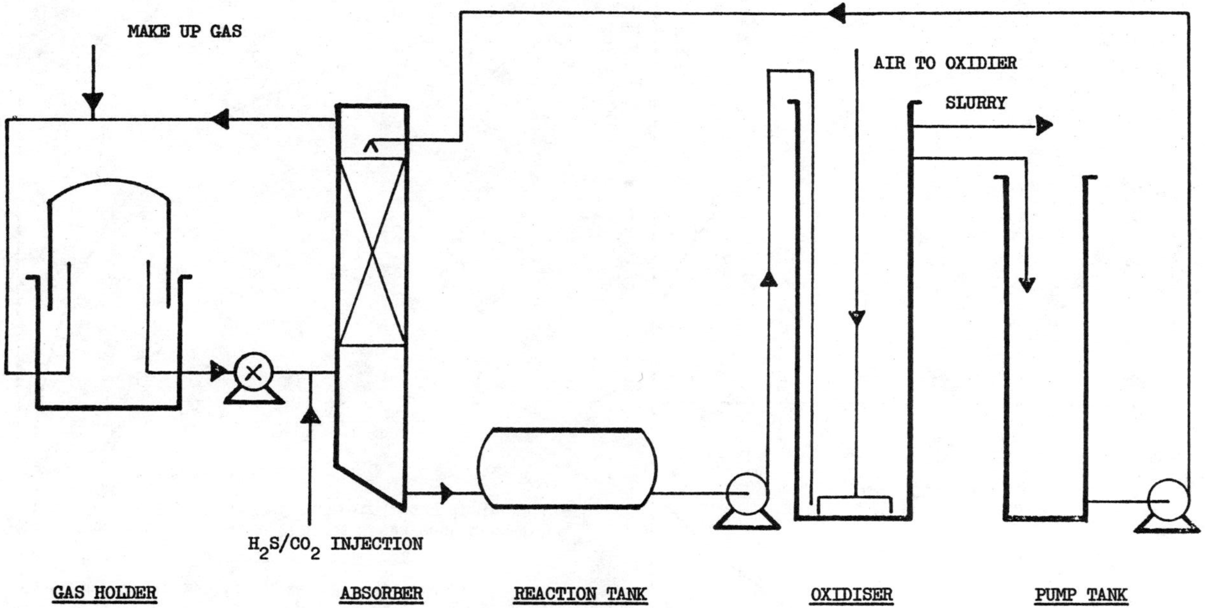

Figure 4. General arrangement of mass transfer pilot plant.

The mass transfer coefficient Kgv has been found to be a function of CO_2 partial pressure described by the function -

$$Kgv = XW^YV^Z$$

Where W = Irrigation Rate

V = Gas Velocity

It has been found that X, Y and Z are each functions of CO_2 partial pressure of the form shown in these graphs.

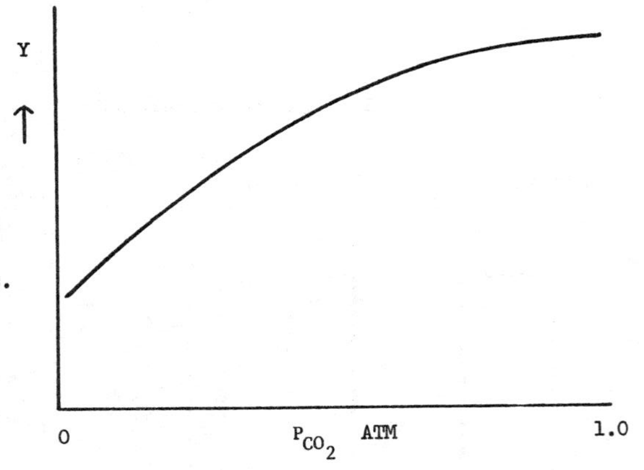

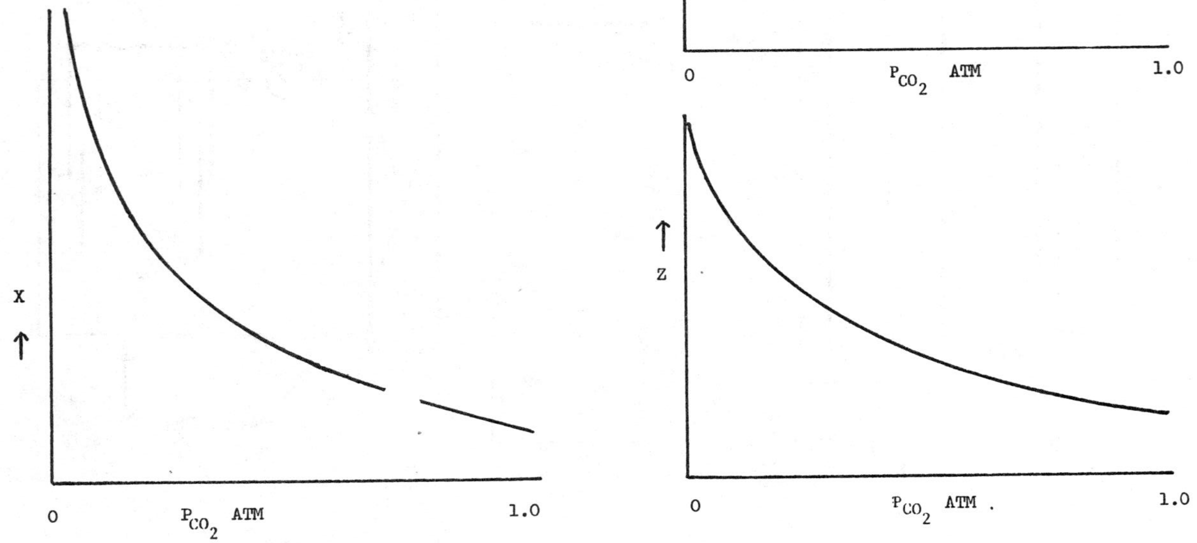

Figure 5. Effect of CO_2 partial pressure on H_2S mass transfer coefficient.

the stage where the reaction becomes the rate limiting step so that the absorption is wholly liquid phase controlled.

It can be seen that estimation of CO_2 effects on the mass transfer coefficient is not a simple factoring calculation, and full consideration of the gas and liquor flow effects must be made to prevent serious errors occurring.

All the data generated in this work have been incorporated into the Stretford Process design program developed by British Gas in order to provide rapid answers to enquiries from licencees.

3.3 Venturi Absorption

The use of a packing material such as the splash plates generally used in Stretford Plants sometimes leads to rather large absorption towers being necessary, particularly at high CO_2 partial pressures. It has been suggested (6,7) that in some instances venturi absorbers are capable of operating at a sufficiently high H_2S removal efficiency to replace packed towers. In the case of high CO_2 partial pressures, it has also been suggested that the absorption rate of CO_2 would be relatively slower than that for H_2S, so that a greater degree of H_2S selectivity would be achieved. In view of these potential advantages, the pilot plant used for atmospheric pressure mass transfer investigation was modified to accept a 200 mm venturi absorber. The plant was operated with a variety of tailpipe arrangements on the venturi, and at low and high (1 atm) CO_2 partial pressure.

Typical operating results are shown in Figure 6, where they are compared with the efficiency of a conventional splash plate packed tower. It can be seen that a packed tower of quite modest dimensions is more efficient at all conditions than the venturi.

The rate of CO_2 pick-up was not less than that in a packed tower as indicated by the change in the pH of the liquor across the venturi. It was also observed that replenishment of the CO_2 from cylinder supplies was not less than in the case of packed tower operation.

It was also found that the different arrangements of tailpipe length and liquor nozzle sizes had very little effect on either H_2S removal efficiency or in reducing CO_2 pick-up. It is accepted that some of these effects may have been contributed to by the gas/liquor separation vessel to which the venturi was attached. However, it is difficult to conceive a venturi system which does not use such a vessel, and so this must be considered an integral

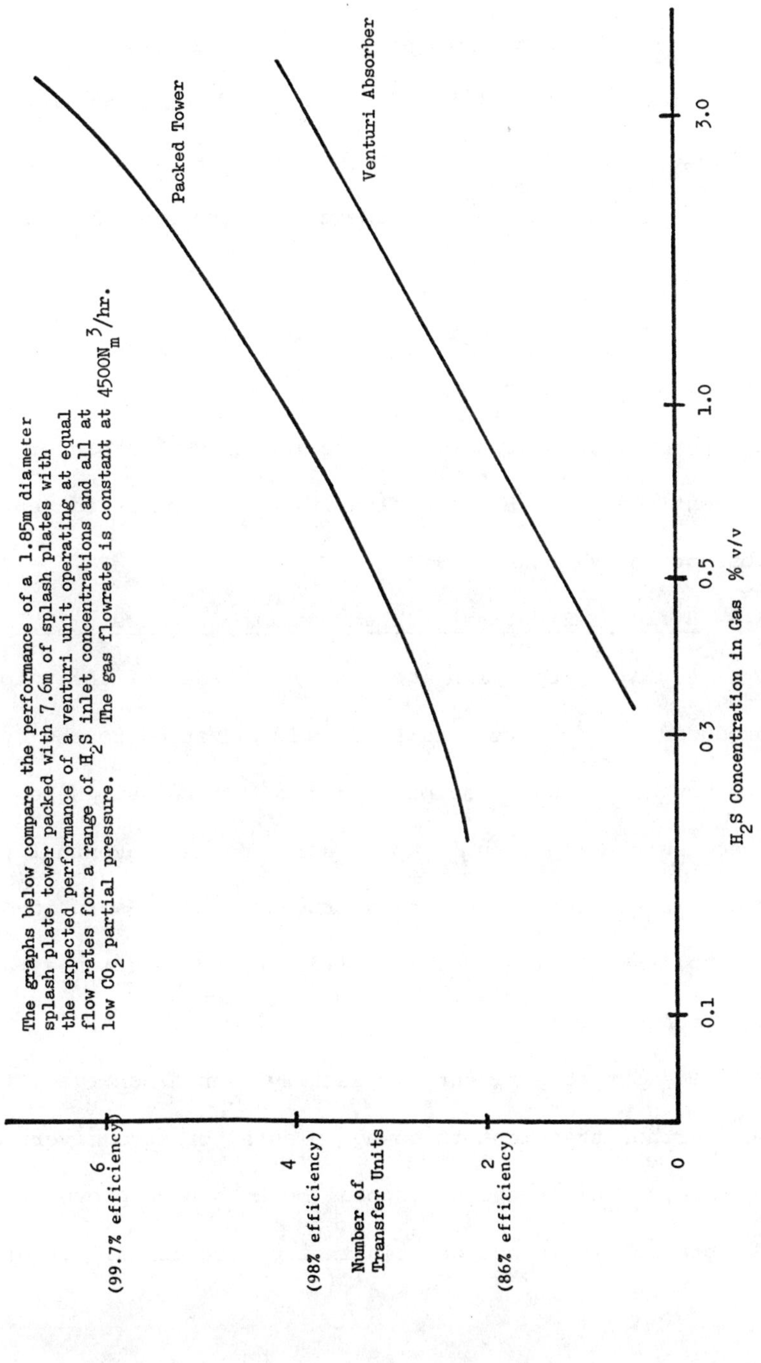

Figure 6. Comparison of Venturi absorption with a packed tower.

part of the system.

The results of this pilot plant work show that venturis are not suitable for high efficiency removal of H_2S, and that their use does not increase H_2S absorption selectivity in the case of operation with high CO_2 partial pressure.

In view of these facts therefore, British Gas would recommend the use of venturis only where lower efficiency H_2S removal is acceptable. Applications with higher concentrations of H_2S are clearly more amenable to venturi absorption as shown in Figure 6.

3.4 Effluent Treatment - The Desalting Process

A small proportion (generally 2 to 5%) of the sulphur produced in the Stretford Process is converted to the by-product salts sodium sulphate and sodium thiosulphate. These salts slowly build up in solution, and eventually a stage is reached where a need arises to purge solutions to control the total dissolved solids in the solution at an acceptable level.

The purged effluent has usually been discharged without further treatment to sewage plants, or into rivers or the sea. Whilst these disposal methods have proved to be acceptable, it is not certain that environmental authorities will continue to allow such practices in the future. In any case the purge of valuable chemicals contained in this effluent is a considerable addition to the running costs of the plants.

British Gas has developed the Desalting Process (8) to treat this effluent, and thus overcome concern about the effect of effluent disposal on Stretford Process operability. In desalting, sodium thiosulphate is converted to sodium sulphate which is then crystallised out and can be further processed to a saleable form of the anhydrous salt. The Desalting Process, which has been patented in the USA, is only suitable for treatment of effluent from plants treating gas with no HCN or very little HCN. In the case of plants with significant quantities of HCN in the gas, high temperature hydrolysis processes have been developed by other companies (6,9), and have been operated successfully on commercial scale Stretford plants.

3.4.1 Description of the Desalting Process. An outline flow sheet of the Process is given in Figure 7.

Sodium thiosulphate is converted to sodium sulphate in an acidification vessel by reaction with concentrated sulphuric acid at high temperature. By sparging the acid in the lower part of the acidifier, the SO_2 produced in the

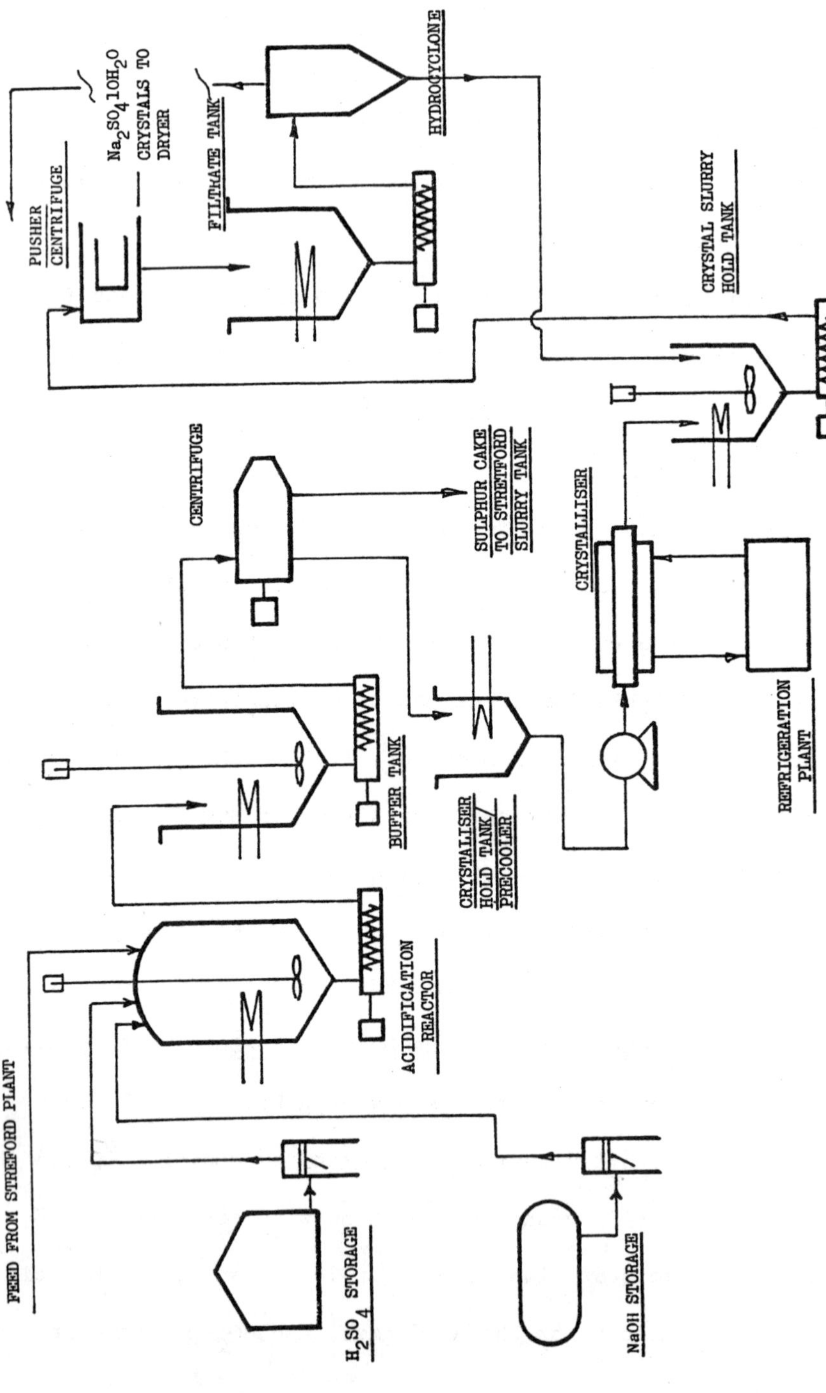

Figure 7. General arrangement of desalting process.

initial acidification reaction is used in further conversion of the thiosulphate by means of the following reactions.

$$Na_2S_2O_3 + H_2SO_4 \longrightarrow Na_2SO_4 + SO_2 + H_2O + S^o$$

$$2Na_2S_2O_3 + SO_2 \longrightarrow 2Na_2SO_4 + 3S^o$$

$$3Na_2S_2O_3 + H_2SO_4 \longrightarrow 3Na_2SO_4 + H_2O + 4S^o$$

The overall reaction will convert approximately 90% of the thiosulphate in about 3 hours with the use of a slight excess of acid. It is obvious that the sodium carbonate content of the solution will be neutralised, thereby increasing alkali consumption; however it is more important that the far more valuable ADA is not destroyed by the acidification process.

The sulphur produced in this reaction is very coarse and granular, and is capable of being separated by fairly simple techniques.

The acidified liquor after pre-cooling and sulphur separation is passed to a crystallisation plant where the sodium sulphate is crystallised out in the decahydrate form at low temperature. The use of high temperature crystallisation to produce anhydrous sulphate would have been preferred, but it was found that ADA was co-crystallised with the sulphate under these conditions.

The sulphate slurry produced in the crystalliser is centrifuged, and the nearly dry cake can either by disposed of in this state or passed to a drying plant to produce a saleable anhydrous product.

The Process has been conceived as one using batch-wise acidification with a continuous crystallisation plant, but there is no reason why the acidification process could not be operated continuously. Kinetic data has been obtained to allow sizing of a continuous process.

3.4.2 Demonstration of the Desalting Process. British Gas has operated the Desalting Process on a pilot plant and has demonstrated all the steps described above to produce sodium sulphate.

The plant consisted of a 500 litre stainless steel acidification vessel fitted with a propeller agitator. Heating and cooling was carried out using a stainless steel coil. The solution to be treated was heated to 90°C, and then concentrated acid added to the vessel through a metering pump.

The acidified solution was filtered using a variety of

separation devices, including centrifuges and vacuum filtration, but since the sulphur produced is granular, much simpler forms of filtration could have sufficed.

Crystallisation was carried out in a scraped surface crystalliser, chilled with glycol from a refrigeration plant. Flow rates up to 100 litres/h were fed to the crystalliser, which produced up to 60 kg/h of decahydrate crystals. A recycle system to encourage crystal growth was developed, and good crystal size was achieved in spite of the recycle pump being considered to be something of a "crystal cruncher".

Various types of centrifuge were used to separate crystals from the slurry, and crystal cake moisture levels as low as 4% were achieved. Some of the crystals were dried to the anhydrous state in a small scale drying plant, and samples were submitted to potential users who considered the material to be suitable for their needs.

Although the operations described were only carried out in a pilot plant, it should be noted that on a commercial scale the Desalting plant would work on a small side stream. In fact the experimental acidifier that was used would be suitable for a 3 tonnes/day sulphur removal Stretford unit and the crystalliser for a 9 tonnes/day plant.

3.4.3 Application of the Desalting Process. The Desalting Process is available to all Stretford Process users with full British Gas process engineering back-up.

The estimated operating costs of the Process are given in Table 3, together with the savings in chemical consumption that would be made by using the Process for a typical application. It is anticipated that the payback time for desalting would be quite short (e.g. for the case described in Table 3, a payback period of about 2 years would be expected).

4. MICROBIOLOGICAL PHENOMENA ASSOCIATED WITH THE STRETFORD PROCESS

4.1 Background

The presence of bacteria in Stretford solution can influence plant performance. Bacteria were first detected in Stretford solution in September 1978, in a sample from an oil refinery in Illinois. Large amounts of uncontrollable light foam rather than the normal sulphur froth were produced, and as the foaming worsened, the plant had to be shut down. The pH had fallen from 9.6 to 7.0. The alkali consumption during a 6 day period had been 25 tonnes of sodium carbonate, and it was calculated that a total of 35 tonnes of sodium thiosulphate had been converted into

Table 3
Costs[1] of Operating Desalting Process Compared with Liquor Purging

		OPERATION WITH DESALTING	OPERATION WITH LIQUOR PURGE
ELECTRICITY	£/day	132	-
L.P. STEAM	£/day	85	-
COOLING WATER	£/day	2	-
SULPHURIC ACID	£/day	140	-
CAUSTIC SODA[2]	£/day	640	460
SODIUM SULPHATE PRODUCTION[3]	tonne/day	8.5	-
SODIUM SULPHATE CREDIT	£/day	425	-
EFFLUENT PURGE	m^3/day	-	29
COST OF CHEMICALS IN PURGE	£/day	-	860
DAILY COST	£/day	574	1320
SAVING MADE BY DESALTING	£/day	746	-
APPROXIMATE CAPITAL COST OF DESALTING PLANT	£	500,000	-

1. Costs estimated for a 54 tonne/day sulphur removal plant treating non cyanide containing gas.

2. Includes cost of caustic replacement required to allow for solubilisation.

3. Costs assume that sodium sulphate is processed to anhydrous form for sale. Processing to decahydrate reduces operating costs but credit from sale is lost.

sodium sulphate. On-site tests by British Gas personnel showed that significant numbers of bacteria (10^6-10^7 per ml) were present.

Subsequently, every plant solution examined at the London Research Station except those treating coke oven gas has been found to be contaminated with bacteria. The work reported in this paper is the result of in-depth investigations that have been carried out by British Gas, and extended to commercial Stretford plants in the past three years. The Stretford Process is by no means unique amongst the wet chemical H_2S removal processes in supporting bacterial growth. The LO-CAT Process which employs iron chelates has also suffered from problems caused by bacteria (10). It is probable that other processes, in which elemental sulphur and thiosulphate are produced, could provide suitable environments for bacterial proliferation.

4.2 Stretford Plant Microbiology

The majority of Stretford plant solutions that we have examined have contained at least 10^6 bacteria/ml with an upper limit of 10^8. Populations in the range of 10^6-10^8/ml are the same as those found in conventional bacterial stirred tank fermentations, and therefore it is not surprising that the bacterial activity in Stretford solutions can be significant.

The bacteria found in Stretford plants include autotrophic sulphur bacteria of the genus *Thiobacillus*, a wide mixture of heterotrophic bacteria, e.g. *Pseudomonas* species, and sulphate-reducing bacteria. Vanadium is toxic to biological cells, and normally concentrations of 1-4 g/l would be expected to prevent bacterial growth. Thus, if strains of thiobacilli and pseudomonads from other sources are introduced into Stretford solution most of them die within 2 days. However, those that survive have adapted to the relatively hostile environment (i.e. high pH and presence of vanadium), and grow less well when they are taken out of this environment.

The following species of thiobacilli have been found in Stretford solutions: *T. thioparus*, *T. neapolitanus* and possibly *T. novellus*. *T. thioparus* and *T. neapolitanus* are the most vigorous. They are obligately autotrophic, i.e. they do not require organic compounds for growth, and obtain the energy they need from the oxidation of inorganic sulphur compounds, e.g. thiosulphate, polythionates, sulphide and elemental sulphur.

The oxidations that thiobacilli undertake vary with the

species concerned and the environmental conditions prevailing. For example, thiosulphate oxidations can proceed according to the following equations:

complete oxidation to sulphate

$$Na_2S_2O_3 + O_2 + H_2O \longrightarrow Na_2SO_4 + H_2SO_4$$

elemental sulphur formation

$$Na_2S_2O_3 + O \longrightarrow Na_2SO_4 + S^o$$

a combination of these reactions

$$5Na_2S_2O_3 + H_2O + 4O_2 \longrightarrow 5Na_2SO_4 + H_2SO_4 + S^o$$

Thus thiobacilli are responsible for the formation of sulphuric acid and elemental sulphur at the expense of thiosulphate. As already stated, the exact stoichiometry will vary with the particular strains of thiobacilli and the environmental conditions. What is clear, however, is that the neutralisation of any sulphuric acid formed in Stretford solutions will lead to the consumption of alkali.

Thiobacilli, which have been the subject of a special study within British Gas, grow slowly in normal laboratory media containing thiosulphate. Colonies take up to 4 days to become visible on agar plates, and are often creamy white due to the deposition of sulphur. Thiobacilli from Stretford solutions grow even more slowly, colonies taking up to 7 days to develop. A more specific medium designated "Stretford Agar" has been developed, and this and one other are routinely used by British Gas to determine the populations of thiobacilli (see Table 4 for details).

The heterotrophs isolated from Stretford solutions are mostly pseudomonads, which are closely related to thiobacilli; however, heterotrophs in contrast to autotrophs require organic compounds for both energy and growth. Organic material is excreted by growing bacteria, and also, when cells die, the entire cell contents become available to heterotrophic bacteria. Bacteria can be killed by autoclaving, and therefore when Stretford solution contaminated with bacteria is recycled from autoclave decanters, it will be enriched with the contents of the cells that have died.

Bacteria carry a net negative charge and often form slimes and aggregates. Some bacteria are surrounded by capsules formed from copious amounts of extracellular mucopolysaccharides and proteins. This capsular slime can break away from the bacteria, and together with proteins released from cells can cause foaming when solutions containing such material are agitated or aerated. In addition, sulphur bacteria release surface active agents

Table 4
Media Used for the Detection of Bacterial Contamination in Stretford Solutions

Starkey's No. 2 Thiosulphate Agar, modified to pH 8.5

	g
KH_2PO_4	4.0
$Na_2S_2O_3.5H_2O$	10.0
$MgSO_4.7H_2O$	0.1
$CaCl_2$	0.1
NH_4Cl	0.1
$FeCl_3.6H_2O$	0.02
$MnSO_4.4H_2O$	0.02
Agar No. 1 (Oxoid)	10.0

Distilled water to 1 litre

Adjust pH to 8.5 with 5M NaOH

Autoclave 10 psi for 20 minutes

Stretford Agar

	g
$Na_2S_2O_3.5H_2O$	10.0
$NaHCO_3$	22.0
Na_2CO_3	0.5
Na_2SO_4	9.0
$NaH_2PO_4.2H_2O$	0.5

Distilled water to 1 litre

Dissolve salts then add,

ADA	2.35
NH_4VO_3	2.29
Agar No. 1 (Oxoid)	15.0

Stir for 15 mins, then heat almost to boiling. Autoclave 10 ps for 10 minutes.

(phospholipids) which enable them to wet elemental sulphur so that they can oxidise it, and these phospholipids are also likely to contribute to foaming.

Even freely suspended microorganisms often occur in aggregates or flocs. Obviously the size of these may be orders of magnitude greater than single organisms (typical size of a sulphur bacterium is 1-2 μm long x 0.4 μm diameter). The tendency for bacteria growing in Stretford solutions to form clumps and to attach to sulphur particles has been confirmed by examination of samples under the microscope.

In the laboratory and in the field, the degree of contamination by viable (living) heterotrophic bacteria is routinely determined by a conventional agar plate technique. A less accurate procedure that is extremely useful for plant operators and has been widely used is the dipslide technique. Suitable dipslides, which can be obtained from a number of manufacturers, consist basically of slides covered with nutrient agar to which a redox indicator (TTC) has been added. The slide is simply dipped into the solution to be tested. The colonies of heterotrophic bacteria that develop after about 2 days are stained red or pink, which aids their counting by the non-microbiologist.

A third group of bacteria that have been isolated from plant solutions are the sulphate-reducing bacteria which occur under densely packed sulphur deposits. They are obligate anaerobes, which means that oxygen is toxic to them. The sulphate-reducers accelerate corrosion by using hydrogen produced at cathodic areas on metal surfaces to reduce sulphate to sulphide, thereby stimulating anodic corrosion and dissolution of iron.

4.3 Commercial Plant Observations

4.3.1 Effects of Bacteria on Plant Operation. The design specifications of Stretford units depend on a number of factors, for example: type of gas, volume of gas, H_2S concentration, H_2S removal duty and CO_2 partial pressure. It follows that no two Stretford plants are exactly the same. Thus the extent to which bacteria have a deleterious effect on Stretford Process operation varies from plant to plant, encompassing a wide spectrum from no detectable impact to maximum inconvenience.

Bacterial contamination of plant solution usually, but not necessarily, leads to one or more of the following consequences:

1. High sulphate to thiosulphate ratio.

2. Increase in demand for alkali.
3. Excessive and uncontrollable foaming in the oxidiser, leading eventually to foaming in the absorber and solution entrainment in the gas.
4. Poor sulphur flotation.
5. Sticky sulphur that tends to aggregate and stick to surfaces.
6. Increased concentration of carbon in sulphur.
7. Increased tendency for sulphur solubilisation.
8. Corrosion.

Some operational difficulties are not easy to recognise; for example, although a sudden increase in alkali consumption would be obvious, an insidious increase might not be recorded at all. Other difficulties however are obvious, e.g. sulphur build-up. If sulphur deposits accumulate to an unacceptable level, experienced operators adopt techniques such as hydraulic shock to release sulphur. If build-up occurs regularly, it may be necessary to redesign some features of the plant e.g. by modifying sprays, or altering packing density or configuration. Also obvious is excessive foaming. A foam is extremely light, and similar to detergent bubbles; it is easily blown away by the wind, and is often accompanied by a scum. On the other hand, a froth should be heavy with sulphur and low in plant solution, but sometimes a light froth low in sulphur and high in plant solution is obtained. A properly-constituted froth is essential for satisfactory performance.

In the context of the Stretford Process, the reaction tank, the oxidiser, and the slurry and balance vessels are merely stirred tanks of widely different dimensions containing baffles or air inputs, although they are comparatively gently agitated tanks compared with conventional biochemical reactors. Chemical and biochemical reactions are similar in that they are affected by mixing rates. The activity of bacteria will vary with the dimensions of the reaction vessel in which they are growing. For aerobic microbial reactions, the principal limiting factor is oxygen and even changes in the dimensions of aerated stirred tanks will affect the rate of the reaction.

Other subtle changes in environment can alter the established bacterial equilibrium. For example, a change in the Claus plant burning-off ratio, so that SO_2 is slipped will decrease the bacterial population because the SO_2 itself acts as a biocide. On the other hand, if there are no biocidal compounds present, the bacteria are able to proliferate and therefore contribute to foaming, and as a consequence the sulphur slurry becomes weaker. In addition, the more solution that passes through the autoclave during sulphur processing, the greater the rate at which bacterial cell contents will be recycled, and the more certain it is that an active bacterial population will be sustained.

Many minor operational difficulties have been resolved with experience, and appropriate remedial practices have become integrated as in-house know-how and accepted as part of normal Stretford plant operation.

4.3.2 Anthraquinone Disulphonic Acid (ADA). Losses of ADA from plant solutions have been reported from time to time, and it has been suggested that such losses are the result of bacterial activity. It is difficult to see how the ADA molecule could be attacked microbiologically, but nevertheless a series of experiments was carried out in order to examine this possibility. Comprehensive tests were carried out with bacteria taken from plants in which ADA losses had been reported, and the incubation medium contained ADA alone or ADA with thiosulphate. None of the bacteria tested was able to utilise the ADA, and it was concluded that ADA losses must be attributed to purely chemical effects.

4.4 Successful and Unsuccessful Approaches to the Control of Bacteria

The more detailed the microbial investigations into the Stretford Process, the more it became apparent that serious bacterial contamination should be eradicated. Some operators had already added biocide in order to cope with badly affected plants. Unfortunately the products used had not been thoroughly tested and evaluated, and not surprisingly the biocide treatments were unsuccessful. Therefore an investigation of biocide effectiveness was started. It soon became apparent that the selection of appropriate biocides was not an easy matter. Although there are hundreds of formulations on the market, the number that are compatible with the Process is small. The successful biocides must conform to the following criteria:

1. Compatibility with Stretford solution.
2. Ability to withstand the redox cycle of the Stretford Process.
3. Effectiveness against Stretford organisms in Stretford solution.
4. Retention of activity after autoclaving.

4.4.1 Laboratory Biocide Tests. After the first screening of approximately 90 products (see Table 5 for biocide types), only 20 qualified for laboratory investigations; when the criteria for selection were applied, only a handful remained for evaluation in the pilot plant.

Laboratory tests were carried out with samples of Stretford solution to which bacteria had been added. Samples were incubated at $30^{o}C$ and examined for surviving bacteria over a 7 day period. Some tests were also carried out on plant solutions from different units. Typical results are shown in Table 6.

4.4.2 Pilot Plant Trials. The 800 litre pilot plant at Stretford was a conventional mild steel unit with a 6 metre sparged oxidiser in which sulphur flotation could be observed. Each biocide run lasted approximately 14 days; the plant was thoroughly cleaned between runs. The solution used for each experiment was of typical composition, and was inoculated with bacteria originally isolated at the London Research Station from a commercial Stretford plant. The appropriate biocide (1000 ppm v/v) was added once the bacteria were established in the plant solution (approximately 2-3 days, 10^{6} bacteria/ml). Samples taken immediately prior to addition, and thereafter

Table 5
Types of Biocides Considered

Biguanides
Diphenyl methane derivatives
Formaldehyde releasing compounds
Glutaraldehyde formulations
Organo-bromine compounds
Phenolics
Quaternary ammonium compounds
Substituted quaternary ammonium compounds
Thiocarbamates
Tri-butyl tin oxide formulations

Table 6
Comparison of the Effectiveness of Biocides Against Stretford Plant Bacteria After 5 Days Incubation

Biocide	Concentration (ppm)	Bacteria/ml		
		Nutrient agar *	Thiosulphate agar ** (pH 8.5)	Stretford agar ***
Control	500	1.4×10^7	5.05×10^7	1.0×10^7
Thiocarbamate biocide	500	6.55×10^6	4.85×10^6	8.4×10^7
	750	1.8×10^6	2.0×10^6	1.6×10^6
	1000	2.1×10^4	7.0×10^5	1.0×10^6
Formaldehyde releasing biocide A	500	Nil	Nil	Nil
	750	Nil	Nil	Nil
	1000	Nil	Nil	Nil
Formaldehyde releasing biocide B	500	Nil	Nil	Nil
	750	Nil	Nil	Nil
	1000	Nil	Nil	Nil

* Initial population 7.0×10^6 bacteria/ml
** " " 1.45×10^7 bacteria/ml
*** " " 1.15×10^7 bacteria/ml

at regular intervals, were tested at Stretford (with dip-slides) for bacterial contamination; duplicate samples were sent to the London Research Station for detailed examination. During certain tests, the detailed microbial examination was made on-site at Stretford. Whenever possible, the rate of loss of biocide was measured directly, but if there was no suitable analytical method the rate of loss was judged by the subsequent growth of bacteria.

Only two biocides were considered suitable for commercial plant trials. These were both formaldehyde-releasing compounds, available commercially.

4.4.3 Commercial Plant Trials.

(1) Plant A. This small plant was built to treat 5.8×10^3 m^3/h of natural gas containing 750 ppm H_2S. Difficulties had been experienced with sulphur flotation and removal, and in mid-1980, alkali consumption had been high and sulphur frothing had become poor. The plant solution was found to be contaminated with bacteria. At this time potentially suitable biocides for commercial plants were being assessed in laboratory and pilot plant trials, and one of the promising biocides was added to Plant A to give a concentration of 750 ppm (v/v) in the

plant solution.

The effects were instantaneous; a head of sulphur froth immediately built up, and excellent sulphur flotation was obtained. Figure 8 shows the extent of the improvement, and the subsequent results. This successful trial demonstrated how impaired the sulphur flotation had been. During the period of regular biocide additions, the rate of sulphur removal increased (i.e. trapped sulphur was gradually released, see Figure 8). Sulphur production eventually exceeded sulphur input. In addition to improved flotation, alkali consumption decreased (see Figure 9). At the annual shut-down, the plant was found to be relatively free of sulphur deposits and the splash plate packing (which during previous shut-downs had to be cleaned) was virtually free of sulphur.

(2) Plant B. An experimental charge of the same biocide (750 ppm v/v) was added to Plant B, which treats refinery gas. However, the rate at which the biocide disappeared was unacceptable so that total sterilisation of the plant would have been prohibitive. But the operators regarded this particular biocide as the best that they had tried, and they decided for cost reasons to control the bacterial population to less than 10^4 bacteria/ml rather

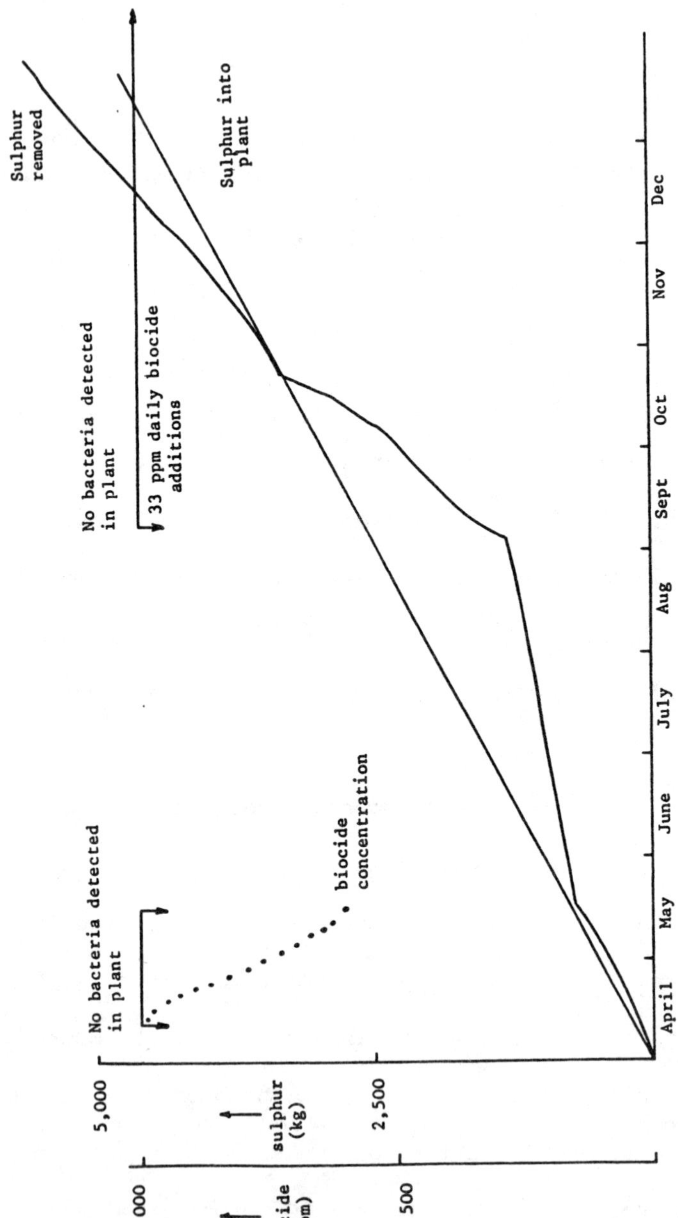

Figure 8. Operating data from Plant A—sulfur input/output balance.

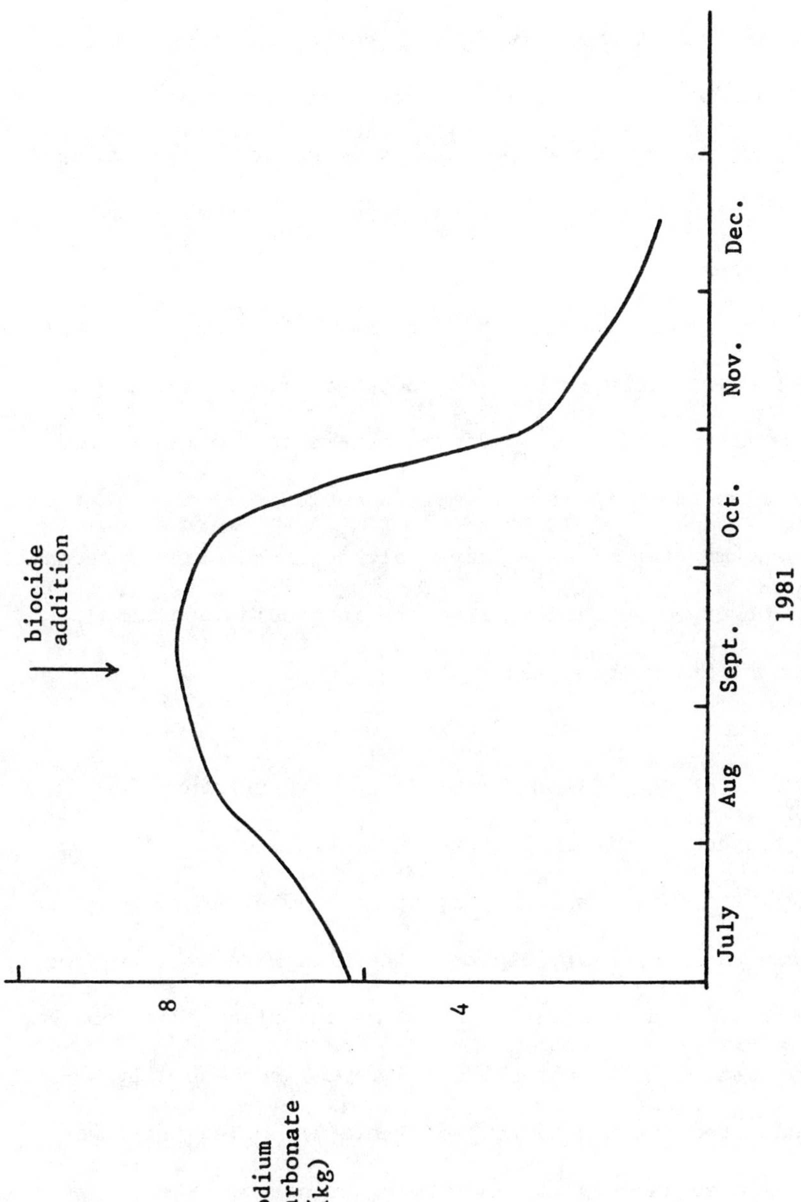

Figure 9. Operating data from Plant A—alkali consumption.

than to eradicate them completely.

(3) Plant C This plant, which removes H_2S from natural gas, had a record of high chemical consumption, sulphur plugging in the absorber, and excessive foaming. It provided an ideal opportunity to scale up the evaluation of biocide treatment from the experience at Plant A. In September 1981, biocide was added (800 ppm v/v), and provision made for daily additions of 100 ppm for a period of 8 days. The problems associated with conducting an experiment at such long distance were considerable, and the results of the trial were inconclusive; the plant was recontaminated after 14 days, although the plant personnel had observed sufficient benefits in biocide addition to justify its routine use.

Prior to the second trial, which British Gas attended for a week, the plant was producing a persistent foam, low in sulphur. Within 3 hours of adding biocide (1000 ppm v/v), the foam was less persistent, and after 24 hours the froth was richer in sulphur. Samples were examined for bacteria; on the fourth day it became clear that some had survived. A further charge (600 ppm) successfully eradicated the survivors (see Table 7). The trial was considered successful, and the plant has operated

Table 7
Biocide Trial at Plant C, Summary of the Microbiological Results

Sample No.	Date Taken	Time	Bacteria/ml		Comments	Amount of sulphur in slurry % w/w
			Nutrient agar	Stretford agar		
12	15.3.82	11.05	2.0×10^7	5.5×10^7	Biocide addition commenced at 11.15 (1000 ppm)	1.8
13	16.3.82	9.45	1.8×10^4	1.75×10^4		1.1
14	17.3.82	9.45	2.6×10^2	2.9×10^2		2.4
15	18.3.82	9.15	7.9×10^2	1.25×10^3	Further biocide added at 11.00 (667 ppm)	2.7
16	19.3.82	9.30	Nil	Nil		1.5
17	20.3.82	9.40	Nil	Nil		3.0

satisfactorily with biocide for over 12 months.

From observations made during this trial, the following procedures for plants undergoing biocide treatment have been formulated.

1. Wherever practicable, any large sulphur deposits (e.g. at the top of the oxidiser weir and sulphur chutes) should be physically removed or broken up and submerged in the Stretford solution. Ideally, areas like those mentioned should be hosed down with a biocide solution.
2. Just before the biocide addition the volume of plant liquor should be at the maximum, so that sulphur deposits and slimes that inevitably accumulate on vessel surfaces are bathed in liquid.

Attention to such detail is well worthwhile.

(4) Plant D. Processing sulphur through a continuous melter was the difficulty in this plant, which treats Claus Process off-gas. Bacteria were detected by both the operator and by British Gas. A mixture of cresylic acids was suggested as a potential biocide, but not by British Gas; this material was used, but the improvement was only

temporary. Glutaraldehyde was then tried; it was effective in killing the bacteria but caused severe foaming. Low concentrations of phenol were then tried and found to reduce contamination, but the amount added had to be continually increased.

A trial with a British Gas biocide (1000 ppm v/v, plus 90 ppm added per day) killed all the bacteria and gave greatly improved operation over a 10 day trial period. Sulphur processing through the continuous melter improved, caustic consumption decreased markedly and foaming ceased. The operators claimed that this biocide was the most effective that they had used. The solution inventory of Plant D is 3.4×10^6 litres and the design capacity for H_2S removal is 59 tonnes of sulphur/day. Therefore the cost of routine treatment with any biocidal agent would be high.

(5) Plant E. This unit had a history of sulphur accumulation. The results of a biocide trial attended by British Gas were inconclusive. The trial was unsatisfactory because the unit suffered a number of unscheduled shut-downs caused by decommissioning of plant upstream from the Stretford unit. Strengthening of the sulphur slurry was evident during the period of biocide treatment, but the inconsistent operation of the plant made it impossible to

demonstrate improved performance. However, further additions of biocide for a longer period have not been ruled out.

4.4.4 Cresylic Acids and Phenols. During the course of our development work, plant operators have made suggestions for different treatments. One often suggested procedure (see Section 4.4.3, Plant D) is the use of cresylic acid-phenol mixtures, principally because the material is a waste-product and cheap. When such preparations have been used, the lowest number of bacteria that we have found is 10^{-5}-10^{6} bacterial/ml. The indications are that although these mixtures do sometimes alleviate effects of bacterial contamination, they are eventually ineffective and can lead to further problems. This is because the bacteria that break down cresols and phenols include pseudomonads, which have been found in Stretford solutions. These organisms use these compounds as a source of carbon and are only killed by very high concentrations. Laboratory tests have confirmed that at least 1% by volume of cresylic acid preparations must be present in the plant solution in order to prevent the growth of Stretford bacteria. Concentrations less than this make matters worse.

4.4.5 Thiocyanate. The difference between coke oven

Stretford solutions and non-coke oven Stretford solutions is that thiocyanate accumulates in the former. A high concentration of thiocyanate inhibits bacterial growth, but thiobacilli can oxidise thiocyanate provided the concentration is less than 5 g/l. Tests undertaken at the London Research Station in 1979 demonstrated that a concentration of 50 g/l NaSCN would kill Stretford bacteria (Table 8). However, 50 g/l is approximately 20% of the maximum total of dissolved salts (including thiosulphate and sulphate) that can be allowed to accumulate in Stretford solution before the solution should be purged. The formation of significant amounts of a variety of sulphur salts in the Stretford Process and similar processes is inevitable because of the complexity of sulphur chemistry. Purging is needed therefore to prevent the precipitation of ADA and vanadate; it also helps to minimise corrosion at high salt concentrations. The presence of thiocyanate would increase the necessity for frequent solution purging, and would contribute to the problems of the disposal of spent solution. For these reasons, the deliberate addition of thiocyanate for the treatment of contaminated plants was not considered a viable option.

Table 8
Effect of Thiocyanate on the Survival of Stretford Plant Bacteria

(i) Thiosulphate agar counts

Sodium thiocyanate (g/1)	Bacteria/ml Time incubated (days) 1	2	6	10
0	1.62×10^7	3.28×10^7	3.98×10^7	2.55×10^7
1.0	1.54×10^7	2.80×10^7	3.78×10^7	3.25×10^7
2.0	1.56×10^7	2.63×10^7	3.43×10^7	4.45×10^7
5.0	1.47×10^7	1.67×10^7	2.65×10^7	4.05×10^7
7.5	1.30×10^7	1.29×10^7	1.92×10^7	4.55×10^7
10.0	9.58×10^6	1.35×10^7	1.36×10^7	3.25×10^7
12.5	9.50×10^6	9.68×10^6	1.28×10^7	2.67×10^7
20.0	4.22×10^6	6.95×10^6	3.88×10^6	1.51×10^6
50.0	1.09×10^5	5.05×10^3	0	0

Initial population = 1.27×10^7 bacteria/ml

(ii) Nutrient agar counts

Sodium thiocyanate (g/1)	Bacteria/ml Time incubated (days) 1	2	6	10
0	1.11×10^7	1.48×10^7	3.56×10^7	1.76×10^7
1.0	8.95×10^6	1.31×10^7	2.83×10^7	3.33×10^7
2.0	8.35×10^6	1.22×10^7	2.50×10^7	3.91×10^7
5.0	8.40×10^6	1.22×10^7	2.71×10^7	2.86×10^7
7.5	8.45×10^6	1.16×10^7	2.83×10^7	4.72×10^7
10.0	8.00×10^6	1.18×10^7	2.41×10^7	3.50×10^7
12.5	7.50×10^6	1.04×10^7	1.11×10^7	1.77×10^7
20.0	4.18×10^6	6.15×10^6	3.65×10^6	1.21×10^6
50.0	4.63×10^6	7.05×10^3	0	0

Initial population = 7.00×10^6 bacteria/ml

4.5 Additives to Stretford Plant Solutions

Flotation agents, antifoams and coagulants are often used by plant operators to encourage smooth running. Additives have seldom been subjected to rigorous scientific investigation, but their use, in our opinion, has evolved gradually because of the effects of bacteria whose presence was not suspected. Information has been passed from one operator to the next, and many procedures have become Stretford folklore.

Diesel oil is the most widely-used flotation agent and it has even been specified as a constituent of the washing solution of one wet chemical sulphur removal process. However, diesel oil can be broken down by bacteria and hence encourages bacterial growth. Moreover, during autoclaving, polysulphonates can be formed from diesel oil and these suppress flotation, particularly when the use of diesel is indiscriminate.

Antifoam agents are added to control excessive foaming in the oxidiser, to prevent foaming in the absorber and to avoid solution entrainment in the gas. Thus minute additions of antifoam can improve frothing, although if overdosing occurs, frothing will be prevented altogether.

When bacterial contamination is effectively controlled

additives are unnecessary. In coke oven Stretford units (which do not contain bacteria, see Section 4.4.5), additions to plant solutions are rarely used, and generally operational experience is far smoother than for non-coke oven units.

4.6 To Treat or Not to Treat With Biocide

The simplest way to prevent or alleviate problems caused by bacteria in Stretford plants is to treat the solution with an effective biocide. However, such treatment is not cheap and may not always be necessary. Some Stretford units may contain bacteria, but the consequences of their activity, if not too severe, are coped with as a matter of routine, and are not necessarily recognised as having been caused by bacteria. For example, a steady but higher than average alkali consumption will result in higher operating costs, but the underlying cause of the alkali consumption (bacterial oxidation of thiosulphate to sulphuric acid) may not be identified. The alkali consumption would drop however, if the bacteria in the plant were to be killed.

Alkali consumption is one of the few consequences of bacterial contamination that can be accurately costed. However, the costs of any general mal-operation which may lead to foaming, solution entrainment, blockages, inability to remove H_2S to design specification etc, or possibly to plant shut-down, are rarely quantified but must be considerable. Furthermore the environmental implications can be serious. It is prudent therefore for plant operators to recognise the adverse effects that can be caused by bacteria, so that appropriate action can be taken.

Any biocide treatment should be carefully monitored for at least 10 days, in order to obtain a fair evaluation. Supervised trials of this length have been completely successful. Shorter trials, or those not conducted as recommended are invariably less successful. After a satisfactory initial trial, a further period of experimentation is required in order to define the amount of biocide that should be added on a daily basis. This is done by allowing the biocide concentration to fall until bacteria are re-established. The whole procedure requires attention to detail and must be monitored.

Bacterial infection of a fresh plant solution or re-infection of an apparently sterile one occurs easily. Many of the component vessels are open to the atmosphere, and can be contaminated by dust and soil particles. Make-up water itself often contains a wide variety of bacteria, including thiobacilli. The recommended practice of maintaining the thiosulphate concentration in new solutions above 20 g/l to prevent ADA loss may also lead to difficulties, because in order to achieve the required concentration of thiosulphate, solution is often taken from an operating plant. Although this may seem reasonable practice because the solution in the working plant is to some extent purged, it is unfortunately an excellent way of introducing bacteria from a contaminated plant to a new one. Moreover, even in biocide treated plants it is possible to find bacteria that are still alive embedded in sulphur slimes. These reservoirs of bacteria can then reinfect the plant solution.

The wide diversity in plant design usually means that each plant is unique, and may need treating in a different way. It should be emphasised, of course, that the successful use of a biocide will solve problems caused by bacteria, but will not overcome any operational difficulties of a chemical engineering origin.

5. ECONOMICS OF THE STRETFORD PROCESS

5.1 Application to the Treatment of Gas From High Pressure Coal Gasification Plants

The use of the Stretford Process for high efficiency removal of H_2S from low pressure sour gases is well established, and the Process must be considered one of the leaders in the field.

British Gas is currently examining the application of the Process to the treatment of gas derived from high pressure coal gasification, and preliminary costings show that it compares very favourably with physical absorption processes.

The sensitivity of Stretford Process economics to H_2S partial pressure and operating pressure differs radically from that of physical processes; Stretford costs will rise with each of these as opposed to a physical process where the reverse would be expected. The Stretford Process is therefore well suited to the treatment of gases with a low H_2S partial pressure. Combined cycle applications are particularly attractive, because there is no very stringent requirement for minor component removal, as is the case for processes employing methanation catalysts.

5.2 Microbiological Aspects

The financial implications of any plant mal-operation are difficult to quantify and vary from plant to plant. Factors that need to be costed include: additional consumption of alkali; manpower needed for physical removal of accumulations of sulphur; the overall effects of shut-down/lost production on upstream processes; and the environmental penalties resulting from the impaired Stretford plant performance. These costs need to be compared with those incurred by routinely applying biocide, which calculations show would increase operating costs by approximately 20%.

An alternative approach that could be adopted involves the maintenance of biocidal concentrations of thiocyanate coupled to the Desalting Process. This combination would overcome the necessity for periodic purging to decrease thiosulphate and sulphate concentrations. The Desalting Process could be operated continuously or semi-continuously, and the thiocyanate concentration could be maintained at approximately 50 g/l to prevent the proliferation of bacteria. The additional operating costs of such a modification to non-coke oven units would be approximately 5½%.

6. CONCLUSIONS

1. The Stretford Process is already well established for low pressure H_2S removal. British Gas has now demonstrated the potential for the Process to be applied to the treatment of high pressure gases. The Process has particular advantages for gas

streams with low H_2S partial pressures.

2. For high efficiency of H_2S removal, packed towers are superior to venturi absorbers. However, venturi absorption is an alternative if lower efficiencies are acceptable.
3. The Desalting Process eradicates the need for solution discharge and produces a saleable by-product.
4. Bacteria can and do grow in Stretford plant solutions. The extent to which bacteria interfere with Process operation varies; in some plants operational difficulties may be encountered, whereas in other plants few deleterious effects may be evident.
5. In plants that contain significant numbers of bacteria, and in which operational difficulties have been experienced, treatment with biocide has resulted in improved performance.
6. Preliminary studies have indicated that the Stretford Process may have a role in the treatment of high pressure gases derived from coal.

ACKNOWLEDGEMENTS

This paper is published by permission of the British Gas Corporation. We wish to thank colleagues for their help in the preparation of this paper, particularly Dr. E.S. Pankhurst, Mr. M.W.G. Tanner, and Mr. H. Crompton. We are grateful for the skill and patience of Mrs. J.D. Morris, Mrs. G.V. King and Mrs. J.B. Mills in typing this paper.

REFERENCES

1. Nicklin, T. and B.H. Holland, "Removal of H_2S from coke oven gas by the Stretford Process," paper presented at European Symposium "Cleaning of coke oven gas," Saarbruken (March, 1963).

2. Nicklin, T., F.C. Riesenfeld and R.P. Vaell, "The application of the Stretford Process to the purification of natural gas," paper presented at I.G.U. 12th World Gas Conference, Nice (1973).

3. Kohl, A.L. and F.C. Riesenfeld, Gas Purification, 2nd edition, pp. 432-441, Gulf Publishing Co. Houston, Texas (1974).

4. Moyes, A.J. and J.S. Wilkinson, "The development of the Stretford Process," paper presented at the Control of Gaseous Sulphur Compound Emission, Salford (1973).

5. Ryder, C. and A.V. Smith, "The application of the Stretford Process to the removal of hydrogen sulphide at high pressure," J. Inst. Gas Eng., 3, 283-300 (1963).

6. Smith, C.R. and B. Mills, "Cost effective improvements to the Holmes-Stretford Process," I. Chem. E. Symposium Series No. 57, Salford (1979).

7. Beavon, D.K. and J.C. Brocoff, "Recent advances in the Beavon Sulphur Removal Process and the Stretford Process," paper presented at The Control of Gaseous Sulphur and Nitrogen Compound Emission, Salford (1976).

8. Farrington, F. and P.S. Clough, U.S. Patent 4360508.

9. Mitachi, K., K. Murakami, and J. Sato, "Thermal decomposition/regeneration," Chem. Eng. Progress 79, (4) 56-61 (1981).

10. Cabodi, A.J., H.R. Van, and L.C. Hardison, "First commercial test is success for catalytic hydrogen sulphide oxidation process," Oil Gas J. 80, (27) 107-110 (1982).

lower limit on the clean gas H_2S content.

The chelated iron solution - ARI-310 - has a demonstrated resistance to often troublesome contaminants such as ammonia, carbon dioxide, cyanide and a variety of hydrocarbons.

The autocirculation configuration is applicable to many gas treating problems which have relative high concentrations of H_2S (over about 5 volume percent at atmospheric pressure). This offers both unusual simplicity and low cost, with efficiencies in the 99.9+ percent range (2).

DESCRIPTION OF THE LO-CAT™ PROCESS

LO-CAT is a liquid phase oxidation process which uses a dilute aqueous solution of iron, held in solution by organic chelating agents. The iron oxidizes the hydrogen sulfide ions by removing electrons which reduces the iron. The iron is then circulated to an oxidizer in which it is regenerated by contact with air for reuse. LO-CAT does not remove CO_2, COS, CS_2 or mercaptans; however, it does remove H_2S to any level required, converting it to elemental sulfur which can be sold or discarded safely (4).

The LO-CAT Process has been under development since 1972 and has been commercialized since 1976. During this period, the process has been installed in a variety of industrial applications including refining and natural gas processing plants and for removing H_2S from carbon dioxide used in commercial beverages.

The basic reactions of the LO-CAT™ Process can be divided into Absorption and Regeneration as follows (5):

ABSORPTION:

$$H_2S(Gas) + H_2O(Aqueous) \rightarrow H_2S(Aqueous) + H_2O(Aqueous) \quad (1)$$

FIRST IONIZATION:

$$H_2S(Aqueous) \rightarrow H^+ + HS^- \quad (2)$$

SECOND IONIZATION:

$$HS^- \rightarrow H^+ + S^= \quad (3)$$

OXIDATION BY METAL IONS(Fe^{+++}):

$$S^= + 2Fe^{+++} \rightarrow S^o\ (Solid) + 2Fe^{++} \quad (4)$$

OVERALL REACTION:

$$H_2S(Gas) + 2Fe^{+++} \rightarrow 2H^+ + S^o + 2Fe^{++} \quad (5)$$

REGENERATION:

ABSORPTION:

$$O_2(Gas) + 2H_2O \rightarrow O_2(Aqueous) + 2H_2O \quad (6)$$

REGENERATION OF METAL IONS:

$$1/2\ O_2 + H_2O + 2Fe^{++} \rightarrow 2(OH)^- + 2Fe^{+++} \quad (7)$$

OVERALL REACTION:

$$1/2\ O_2(Gas) + H_2O + 2Fe^{++} \rightarrow 2(OH^-) + 2Fe^{+++} \quad (8)$$

Adding equations (5) and (8) yields the modified Claus reaction:

$$H_2S + 1/2\ O_2 \rightarrow H_2O + S^o \qquad (9)$$

In the overall reaction, the metal ions serve to transport electrons from the absorption side of the reaction to the regeneration side, and it is necessary to supply at least two ferric ions per atom of sulfur produced. In this sense, the ferric ions are a reagent. However, they are not consumed in the overall reaction and serve as a catalyst for the reaction of H_2S and oxygen (3).

In addition to the sulfur-forming reaction, thiosulfate ions are also formed whenever sulfide ions and oxygen are in contact with each other. This can occur in liquid oxidation processes whenever sulfide ions are transferred from the absorber to the oxidizer or when dissolved oxygen is transferred from the oxidizer to the absorber. These byproduct reactions are minimized in the LO-CAT™ Process. Due to the high reaction rates which occur in the LO-CAT absorber, sulfide ions are <u>not</u> present in the solution entering the oxidizer. In addition, the LO-CAT Process operates at very low dissolved oxygen concentrations (1ppm).

Neither ferrous or ferric ions are stable in aqueous solutions at neutral or alkaline pH levels and ordinarily will precipitate as either ferric hydroxide or ferrous hydroxide as follows:

$$Fe^{+++} + 3(OH)^- \rightarrow Fe(OH)3 \downarrow \quad (10)$$

or

$$Fe^{++} + 2(OH)^- \rightarrow Fe(OH)_2 \downarrow \quad (11)$$

Precipitation is prevented by complexing the iron with organic chelates (claw-like compounds) which are capable of holding both forms of iron in solution. Two types of chelates are required to make the solution stable over the entire pH range. Type A chelates, such as Ethylene diamine tetraacetic Acid (EDTA) are powerful chelating agents at low pH levels. Above 8 pH, Type B chelates consisting of polyhydroxylated sugars function effectively, and the combination makes the catalyst stable at any pH (5).

Applications for LO-CAT™

Applications for the LO-CAT process may be generally divided into two types:

- Applications where hydrogen sulfide is present under aerobic conditions.
- Applications where hydrogen sulfide is present under anerobic conditions.

The aerobic form of the process is very simple, and likely to show great economic advantages over competing processes, such as incineration or chemical oxidation of H_2S. However, there is usually less potential for recovering useful byproduct sulfur in these applications.

The process application for which LO-CAT™ was originally developed involves the emission of hydrogen sulfide and carbon disulfide from the Xanthate Process used in the manufacture of cellulose. The basic chemical reactions involved in the generation of the hydrogen sulfide and carbon disulfide are as follows:

Digestion

$$\text{Cellulose} + CS_2 \rightleftarrows \text{Cellulose: } CS_2$$

(Xanthate)

Regeneration

$$\text{Cellulose: } CS_2 + H_2O \rightleftarrows \text{Cellulose} + CS_2 + H_2S$$

Hydrogen sulfide and carbon disulfide are both released as gases from the acidic solution, and carried away from the forming or spinning machines by large volumes of ventilating air.

Very large ventilating air volumes are typical of these applications. Gas flows from 100,000 to several million standard cubic feet per minute are common. Hydrogen sulfide levels may be as low as several parts per million, or as high as several hundred. Carbon disulfide concentrations are usually several times as high as the hydrogen sulfide concentration, but the lower odor threshold associated with the carbon disulfide tends to make this a secondary problem. Figure 1 is a photograph of the large scrubbers used on a Xanthate process plant and Figure 2 is a simplied flow scheme of a LO-CAT™ unit for this application (1).

Sewage treatment plants and sewage lift pump stations have chronic problems with the anerobic generation of hydrogen sulfide in transfer lines and sumps, which contaminates the ventilating air inside the plant or pumping station. The solution for in-plant safety and comfort is to ventilate the enclosure or sump at a high rate, which leads to local odor nuisance problems. Ordinarily, the rate of hydrogen sulfide generation for a lift pump station is relatively low and may be as little as one pound per day. The principal design problem in treating the ventilating air from these stations lies in the assessment of the ventilation rate required for good odor control and safety. Variations in rate are likely to be encountered due to tempo-

Figure 1. Large Xanthate plant H_2S scrubbers.

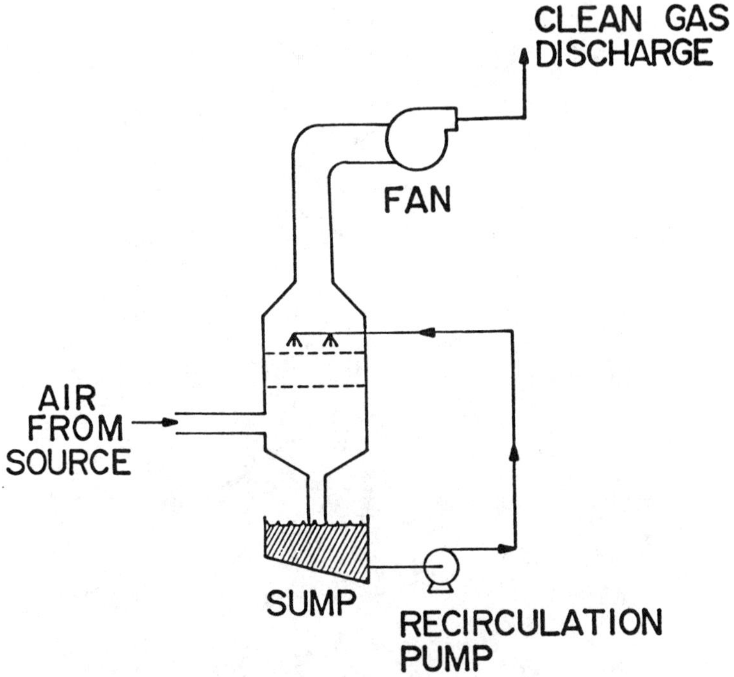

Figure 2. Simple spray chamber absorber.

rary ventilation for safety when workers must enter wet wells or other areas in which hydrogen sulfide generation is normally confined in a small ventilating gas stream.

The control equipment for lift pump station operation is very simple, and the sulfur may be allowed to accumulate within the system for fairly long periods of time before there is any necessity to remove it. Figure 3 is a sketch of a simple lift pump station scrubbing unit. Note the ventilating fan is arranged to keep the station under slight negative pressure. This fan is capable of delivering sufficient air to the scrubber to maintain good ventilation of the station when access doors are left open for worker comfort. The only attention required for such systems involves periodically checking the water makeup control valve (which provides additional water to compensate for evaporation losses into the ventilating air stream), and checking for solid sulfur buildup and catalyst activity loss. These latter items may be attended to on a monthly schedule.

Paper mills have numerous points where atmospheric emissions of hydrocarbon sulfide may be capable of causing a nuisance. These involve too little hydrogen sulfide to warrant modification of the process. Gas volumes for such

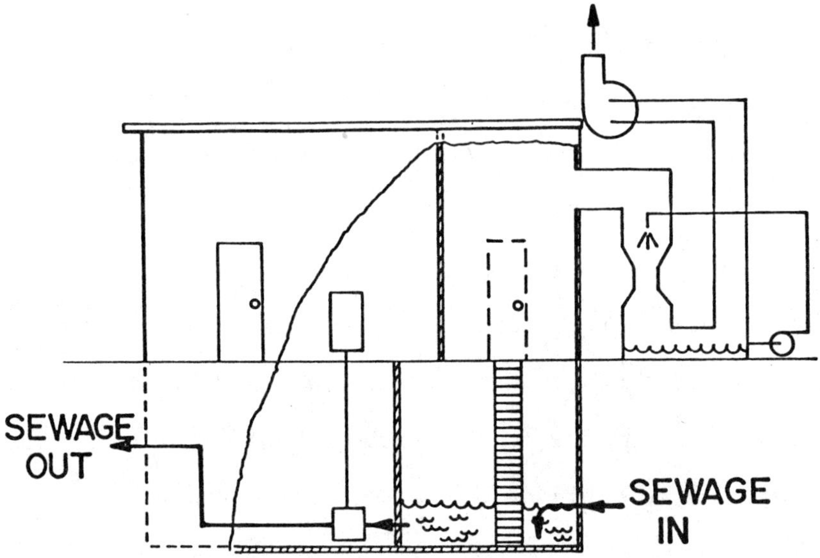

Figure 3. Simple system for lift pump station.

applications are likely to be intermediate between the large Xanthate process gas flows and the small sewage lift pump stations. Sulfur production is likely to be sufficiently large to require mechanical separation of the sulfur to minimize solution losses. Provision for autoclaving of the sulfur in order to produce a molten product for feed to sulfur burner used in making sodium sulfide may be also required.

Petroleum refineries often experience difficulty with hydrogen sulfide emission from tanks which store sour crude oil or from sumps and drains where wash water is used to remove sodium and ammonium sulfide salts from coolers and condensers. Storage tanks and sour water receiver vents also generate sporadic emissions of air contaminated with hydrogen sulfide. These vent streams are irregular in nature and ordinarily occur at low pressure and are not suitable for compression into the refinery fuel system. For these cases a simple LO-CAT™ reactor is an ideal solution to the problem. The water induced venturi scrubber is well suited for small gas flows and may be applied as shown in Figure 4. For very small gas flow rates, a simple non-circulating scrubber provides contact as well as mixing solely by action of the gas flow (1).

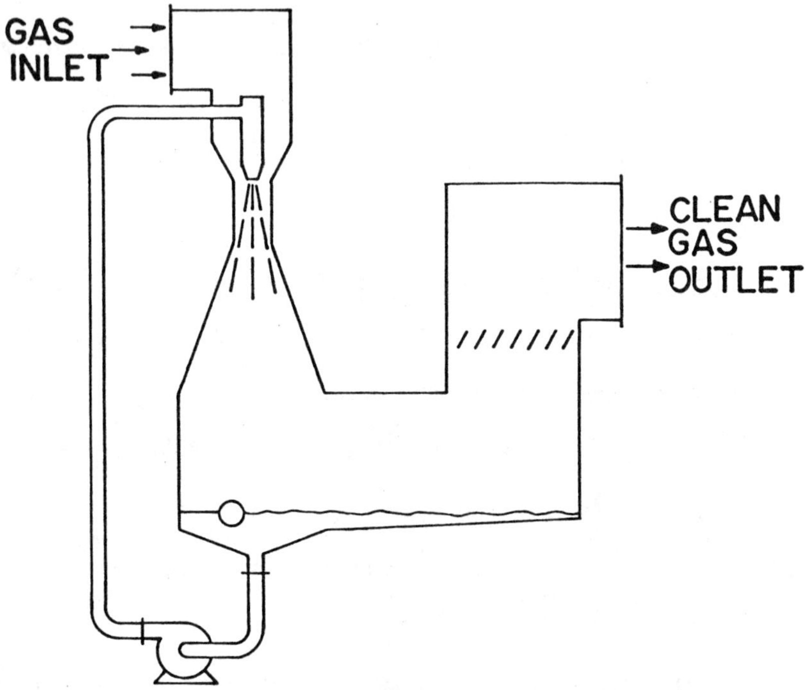

Figure 4. Water induced Venturi absorber.

Hydrogen sulfide is generated under anaerobic process conditions in many energy production processes and in some manufacturing plants. Examples of such processes for which LO-CAT™ has been selected to remove H_2S to meet product purity requirements or to meet exacting air pollution control standards include:

- Natural gas sweetening to pipeline specifications
- Natural gas amine plant acid gas cleanup
- Associated gas treating
- Enhanced oil recovery
- Shale oil retorting
- Synthetic fuel gas production
- Refinery fuel gas treating
- Sour water stripper effluent cleanup
- Claus sulfur plant tail gas cleanup
- Steel mill coke oven gas
- Sulfonation process effluent cleanup
- CO_2 cleanup for beverage use
- Lubricating oil additive manufacture

Natural gas sweetening processes are used to remove H_2S from natural gas in order to meet pipeline specifications, usually 1/4 grain per 100 SCF or about 4.3 ppm. Much of the natural gas produced in the United States is sour, with H_2S levels from a few ppm to as much as 20 volume percent.

LO-CAT™ can be used to sweeten natural gas selectively, removing only the H_2S and not the CO_2 or any other components. A conventional flow scheme is shown in Figure 5. This works well at low pressures, but leads to some transfer of hydrocarbons to the oxidizer at high pressure. Figure 6 is a mechanism for avoiding this, and simplifying the process at high pressure (3).

Alternatively, LO-CAT can be used to treat amine scrubber acid gas which has been removed from wellhead or pipeline gas regardless of the ratio of CO_2 to H_2S. When the CO_2 content is very high, the flow scheme shown in Figure 7 is used. Figure 8 shows the same system but modified to keep oxygen out of the product gas so it can be used for miscible phase EOR, or sold for beverage use.

Figure 9 illustrates a particularly inexpensive arrangement for low CO_2, high H_2S applications to amine regenerator gas (4). This patented autocirculation design:

- Is lower in capital cost than a simple Claus unit

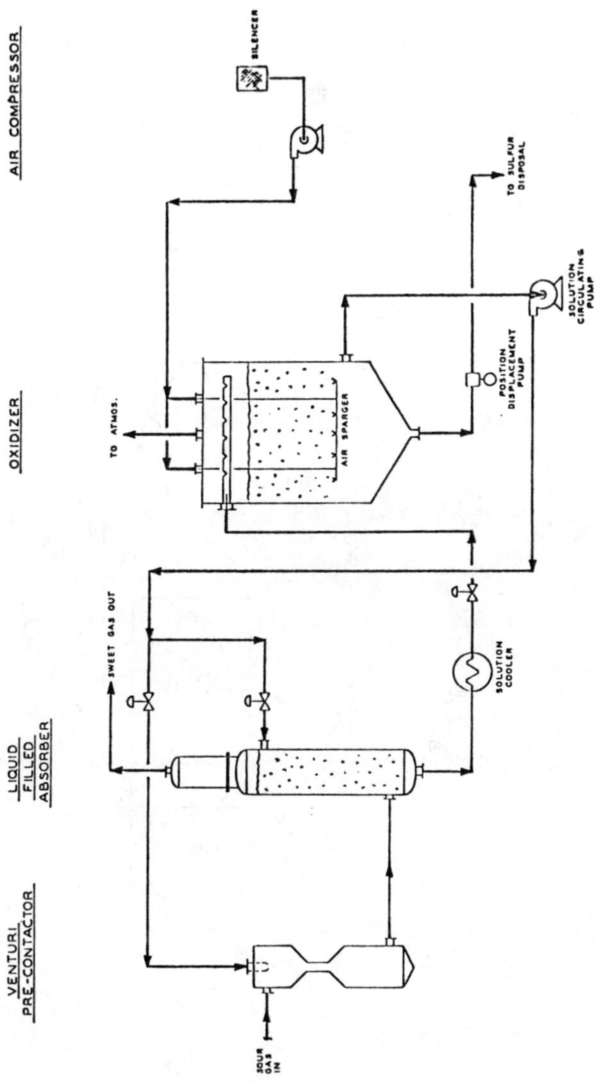

Figure 5. Conventional LO-CAT™ flow scheme.

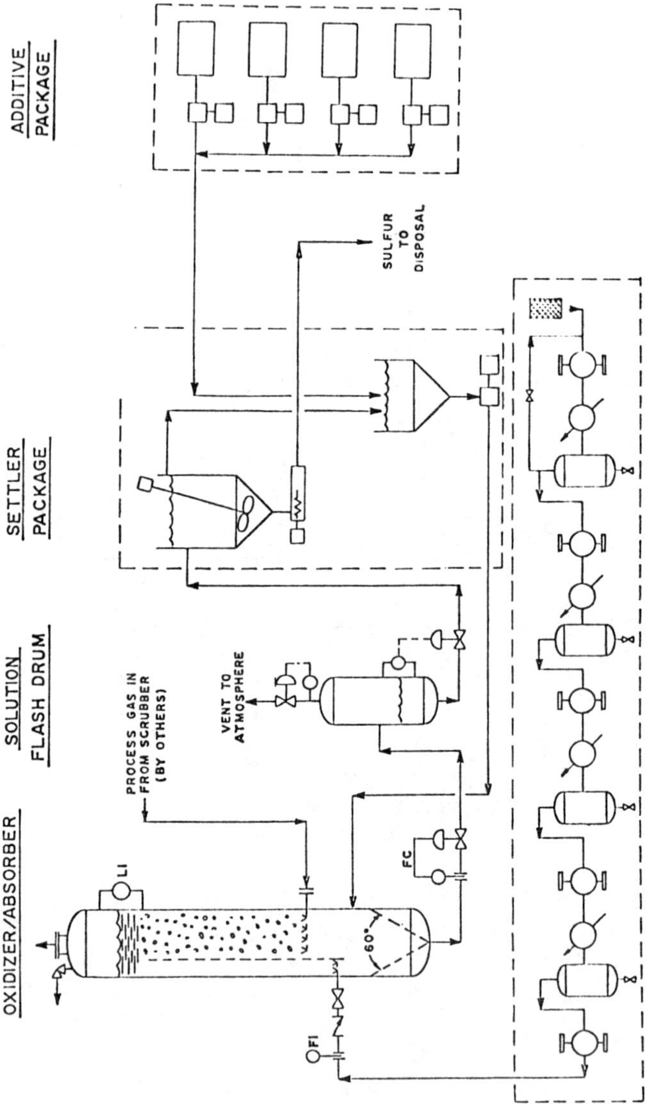

Figure 6. High-pressure autocirculation flow scheme.

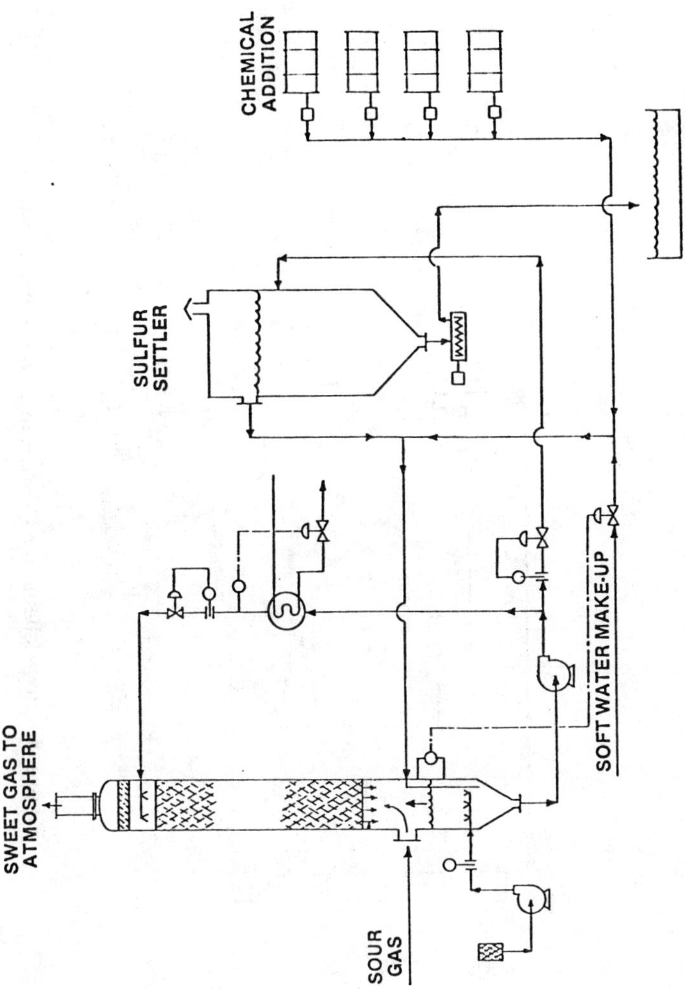

Figure 7. Flow scheme for CO_2 treating unit.

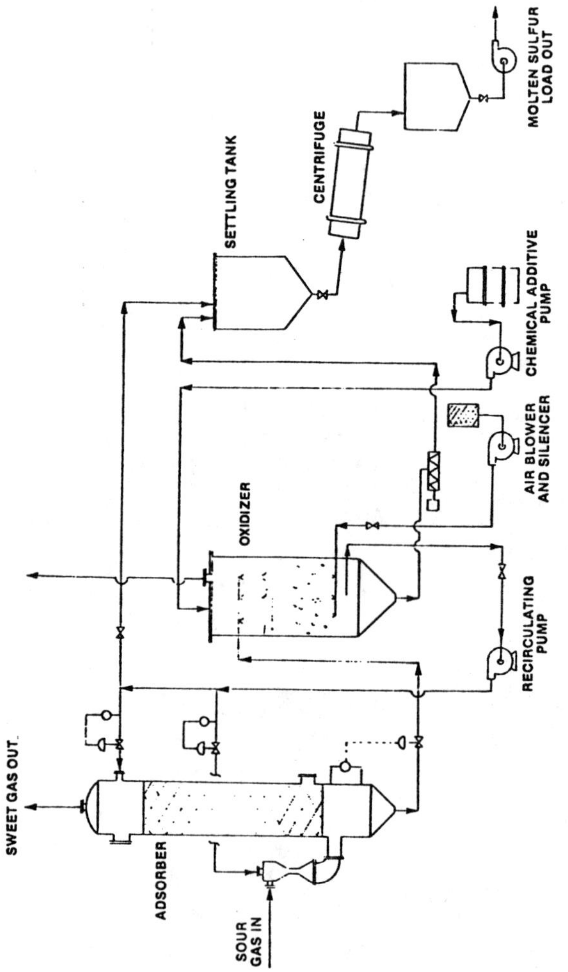

Figure 8. Flow scheme for CO_2 treating with separate oxidizer.

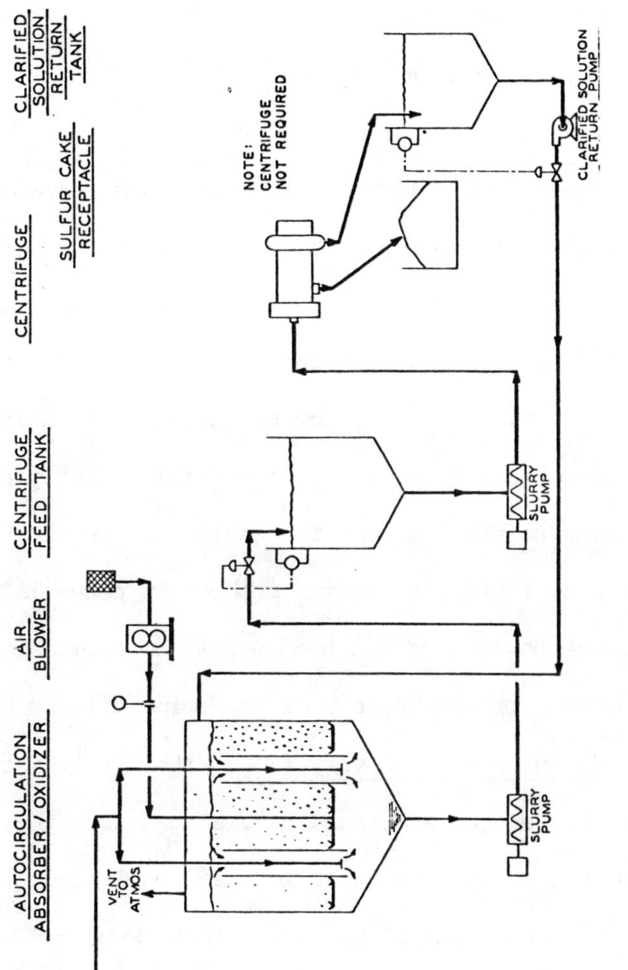

Figure 9. Autocirculation unit flow scheme.

- Provides complete air pollution control without incineration
- Has infinite turndown

These applications of LO-CAT™ to the production of natural gas or to the treatment of associated gas from oil wells has been accelerated by the need for improved control of SO_2 emissions into the atmosphere. Whereas previous gas treating practice involved the removal of H_2S from the gas stream by regenerable amine scrubbing and flaring the acid gas for installations with less than 5 or 10 LT/D of sulfur production potential, good practice now includes a LO-CAT unit to eliminate SO_2 emissions, or replacement of the amine unit by LO-CAT.

Enhanced oil recovery (EOR) with CO_2 is a special case related to associated natural gas treating. CO_2 at supercritical pressures is used to displace liquid hydrocarbons from oil bearing strata. The H_2S need not be removed from CO_2 before injection, but it is often removed for safety or corrosion prevention. However, the gas produced along with the oil must be treated for CO_2 removal, and this presents a unique and challenging problem as the gas flow increases with CO_2 breakthrough.

The flow may increase by a factor of 10 over the life of a project, as the CO_2 content of the gas increases from 20% to 90% (5). This application is ideally suited to the LO-CAT™ Process characteristics.

Shale oil retorting produces gases rich in CO, hydrocarbons, ammonia and H_2S from typical western shales. The anerobic LO-CAT as shown in Figure 8 was selected for the first in situ oil shale retorting operation by Geokinetics, Inc. in Vernal, Utah in 1983.

Refinery full gas treating units have been in operation since 1981. These utilize liquid-filled absorber vessels and atmospheric pressure oxidizers, as shown in Figure 5, and can meet any specification for refinery fuel gas. The LO-CAT approach is particularly advantageous where fuel gas flows or compositions are too variable to permit use of a Claus sulfur plant, or where air pollution emission standards are very stringent (6).

Sour water stripper effluent in oil refineries is difficult to handle by conventional means because of the high ammonia content. The LO-CAT process is tolerant of ammonia and has been used to treat sour water stripper gas since 1981 (6). The autocirculation flow scheme in Figure 9 is useful in this situation.

Claus sulfur plants are used in most applications where amine scrubbers are used to treat gas streams for the removal of 10 LT/D to hundreds of LT/D of sulfur and the ratio of CO_2 to H_2S is not too high. These plants are economical to operate and produce good quality sulfur. However, they are limited in efficiency by chemical equilibrium and the vapor pressure of sulfur to efficiencies in the 95-98% range. LO-CAT™ may be used as an after-treatment for Claus plants after the non-H_2S sulfur-bearing compounds are hydrogenated to H_2S, and the excess moisture is removed by condensation. Overall efficiencies of 99.99% can be achieved with conventional LO-CAT units, and incineration of the Claus tail gas is not required.

Steel mill coke ovens produce a hydrocarbon and CO rich gas which is usually contaminated with substantial amounts of NH_3, HCN and H_2S. The LO-CAT Process can be used to treat the coke oven gas after removal of the tars. In many cases, the ammonia and cyanide can be removed in the LO-CAT along with the H_2S. Large units may require reductive incineration systems to decompose cyanide compounds and prevent buildup to high concentrations in the catalyst solution. The LO-CAT process appears to be more tolerant of both ammonia and HCN than Stretford.

SUMMARY

The LO-CAT™ Process is a relatively new approach to removal of H_2S from gas streams, but it has been demonstrated in a wide variety of industrial processes such as petroleum refining, natural gas production, shale oil retorting, synthetic fuel production, chemical processing and sewage treatment.

Superior catalytic activity and non-toxicity are significant advantages of the catalyst over older processes such as Stretford, Takahax, etc.

Both aerobic and anerobic gas streams can be treated, but a much broader range of anerobic applications have been treated to date.

REFERENCES

1. Hardison, L.C. and L.R. Steenberg, "A New Process for H_2S Odor Control", Presented at The 79th National Meeting of The American Institute of Chemical Engineers, Houston, Texas (March 16-20, 1975).

2. Thompson, Ralph B., U.S. Patent No. 4,189,462, "Catalytic Removal of Hydrogen Sulfide From Gases," (February 19, 1980).

3. Hardison, L.C., "Applications of the LO-CAT™ Process to Sweetening Natural Gas", 34th Annual Gas Conditioning Conference, University of Oklahoma, Norman, Oklahoma (March 5-7, 1984).

4. Hardison, L.C., "Two Ways LO-CAT™ Can Help Gas Producers", South Texas Regional Meeting of the Gas Processors Association (February 23, 1984).

5. Nagl, Gary J., "Applications of the LO-CAT™ Process to Enhanced Oil Recovery", New Orleans Oil Show and Conference (May 3, 1984).

6. Cabodi, A.J., H.R. Van and L.C. Hardison, "First Commercial Test is Success for Catalytic Hydrogen Sulfide Oxidation Process", Oil & Gas Journal, (July 5, 1982).

25

Electrochemical Flue-Gas Desulfurization

Kevin Scott
Jack Winnick
Georgia Institute of Technology

ABSTRACT

An electrochemical membrane cell has previously been found to remove and concentrate the sulfur oxides in simulated flue gas. The high operating temperatures ($\geq 500^{o}C$) needed for this early device make application to conventional power plants difficult. New cell components are described which are being tested for cell operation at signifcantly lower temperatures ($\geq 150^{o}C$) . Preliminary results are presented.

INTRODUCTION

Increased use of fossil fuels in the industrialized nations has resulted in increased emissions of SO_x and NO_x. These gases are transformed into acids in the atmosphere. Data from the eastern United States indcate that between sixty and seventy percent of the acidity is due to sulfuric acid [1].

Present flue gas desulfurization methods generally involve absorption in an aqueous medium. Regenerable systems producing sulfur or sulfuric acid often produce a sludge which must be disposed of. The absorption stage requires that the flue gas be reheated before release to the atmosphere. An alternative which prodcues no sludge and requires no reheat step is a molten salt electrochemical membrane concentration cell*. In an ideal concentrator the reaction at one electrode is the reverse of that at the other electrode. Though there is no net overall reaction, mass can be transfered from a region of low concentration to a region of high concentration by means of the electrochemical potential driving force. A similar process has been used to remove CO_2 from spacecraft cabin atmospheres.

Successful cell operation with a ceramic membrane containing sulfate electrolyte has been obtained at temperatures above 500°C. However, compatability with conventional power plants requires lower operating temperatures. Bisulfate and pyrosulfate electrolytes and

* U.S. Patent 4246081

new ceramic electrodes have been tested, and preliminary results show that these components may be usable as low as 150°C.

THEORY

In this electrochemical SO_x concentrator, application of a small potential across a membrane provides the driving force to selectively remove the SO_x from the contaminated flue gas fed to the cathode compartment.

Molten electrolyte is held by capillary forces in the interstices of the porous ceramic membrane. The anode and cathode, both porous conductive ceramics, are in intimate contact with both faces of the ceramic membrane. During cell operation a small fraction of the liquid electrolyte wets the porous electrodes so as to minimize the interfacial component of the total system free energy. Electrochemical reactions occur at the solid-liquid-gas interface in the electrodes. At the cathode, gas phase diffusion of the reactants from the bulk to these active sites must first occur, and is expected to be current limiting.

Also important is the gas phase equilibrium between

SO_2 and SO_3:

$$SO_2 + \frac{1}{2} O_2 = SO_3 \tag{1}$$

Though SO_3 is thermodynamically favored at temperatures below 700°C, the kinetics of reaction (1) are extremely slow and a catalyst such as Pt or V_2O_5 is required to obtain significant conversion to SO_3.

The overall cathodic reaction in a sulfate melt is

$$SO_3 + \frac{1}{2} O_2 + 2e = SO_4^{2-} \tag{2}$$

The SO_4^{2-} ions then migrate across the membrane to the anode where the reverse reaction takes place:

$$SO_4^{2-} = SO_3 + \frac{1}{2} O_2 + 2e \tag{3}$$

To complete the process, the anodic products diffuse from the electrode pores and are swept away with nitrogen or air, forming a gas stream with a high SO_x concentration.

Cell performance is evaluated by examining the current density (superficial area basis) at the electrodes. This will be dependent on the gas diffusion rates, the overpotential, and the active area available for reaction.

The rate of gas diffusion is a function of the electrode porosity, temperature, and gas flow rate, and can be calculated by standard mass-transfer procedures. The mass flux, which is directly related to the current density, is proportional to the concentration gradient at the electrode. When the reaction is occurring at the maximum possible rate the concentration of reactant at the electrode approaches zero. The current flowing under these conditions is known as the limiting current.

Under equilibrium conditions the elecrode potential is given by the Nernst Equation:

$$E = E^{o'} + \frac{RT}{nF} \ln\left(\frac{[O]}{[R]}\right) \qquad (4)$$

where:

$E^{o'}$ = the formal potential

[O] = the concentration of oxidized species

[R] = the concentration of reduced species

An electrochemical driving force is required to cause deviation from equilibrium, or current flow. This driving force is the overpotential, η, and is given by:

$$\eta = E - E_{eq} \qquad (5)$$

Under conditions where mass transport can be neglected the steady state Force-Flux equation relating the current density to the overpotential is expected to follow the Butler-Volmer equation:

$$i = i_o \, [(\exp\,(\alpha_a \, f\eta) - \exp\,(-\alpha_c \, f\eta)] \qquad (6)$$

where i is the faradaic current density, i_o is the exchange current density (analogous to the reaction rate constant in chemical kinetics), α_a and α_c are the anodic and cathodic transfer coefficients (related to the symmetry of the energy barrier for the electrochemical reaction) and f is given by

$$f = F/RT \qquad (7)$$

where F is Faraday's constant, R is the universal gas constant and T is absolute temperature.

At high overpotentials (above ± 90mV) one of the exponential terms in the Butler-Volmer equation becomes negligible and Equation (6) reduces to:

$$\ln i = \ln i_o - \alpha_c f\eta \tag{8}$$

or

$$\ln i = \ln i_o + \alpha_a f\eta \tag{9}$$

depending on the type of polarization. The intercept of an experimental plot of ln i versus η gives the exchange current i_o while the slope yields either α_a or α_c.

Treatment of porous electrodes is complicated [2] due to the unknown current density distribution. It is difficult to obtain meaningful values for fundamental electrochemical parameters. However, these are crucial for commercial cell design specifications. This information must be gained from experiments involving planar electrodes immersed in free electrolyte. Steady-state experiments and standard transient electrochemical techniques can provide information on the reactions occurring, limiting currents, transfer coefficients and exchange current densities.

A typical commercial cell would utilize porous electrodes with an actual to superficial area ratio of approximately 300. Therefore, prediction of full-cell current densities would require this scale-up factor on those obtained with planar electrodes.

BACKGROUND

Previous work [3] has focused on the ternary eutectic (0.79 Li, 0.12 K, 0.09 $Na)_2 SO_4$ as the electrolyte. This mixture melts at 512°C and is stable with low SO_2 and SO_3 vapor pressures [4].

Salzaro and Newman [5] investigated the cathodic reaction in a sulfate melt under an atmosphere of SO_2 and O_2 and found that the electrode potential did not respond to the SO_2 dirrectly but rather to the SO_3 resulting from the oxidation of SO_2.

$$SO_2 + \frac{1}{2} O_2 = SO_3 \qquad (10)$$

$$SO_3 + \frac{1}{2} O_2 + 2e = SO_4^{2-} \qquad (11)$$

This sequence gives an overall reaction of

$$SO_2 + O_2 + 2e = SO_4^{2-} \qquad (12)$$

The reaction mechanism is unknown though a number of possibilities have been postulated [6-8]. In a sulfate melt the anodic reaction has been shown to be [9,10]

$$SO_4^{2-} = SO_3 + \frac{1}{2}O_2 + 2e \qquad (13)$$

Earlier tests in this laboratory with a small test cell (Figure 1) were done using a sulfate melt. The cathode and anode gases were introduced and removed through alumina tubes cemented into the ceramic housings. Sandwiched between the electrodes was a ceramic membrane across which the transfer of SO_x occurred. The assembled cell was held together inside a custom-made furnace by means of a pneumatic ram.

Typical SO_2 removal efficiencies (percent of the SO_2 to the cathode removed at a given current density and flow rate) for experiments with gold and $Li_2O.9Cr_2O_3$ electrodes [11] are shown in Figure 2.

The cathode gas contains 0.3% SO_2, 3% O_2, 15% CO_2, and the balance N_2. The cell's polarization

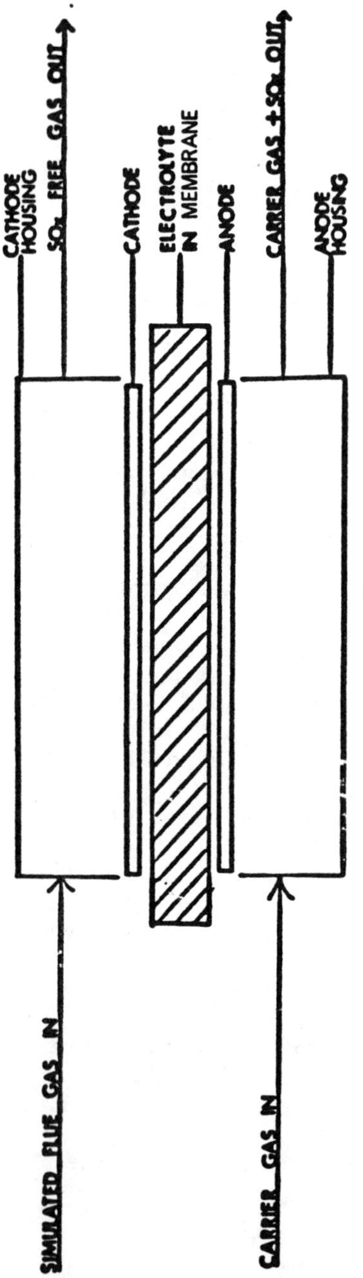

Figure 1. Test cell schematic.

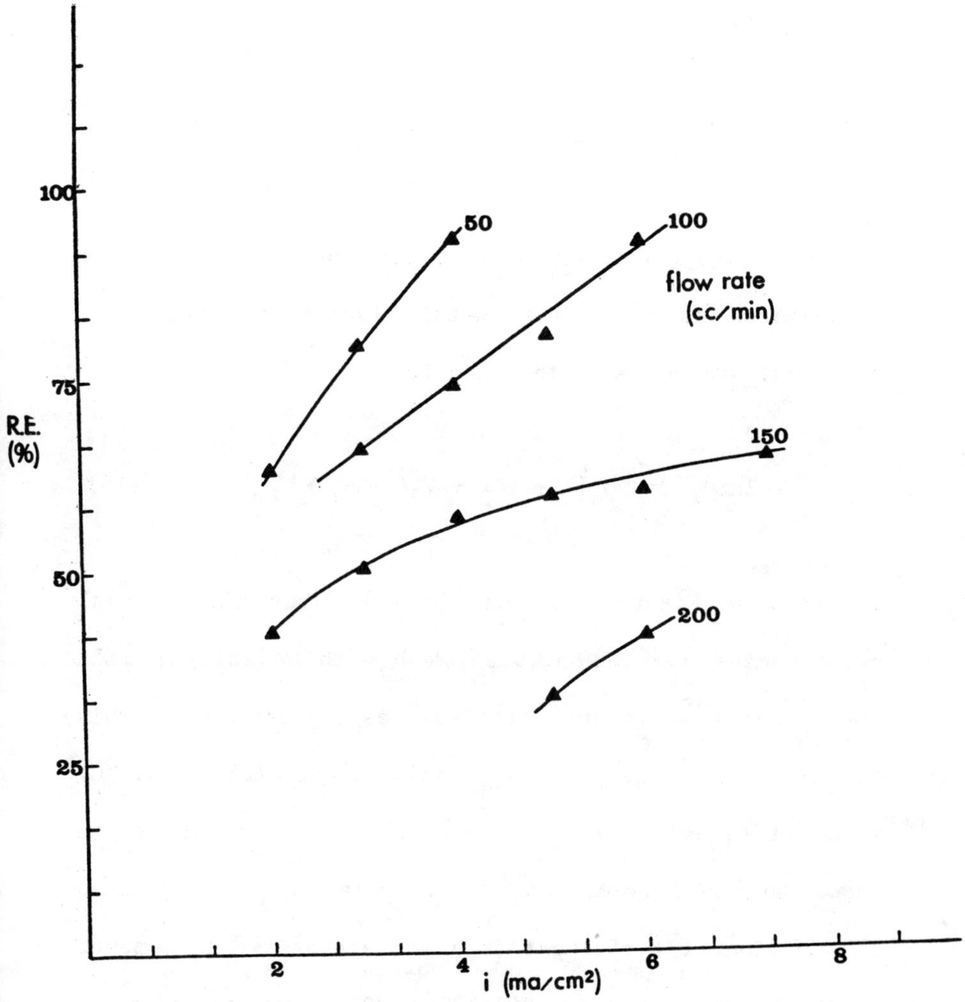

Figure 2. Removal efficiency vs. current density.

characteristics are shown in Figure 3.

The cell is insensitive to carbon dioxide as it is inhibited from reaction due to the equilibrium between sulfate and carbonate which lies far to the right ($K \approx 10^{12}$);

$$CO_3^{2-} + SO_2 + \frac{1}{2} O_2 = SO_4^{2-} + CO_2 \tag{14}$$

Any water vapor present in real flue gas will not affect performance as any alkali metal bisulfates formed will decompose according to the reaction:

$$2HSO_4^{-} = SO_4^{2-} + SO_2 + H_2O + \frac{1}{2} O_2 \tag{15}$$

Though from Figure 2 it can be seen that high removal efficiencies can be accomplished with a sulfate melt there are two major problems associated with this electrolyte. The high operating temperature (>512°C) required to maintain the electrolyte in the molten state leads to rapid corrosion of any metal cell components. In addition, the effluent from the economizer in a power plant is typically at 250-330°C [12] and therefore a

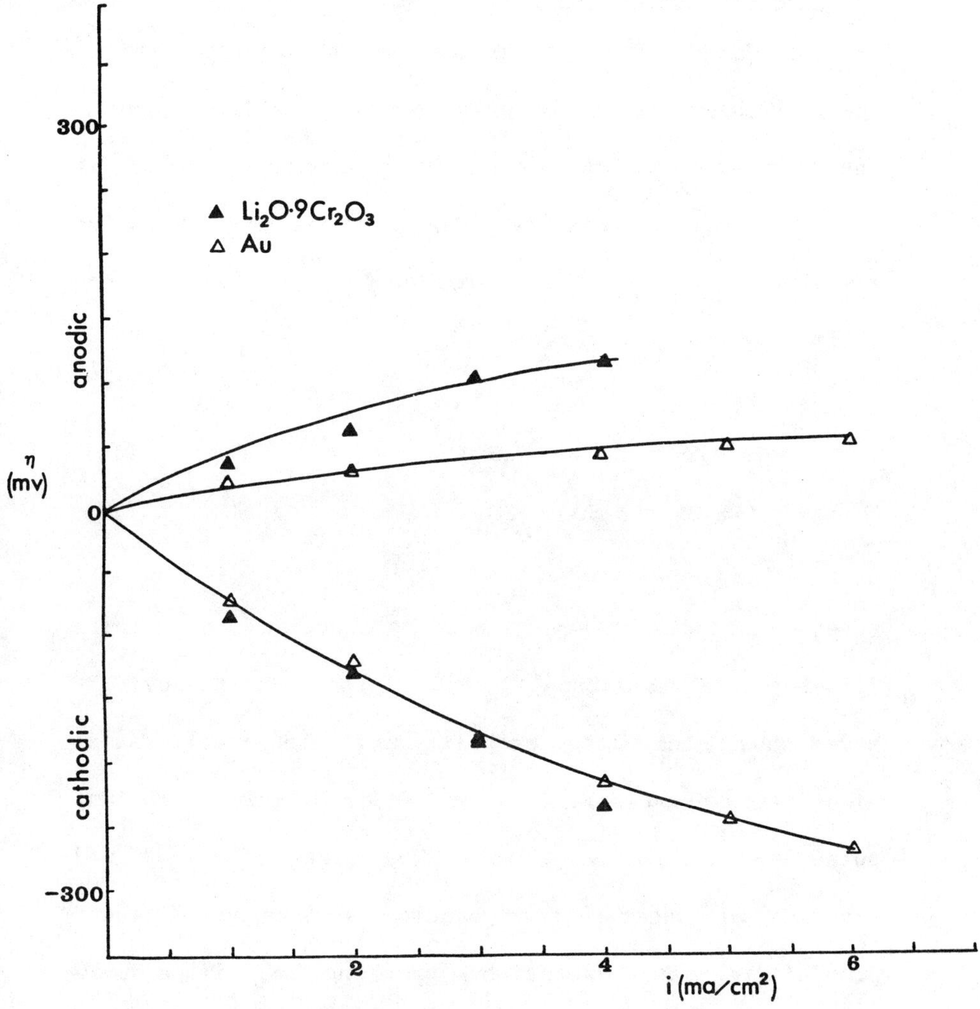

Figure 3. Overpotential vs. current density.

regenerative heat exchange section would be necessary thereby increasing operating costs.

LOWER TEMPERATURE OPERATION

Bisulfate or pyrosulfate melts may provide a feasible lower melting electrolyte alternative. One of the main problems in working with bisulfate systems is their thermal instability. Loss of water and sulfur trioxide lead to the formation of pyrosulfate and ultimately sulfate, thus:

$$2MHSO_4 = M_2S_2O_7 + H_2O \qquad (16)$$

$$M_2S_2O_7 = M_2SO_4 + SO_3 \qquad (17)$$

Several studies have been conducted on these decomposition reactions [13-15]. Spitsyn and Meerov [15] have shown that the stability of the alkali metal bisulfates increases in the order lithium, sodium, potassium, cesium, rubidium. The salts of cesium and rubidium are unattractive candidates because of their prohibitive cost. Recent work by Cheng [16] with a 50-50 mole % Na/K bisulfate melt (m.p $\approx 125^{\circ}C$) indicates that with a hydrated flue gas stream the overall cathodic reaction is

$$2SO_2 + 3/2\ O_2 + 2e = S_2O_7^{2-} \tag{18}$$

The anodic reaction has been studied by Arvia et al. [17,18] and De Tacconi et al. [19,20]. They suggest the following

$$2HSO_4^- = H_2SO_4 + SO_3 + \tfrac{1}{2}\,O_2 + 2e^- \tag{19}$$

though the exact mechanism is unknown.

There is some disagreement in the literature on the order of stability of the alkali metal pyrosulfates. De Vries and Gellings [21] have shown that there is a wide variety of data on the decomposition temperatures of dry, powdered potassium- and sodium-pyrosulfates. From their results it appears that both the initial decomposition temperature and the decomposition range over which the pyrosulfates convert into the normal sulfates depend on the source of the starting material and the heating rates used. In any case, it is generally accepted [22,23] that the thermal stability increases in the order of increasing cation size. Potassium pyrosulfate (m.p. 300°C) is the most attractive because of its availability and low cost.

The exact electrode reactions in a pyrosulfate melt have not been detailed. One possibility is that the reactions are analogous to those occurring in a sulfate melt (12,13), namely

$$2SO_3 + \tfrac{1}{2}O_2 + 2e = S_2O_7^{2-} \qquad (20)$$

at the cathode and the reverse at the anode. However, Fang and Rapp [24] demonstrated that at 900°C in a Na_2SO_4 melt under an atmosphere containing SO_2, SO_3 and O_2 the predominant electrochemical reaction at the cathode is

$$S_2O_7^{2-} + e = SO_4^{2-} + SO_3^{-} \qquad (21)$$

Fundamental studies are required to determine the reactions taking place, exchange current densities, transfer coefficients, and polarization characteristics. With this information and by applying suitable scale-up factors it will be possible to estimate the viability of a commercial cell operating with these lower melting electrolyte alternatives.

EXPERIMENTAL

Free electrolyte studies were conducted on both electrolytes. The stability of each electrolyte was monitored by observing any precipitate formation (analyzed later by x-ray powder diffraction) and any decrease in melt conductivity as each experimental run progressed.

The cell consists of a pyrex glass cell containing planar gold electrodes in free electrolyte (Figure 4). During operation the gases flow through the glass inlet tubes into the melt, saturating it in the vicinity of the respective electrodes, and then exit through the glass exhaust tubes. Electrical connection to the electrodes is produced by gold wires running through the inlet tubing.

The reference electrode is composed of a 0.010 inch gold wire in a Pyrex tube made into a Luggin-Haber capillary tip close to the working electrode. In some experiments a Ag/Ag^+ reference consisting of a silver wire dipping into a nitrate melt contained in a beta-

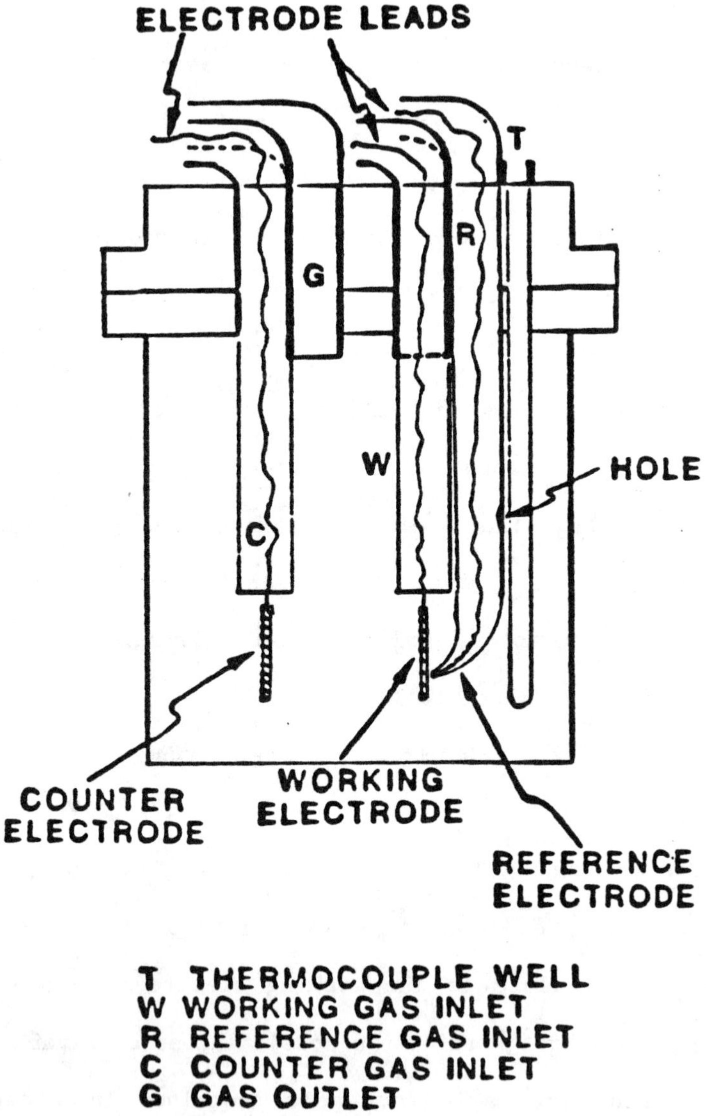

Figure 4. Experimental cell.

alumina tube was utilized.

The entire cell fits into a custom-made furnace and operating temperatures may be controlled accurately.

RESULTS AND DISCUSSION

Experiments with the bisulfate melt at 205°C show that under atmospheres containing H_2O, SO_3, SO_2, O_2 and N_2 the rate of thermal decomposition is slow and only small quantities of pyrosulfate are formed over a period of several days.

Tests with the pyrosulfate melt indicate, as expected, that the decomposition reaction (17) occurs faster with lower SO_3 partial pressures above the melt. Addition of small amounts of V_2O_5 thermally stabilize the melt. In addition the V_2O_5 lowers the electrolyte melting point and probably catalyzes the gas phase oxidation of SO_2 to SO_3. Pyrosulfate melts containing 10% V_2O_5 remain higly conductive for periods in excess of two weeks at 300°C.

Steady state results of ln i vs η are shown in Figure 5 for the bisulfate and pyrosulfate (containing 10% V_2O_5) systems. The working gas is 10% SO_2, 20% O_2, and 70% N_2. The current obtained in the pyrosulfate

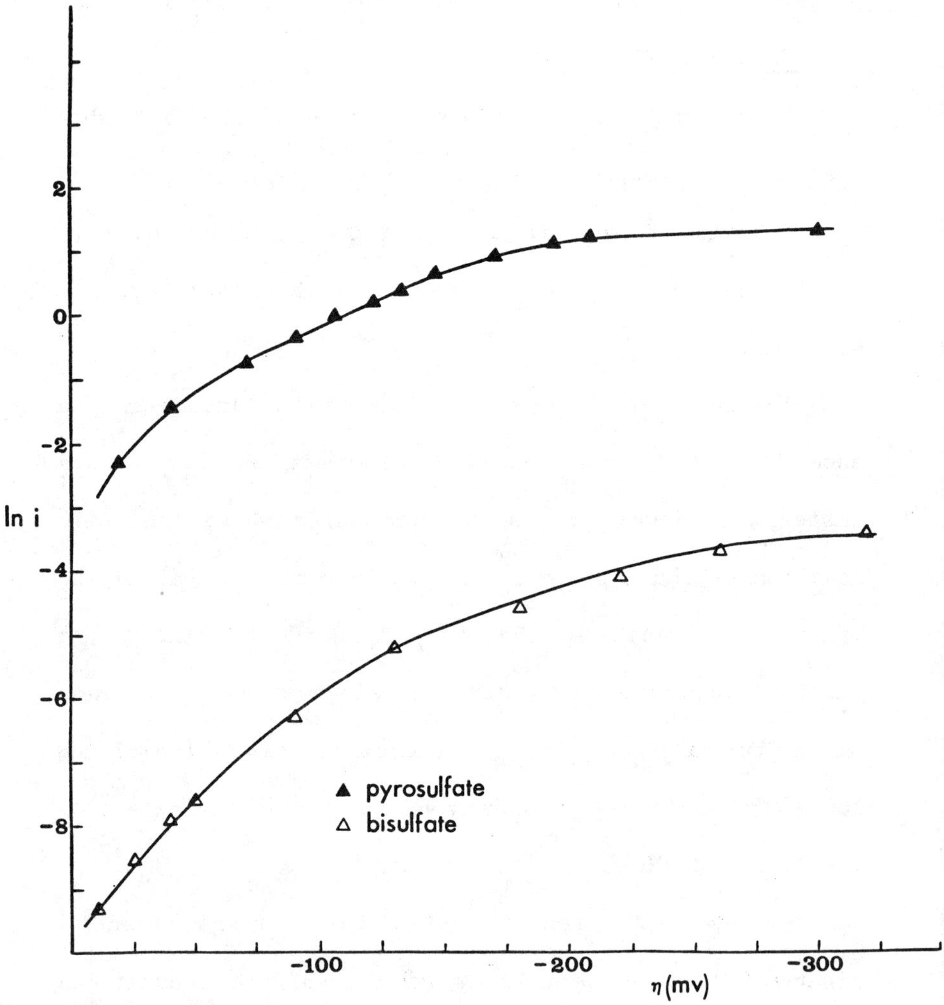

Figure 5. Steady state current density vs. overpotential.

system is much higher at any given overpotential. This is patly due to a temperature effect since the bisulfate is at 205°C while the pyrosulfate is at 325°C. Another reason is that the pyrosulfate might be the electroactive species while the bisulfate itself is not being directly reduced. The results of varying the gas composition to the melts support this theory. With the bisulfate system the current is dependent on the SO_2 partial pressure indicating that this reactant is supplied by the gas phase. In addition the shape of the steady state bisulfate polarization curve is tyical of a diffusion controlled process. On the other hand the current in the pyrosulfate system is practically independent of gas composition. The following scheme for the reactions at the cathode is proposed

$$2SO_3 + 2SO_4^{2-} = 2S_2O_7^{2-} \quad (22)$$

$$S_2O_7^{2-} + e = SO_4^{2-} + SO_3^{-} \quad (23)$$

followed by a further electrochemical reaction

$$SO_3^{-} + e = SO_3^{2-} \quad (24)$$

Since SO_3^{2-} is unstable, further chemical reactions probably take place:

$$SO_3^{2-} = SO_2 + O^{2-} \quad (25)$$

$$SO_2 + 1/2\ O_2 = SO_3 \quad (26)$$

$$SO_3 + O^{2-} = SO_4^{2-} \quad (27)$$

The sum of these reactions gives the overall cathodic reaction:

$$2SO_3 + 1/2O_2 + 2e = S_2O_7^{2-} \quad 28)$$

At steady state the melt would consist of a mixture of sulfate and pyrosulfate. Successful operation would require that the mixture remain molten at the operating temperature employed.

In steady state experiments only part of the applied overpotential drives the charge transfer reaction while part drives the mass transfer process. To eliminate mass transfer effects transient measurements are required. Potential steps of increasing magnitude were applied and the resulting current decay as a function of time was monitored. From these measurements the pure Faradaic current, with no mass transfer efffects, i(0) can be derived. The results for both systems are shown in Figure 6.

The exchange current density, i_o, is obtained from the intercepts of the straight lines in Figure 6. For the pyrosulfate $i_o = 6.7\ mA/cm^2$, while for the bisulfate $i_o = 0.43\ mA/cm^2$. These exchange current densities are not diffusion limited but are related to the charge-transfer reaction.

The average value for the transfer coefficients calculated from the slope of ln i(0) vs η of a number of data sets are shown in Table 1.

Now that values for the transfer coefficients and exchange current densities in each electrolye have been calculated, an estimate of the current at a given overpotential may be obtained by substitution into Equation (6). Currents to be expected in a full cell with porous electrodes will be a factor of 300 greater. The results of these calculations for overpotentials of 0.2 V and 0.5 V are shown in Table 2. Note that these

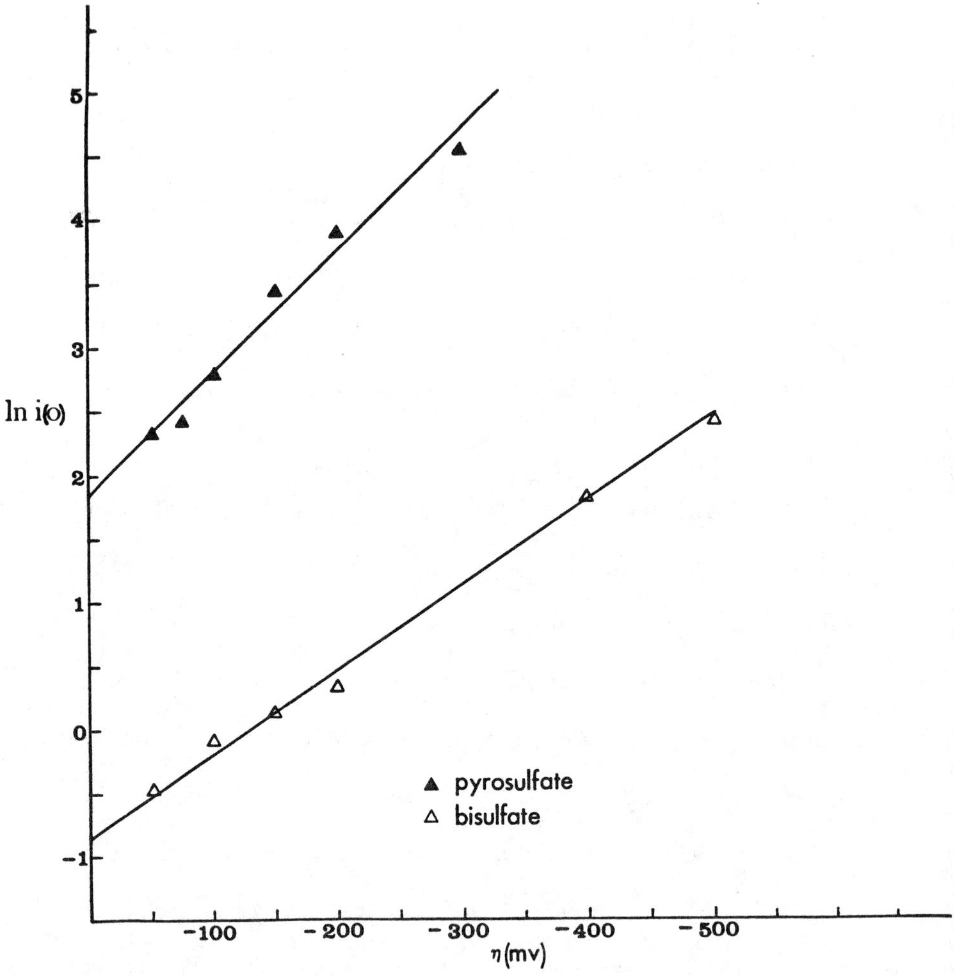

Figure 6. Faradaic current density vs. overpotential.

Table 1
Transfer Coefficients

System	α_c	α_a
Bisulfate	0.4 ± 0.1	0.3 ± 0.1
Pyrosulfate	0.5 ± 0.1	0.4 ± 0.1

Table 2
Predicted Current Densities Based On No Mass Transfer Effects

Overpotential (V)	System	Predicted Current Density (No Mass Transfer Effects) mA/cm^2
0.2	Bisulfate	5.35×10^2
0.2	Pyrosulfate	9.21×10^3
0.5	Bisulfate	4.92×10^3
0.5	Pyrosulfate	9.75×10^4

currents are based on kinetic performance with no mass transfer limitations.

Standard mass transfer methods may be used to calculate the limiting current imposed by transport of SO_x to the cathode of a test cell. The Reynolds analogy for mass transfer can be applied:

$$Re = \frac{2d_w \mu\rho}{\mu} \tag{29}$$

$$1/f^{1/2} = 4 \log (Re \cdot f^{1/2}) - 0.4 \tag{30}$$

$$f = \frac{d_w g_c}{\rho u^2} \frac{\Delta P}{\Delta l} \tag{31}$$

$$St = \frac{k_m}{\mu} = \frac{f}{2} \tag{32}$$

where

Re = Reynolds number

d_w = plate spacing

u = gas velocity

ρ = gas density

f = friction factor

$\frac{\Delta P}{\Delta l}$ = pressure drop per length of plate

St = Stanton number

k_m = gas phase mass transfer coefficient

An estimate for the Reynolds number for the pyrosulfate and bisulfate systems can be calculated from Equation (29) by assuming a gas velocity of 24 m/s and a plate spacing of 0.5 cm. Then the friction factors may be estimated by trial and error from Equation (30). Substitution into Equation (31) yields a pressure drop of approximately 2 inches of water over a 4 foot plate length. Equation (32) is next used to calculate k_m, the gas phase mass transfer coefficient. Assuming that mass transfer control exists in the gas flow channel adjacent to the cathode, the overall mass transfer coefficient, k_{ov}, will be nearly equal to k_m. Therefore the limiting current, i_1, can be predicted using the following equation

$$i_1 = nFk_{ov}C^* \quad \Delta x \tag{33}$$

where

n	=	number of electrons transferred per mole of reaction
F	=	Faradays constant
C^*	=	molar density of the flue gas
Δx	=	log mean difference in mole fraction of SO_2 between the bulk and the electrode surface from the inlet to the outlet.

The result is a limiting current of 29 mA/cm^2 for the bisulfate system and 26 mA/cm^2 for the pyrosulfate. These values are orders of magnitude lower than those predicted by the Butler-Volmer equation. This means that the electrode reaction is a much faster step in the overall process than mass transfer to the cathode. It should be noted that, as far as SO_x removal rates are concerned, current densities of 29 mA/cm^2 and 26 mA/cm^2 in the bisulfate and pyrosulfate systems are equivalent to current densities of 58 mA/cm^2 and 52 mA/cm^2 in the sulfate system. This is because of the number of moles of SO_x removed per mole of electrons (1/2 for sulfate, 1 for bisulfate and pyrosuulfate).

In a commercial unit the current density must be maintained below the mass transfer limit to prevent undesired reactions from occurring. It has been shown [25] that with operating currents of 50 mA/cm^2 the economics of a unit (employing a sulfate electrolyte for a 500 mW plant are extremely favorable. Since our results show that equivalent current densities can be obtained while at potentials under 1.0 V, the bisulfate and pyrosulfate electrolytes appear to be bright prospects as components of the molten salt electrochemical flue gas desulfurization cell.

REFERENCES

1. Glass, N. R., G.E. Glass and P. J. Rennie, "Effects of Acid Precipitation", Environ. Sci. Tech., 13, no. 11, 1350 (1979).

2. Newman, J. S. and C. W. Tobias, J. Electrochem. Soc., 109, 1183 (1962).

3. Townley, D., Ph.D. Thesis, "Concentration of Sulfur Oxides", Georgia Tech (1981).

4. Barin, I., and O. Knacke, "Thermochemical Properties of Inorganic Substances", Springer verlag, Berlin (1973).

5. Salzano, F. J., and L. Newman, J. of the Electrochem. Soc., 119, 1273 (1972).

6. Burrows, B. W. and G. J. Hills, Electrochim. Acta., 15, 495 (1970).

7. Wrench, D. M. and D. Inman, Electrochim. Acta., 12, 1601 (1967).

8. Wrench, D. M. and D. Inmna, J. Electroanal. Chem., 17, 319 (1968).

9. Johnson, K. E. and H. A. Lastiner, J. of the Electrochem. Soc., 110, 314 (1963).

10. Liu, C. H., J. P. Chem., 66, 164 (1962).

11. Townley, D. and J. Winnick, Electrochim. Acta., 28, 389 (1983).

12. Potter, P. J., "Power Plant Theory and Design", The Ronald Press Company, New York (1959).

13. Ingram, M. D., J. A. Duffy and S. M. Forbes, J. of App. Electrochem., 1, 53 (1971).

14. Cambi, L. and G. Bozza, Ann. Chim. Apl., 13, 221 (1923).

15. Spitsyn, V. I. and M. A. Meerov, J. of Gen. Chem. of U.S.S.R., 20, 959 (1947).

16. Cheng, A., Master's Thesis, "The Electrochemical Kinetics of Sulfur Dioxide Reactions in Molten Bisulfates", Georgia Tech (1983).

17. Arvia, A. J., A. J. Calandra and H. A. Videla, Electrochimica Acta, 10, 33 (1965).

18. Arvia, A. J., A. J. Calandra and H. A. Videla, Electrochimica Acta, 13, 167 (1968).

19. De Tacconi, N. R., A. J. Calandra and A. J. Arvia, J. Electroanal. Chem., 51, 25 (1974).

20. De Tacconi, N. R., A. J. Calandra and A. J. Arvia, J. Electroanal. Chem., 57, 325 (1974).

21. De Vries, K. J. and P. J. Gellings, J. of Inorganic Chem., 31, 1307 (1969).

22. Flood, H. and T. Forland, Acta Chemica Scandinavia, 1, 781 (1947).

23. Spitsyn, V. I. and M. A. Meerov, J. of Gen. Chem. of U.S.S.R., 22, 963 (1952).

24. Fang, W. C. and R. A. Rapp, J. Electrochem. Soc., 130, 2335 (1983).

25. Townley, D. and J. Winnick, I. and E. C. Proc. Des. and Dev., 20, 435 (1981).

26

The Chevron WWT Process—An Economical Way to Treat Sour Water from Refineries and Synthetic Fuel Plants

James P. Leonard
Nicholas J. Haritatos
Process Design Division,
Chevron Research Company

David V. Law
Patent, Trademark & Licensing Division,
Chevron Research Company

In recent years many oil companies have installed refinery facilities that allow them to process cheaper, heavier crudes which contain large amounts of contaminants such as sulfur, nitrogen, and metals. To handle these heavy crudes while meeting the demand for light products requires more treating and conversion facilities. These facilities will generate large amounts of sour water, containing hydrogen sulfide and ammonia, that must be treated before being discharged. One way of handling this water is to strip out the hydrogen sulfide and ammonia and feed the gas stream to a Claus Sulfur Recovery Unit (SRU). However, ammonia fed to a SRU can cause operating problems such as catalyst deactivation, equipment plugging, and lower sulfur recoveries. In addition, the investment costs for sulfur and tail gas units are higher if they are designed to handle NH_3 in the feed. So it is advantageous to the refiner if the ammonia and hydrogen sulfide in the sour water are recovered separately.

Chevron has a commercially proven process for economically separating and recovering ammonia and hydrogen sulfide from refinery sour water. This is the Chevron Waste Water Treating (WWT) Process. In addition to refinery sour water, the WWT process can also easily handle the high ammonia sour waters that will be produced in shale oil and tar sands upgrading plants. It is also capable of treating the sour water from shale oil retorting plants. This water will typically contain high amounts of carbon dioxide in addition to ammonia and hydrogen sulfide.

The WWT process separates and recovers ammonia and acid gas (hydrogen sulfide and carbon dioxide) while producing a stripped water suitable for reuse or discharge. The ammonia can be produced as a saleable by-product in either the anhydrous or aqueous form. The acid gas produced is suitable for feeding to a conventional SRU or a sulfuric acid plant. Removing ammonia from the hydrogen sulfide eliminates the many operating problems common to ammonia burning SRU plants.

Stripped water from WWT plants has been successfully reused as injection water in ISOCRACKING plants (the Chevron licensed hydrocracking process). One refinery has used WWT water as injection water for over 12 years without ever having a plant shutdown due to failure of the WWT plant. WWT stripped water has also been successfully reused as crude unit desalter water, cooling tower makeup, and process wash water. The reuse of WWT water is particularly attractive because it reduces net makeup water requirements and can be an important step toward reducing waste water discharge rates to treating facilities and to receiving waters.

WWT FEEDS

WWT plant feeds consist primarily of ammonia and hydrogen sulfide dissolved in water, with traces of carbon dioxide, phenol, cyanides, and other contaminants present. WWT plants have been typically designed for sour water with ammonia concentrations between 0.2 wt % and 4.5 wt % and hydrogen sulfide concentrations between

0.3 wt % and 8.0 wt %. However, the process is not limited to feeds in this composition range; it can handle water with high ratios of ammonia to hydrogen sulfide and water with high carbon dioxide content.

WWT units have successfully processed sour water from the following refinery sources:

1. Chevron Isocrackers and Hydrotreaters
2. Other Hydroprocessing Units
3. FCC Units
4. Cokers - Both Delayed and Fluid
5. Crude Units
6. Amine Plants
7. Flare Seals

PROCESS DESCRIPTION

The Chevron WWT process consists of four main processing steps:

1. Degassing and feed storage.
2. Acid gas stripping.
3. Ammonia stripping.
4. Ammonia purification and liquefaction.

Figure 1 shows a simplified WWT flow diagram.

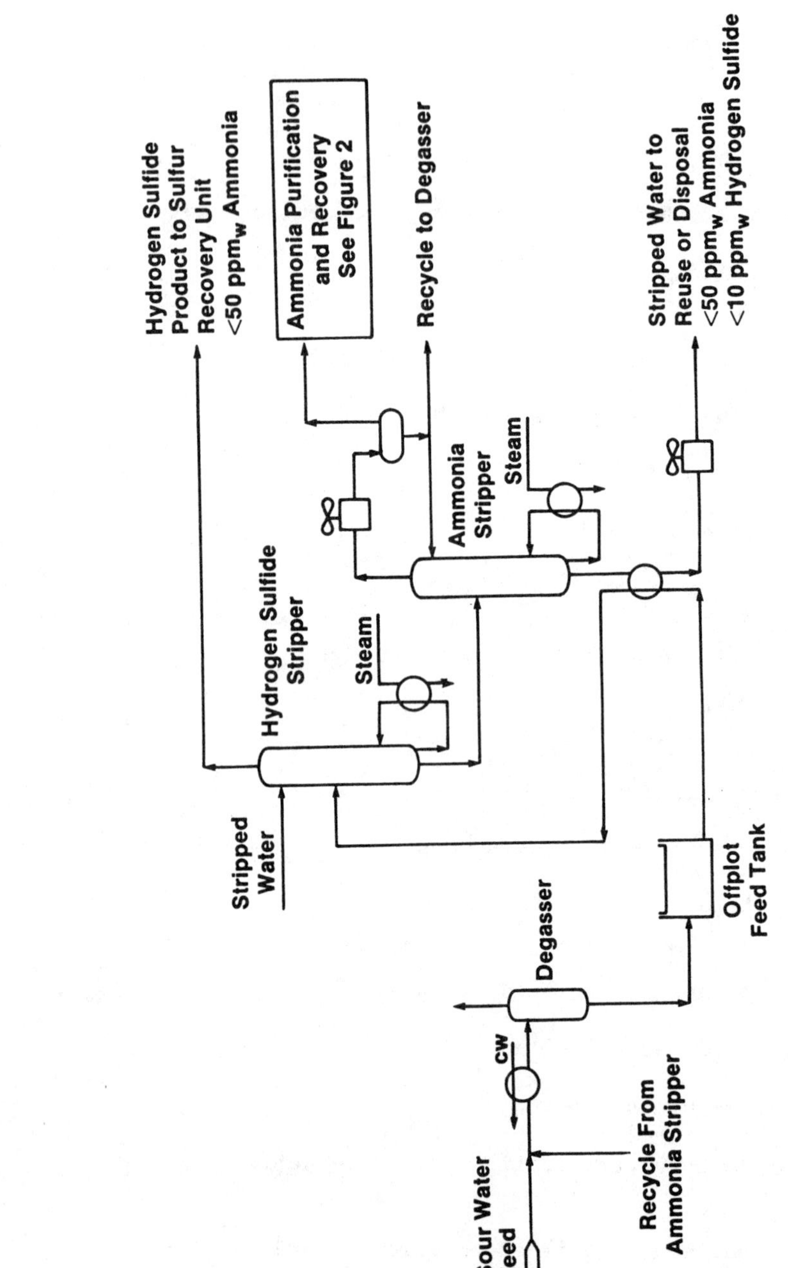

Figure 1. The Chevron WWT Process.

The sour water feed to the plant is combined with a recycle stream from the ammonia stripper, cooled and fed to a degasser where dissolved hydrogen, methane, and other light hydrocarbons are removed. The recycle stream is rich in ammonia, which helps keep acid gases in solution in the degasser. Thus, the release of acid gas and possible air pollution are minimized. The degassed sour water is pumped to an offplot storage tank which serves to dampen flow rate and composition changes. It also provides the opportunity to remove entrained oil and solids.

From the feed tank, the degassed sour water feed is pumped to the WWT unit, where it is heated by feed-bottoms exchange and fed to the acid gas or hydrogen sulfide stripper. This stripper is a steam-reboiled distillation column. The hydrogen sulfide, which is stripped overhead, is of high purity--an excellent feed for a sulfur or sulfuric acid plant. It contains negligible ammonia, less than 50 ppm, and very little hydrocarbon since the plant feed has been degassed. However, it does contain any carbon dioxide that is present in the feed. The hydrogen sulfide is available at about 100 psig and 100°F and will be saturated with water vapor. Typical product specifications are shown in Table 1.

The bottoms from the hydrogen sulfide stripper, containing all the ammonia in the feed and some hydrogen sulfide, is fed directly to the ammonia stripper. In this column, which is a steam-reboiled refluxed distillation column, essentially all the ammonia and hydrogen sulfide are stripped from the water. The water leaves as the column bottoms stream. After exchanging heat with the feed to the hydrogen sulfide stripper, this stripped water is cooled and sent offplot for reuse or treating. The stripped water contains less than 50 ppm of free ammonia and less than 10 ppm of free hydrogen sulfide. The stripped water will also contain traces of phenols and salts which entered with the feed. If the feed contains acidic compounds that "fix" ammonia, the fixed ammonia can be released and then stripped off.

The ammonia and hydrogen sulfide stripped from the water in the ammonia stripper are passed through an overhead condenser and are partially condensed. Most of the liquid is used as column reflux and a portion is recycled to the degasser and offplot feed tank. The vapor product

Table 1
Typical Product Inspections
Chevron Waste Water Treating Process

	Hydrogen Sulfide (Acid Gas)	Stripped Water	Anhydrous Ammonia
Form	Gas	Liquid	Liquid
Ammonia, ppm	<50	<50	-
Hydrogen Sulfide, ppm	-	<10	<5
Water, Wt %	0.1	-	0.4
Temperature, °C (°F)	38 (100)	38-93 (100-200)[1]	38 (100)
Pressure, kPa (psia)	172-791 (25-115)[1]	-[1]	1479 (215)

[1]As required.

is an ammonia-rich gas, which may be handled in a variety of ways, as shown in Figure 2.

For small plants where ammonia recovery is not desired or economic, the overhead product may be incinerated, either directly off the reflux drum or after being scrubbed with water to reduce the hydrogen sulfide content to 1500 ppm or less. However, in most cases, the desirable and economic choice is to purify the gas and produce either anhydrous or aqueous ammonia suitable for sales or further processing.

For production of anhydrous ammonia, the gas is passed through a two-stage scrubbing system to remove hydrogen sulfide and is then liquefied to produce the anhydrous ammonia. The hydrogen sulfide content of the ammonia is typically less than 5 ppm. For production of aqueous ammonia, a one- or two-stage scrubber may be used to remove the hydrogen sulfide; and the ammonia gas is then dissolved in water to yield the desired product grade.

Most plants that produce liquid anhydrous ammonia do it by compressing the ammonia vapor to 1479 kPa (215 psia) and then condensing it with cooling water. However, some recent WWT plants have used ammonia

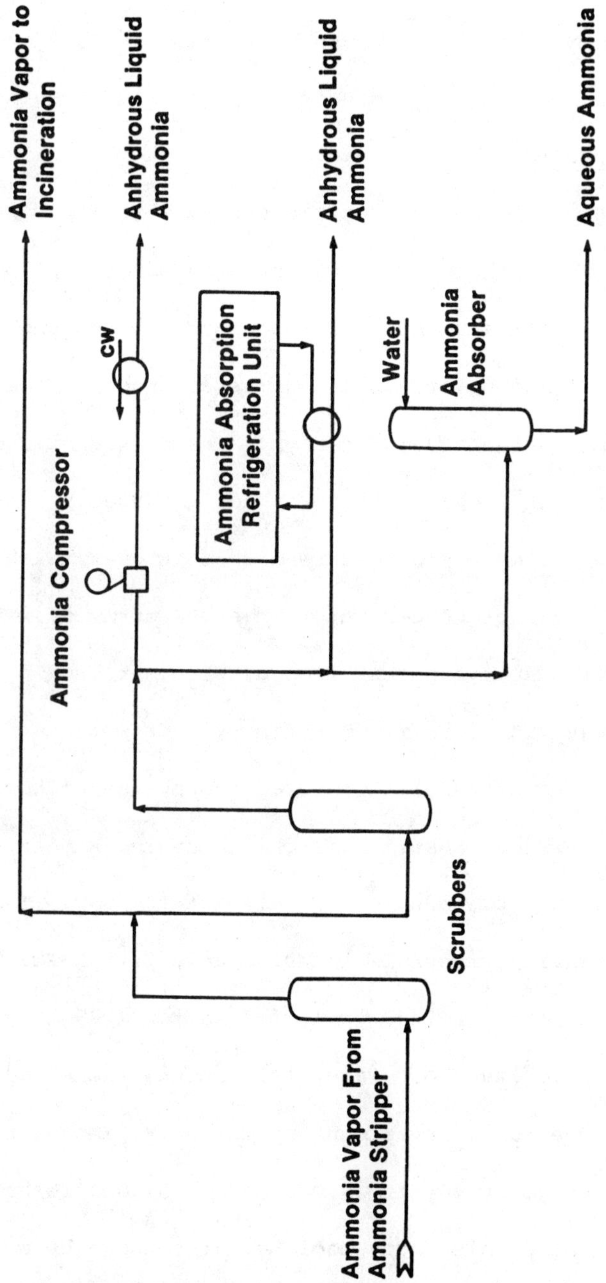

Figure 2. The Chevron WWT Process—ammonia recovery options.

absorption refrigeration units to condense the ammonia. By utilizing waste heat, these units reduce plant operating costs. They also eliminate compressor maintenance.

WWT FLEXIBILITY

The WWT process is flexible and versatile. For instance, where there are large volumes of dilute sour water, it is often economic to use the three-column WWT process. This consists of a sour water preconcentrator followed by a standard WWT design. The dilute streams are sent to the preconcentrator which produces a concentrated sour water overhead stream that is sent to the WWT feed tank. The preconcentrator produces stripped water of the same quality as that from a WWT plant.

In many cases, existing refinery sour water strippers (SWS) can be easily converted to sour water concentrators. This can result in significant investment and utility savings by reducing the size of the WWT. Table 2 shows the savings possible in using an existing SWS to concentrate 91 m^3/hr (400 gpm) of feed to 22.75 m^3/hr (100 gpm). For the sour water feed shown, the effect is to reduce the total investment by 12% or $1 MM and reduce the steam usage by 40% or 16,000 kg/hr (35,000 lb/hr). Preconcentration also cuts cooling water usage by more

Table 2
Investment and Utility Estimates
Preconcentration of WWT Feed

Sour Water Feed Flow Rate, m^3/Hr (gpm)	91 (400)	
	Weight Fraction	Metric Tons/Day
Ammonia	0.011	24
Hydrogen Sulfide	0.022	48
	Stand-Alone WWT	Preconcentrator Followed by WWT
WWT Size, m^3/Hr (gpm) of Feed	91 (400)	22.75 (100)
Investment, $MM[1]	8.4	7.4[2]
50 psig Steam, kg/Hr (Lb/Hr)	-	12,200 (27,000)
150 psig Steam, kg/Hr (Lb/Hr)	40,400 (89,000)	12,200 (27,000)
Cooling Water, m^3/Hr (gpm)	1,300 (5,700)	450 (2,000)
Power, kwh	310	250

[1] Onplot investment only, Third Quarter 1984, U.S. West Coast site.

[2] Investment is for reuse of an existing 75 m^3/hr (330 gpm) sour water stripper and its heat exchangers as a preconcentrator, followed by a new WWT plant.

than half. The three-column WWT process using existing SWS columns has been used successfully in all three of Chevron's WWT plants as well as in several licensee plants.

The WWT process has the flexibility to be implemented in two phases, which in certain situations may be desired. This can be achieved by installing only the ammonia stripper column of the WWT and operating it initially as a SWS. Then, when the situation and economics dictate, it can be integrated into a WWT plant. The design of the WWT plant is completed by addition of a hydrogen sulfide stripper, appropriate ammonia purification equipment, and other auxiliary equipment.

WHY BUILD A WWT?

The two major reasons for building a WWT instead of a SWS followed by an ammonia-burning SRU are the economics of ammonia recovery and improved SRU operation.

The economics of the WWT process are strongly dependent on the amount of ammonia present in the feed. The more ammonia in the sour water, the more incentive there is to recover it as a valuable by-product. Table 3 shows the economics of a WWT that was built recently. In this case it was more economic to recover the ammonia in a WWT

Table 3
Economic Comparison - WWT Versus Sour Water Stripper

Sour Water Rate, m^3/Hr (gpm)	65 (285)	
	Weight Fraction	Metric Tons/Day
Ammonia	0.031	47
Hydrogen Sulfide	0.062	94
	WWT	SWS
Investment, $[1]	9,000,000	10,000,000
Net Operating Costs, $/Yr[2]		
Utilities[3]	2,900,000	1,500,000
Steam Credit[4]	-	(1,400,000)
Maintenance[5]	255,000	285,000
Taxes, Insurance, Labor, Overhead[6]	495,000	545,000
Ammonia Credit at $165/ Metric Ton	(2,600,000)	-
Total, $/Yr	1,050,000	930,000

[1]Onplot investment only, Third Quarter 1984, U.S. West Coast site. SWS investment includes the incremental investment for the sulfur and tail gas plants to enable them to handle ammonia.
[2]90% operating factor. SWS operating costs include estimates for the incremental SRU operating costs due to the ammonia in the SRU feed.
[3]Steam at $4.75/455 kg (1000 lb), power at 5.5¢/kwh, cooling water at 15¢/3.8 m^3 (1000 gal.).
[4]Credit is for the extra steam produced in the SRU due to the combustion of the ammonia.
[5]Maintenance costs are estimates based on 3% of the investment per year.
[6]Taxes, insurance, and overhead based on 5% of the investment per year. Labor is half a shift position.

than to strip it in a SWS and then burn it in an ammonia-burning SRU. The one-million dollar lower investment for the WWT case compared to the SWS case more than makes up for the slightly lower operating costs for the SWS.

However, even in plants where the economics of ammonia recovery do not appear favorable, the WWT process can still be justified based on improved SRU operation. Separation of ammonia from hydrogen sulfide in the WWT has the following desirable effects on the SRU:

1. Eliminates ammonia-related problems in SRU's such as:
    a. Lower operating factor.
    b. Plugging in condensers and seals caused by ammonia salts.
    c. More rapid catalyst deactivation.
2. Increases overall sulfur recoveries.
3. Reduces the size of SRU's and tail gas units because ammonia and the air needed to burn it act as diluents.

Because of these effects, a WWT plant is an excellent way to debottleneck an existing SRU by removing the ammonia from the acid gas feed.

Reference 1 shows an example of these effects for a typical refinery SRU that gets 70% of its acid gas feed from an amine plant with the rest coming from a SWS. The gas from the SWS contains 24 metric tons/day of ammonia. Removing the ammonia from the SRU reduces the SRU feed and combustion air flow rates by about 45%, thus reducing the size and cost of the SRU. The tail gas flow rate from the SRU is reduced by about 55%. Removing the ammonia diluent from the SRU can also increase the sulfur recovery by as much as 2%.

OPERATING EXPERIENCE

The WWT process and its product specifications have been commercially proven in over 100 plant years of experience. Table 4 gives a list of the 14 operating WWT plants ranging in capacity from 4.5 m^3/hr (20 gpm) to 143 m^3/hr (630 gpm). In the last year alone, three new WWT units have come onstream. Figure 3 is a picture of a typical WWT plant.

Chevron Research provides a variety of services to WWT process licensees such as preparation of the process design, review of the contractor's mechanical engineering package, preparation of process manuals, and assistance with operator training and startup.

Table 4
Chevron WWT Plants

	Location	Capacity, gpm	Startup Date
Onstream			
Chevron	Richmond, CA	230	1966
	El Segundo, CA	260	1970
	Pascagoula, MS	285	1983
	Richmond, CA	160	1984
Fuji Oil Company	Sodegaura, Japan	23	1968
	Sodegaura, Japan	70	1976
Kuwait National Petroleum	Shuaiba, Kuwait	140	1968
Nippon Petroleum Refinery	Negishi, Japan	93	1969
Idemitsu Kosan	Himeji, Japan	125	1970
Mobil Oil Corporation	Beaumont, TX	630	1975
Petro-Canada	Montreal, Canada	250	1978
Syncrude Canada	Alberta, Canada	300	1978
Getty Refining	Delaware City, DE	97	1979
TOSCO Corporation	Martinez, CA	280	1983
Engineering and Construction			
Unannounced	Middle East	175	1988

Figure 3. The waste water treating plant at Chevron's El Segundo, California, refinery.

SUMMARY

The Chevron WWT process is being used to economically treat sour water produced in oil refineries. It is a proven process, with 14 plants in operation. Products from a WWT plant are stripped water which is suitable for reuse or discharge, a high quality hydrogen sulfide stream suitable for feeding a SRU, and an anhydrous ammonia stream which can be sold. One particular advantage of the WWT process is that it removes ammonia from the hydrogen sulfide fed to a SRU. This improves the operation of the SRU because it eliminates plugging, reduces the deactivation rate of the catalyst, and increases sulfur recovery in the plant. The WWT process is also very versatile. It is possible to build a SWS first and then use this column in a future WWT.

REFERENCE

1. Goar, B. Gene, Hydrocarbon Processing, 129, (July, 1974).

27

Sour Water Equilibria Extended to High-Temperatures and with Inerts Present

Grant M. Wilson
President, Wiltec Research Co., Inc.

Paul C. Gillespie
Project Leader, Wiltec Research Co., Inc.

Jonathan L. Owens
Project Leader, Wiltec Research Co., Inc.

INTRODUCTION

Undesirable sulfur, nitrogen, and oxygen compounds are often encountered in commercial gas production and treating facilities processing either natural or substitute gas. Water also occurs as condensate in the gas streams or water is brought in contact with these gas streams in various process steps. As a result, aqueous waste streams are produced in which undesirable compounds are present. The concentrations of these compounds in the aqueous waste streams depend on their equilibrium concentrations in gas streams from which the aqueous streams are derived.

Present and future environmental control restrictions dictate that the concentrations of these undesirable compounds be reduced to very low levels before these streams exit the plant boundary. This implies that methods must be developed to control the concentrations of the undesirable compounds. One method currently used in refineries is a steam stripper where these trace compounds are steam distilled and then concentrated. Similar processes will be necessary in new and existing gas production and treating facilities.

In order to assist in the design of these processes new equilibrium data have been measured for sour water systems containing ammonia, carbon dioxide, and hydrogen sulfide at temperatures from 200 to 500°F. Additional equilibrium measurements on these sour water systems with the inert gases carbon monoxide, nitrogen, methane and hydrogen added have also been made from 100 to 400°F.

These new measurements will extend the upper temperature limit of sour water equilibrium data to temperatures high enough to be used for design purposes for proposed water treating unit operations in the synthetic fuels and gas processing industries. They will also enable us to better understand the effect of the inert components on the volatility of the acid gas compounds as well as the effect of NH_3, CO_2, and H_2S on the Henry's constants of the inert gases present. These data will be important in the design of high pressure separators.

EXPERIMENTAL APPARATUS AND PROCEDURE

These measurements were made by means of rocked static equilibrium cells which were heated electrically and insulated with fiberglass. Pressures were measured by means of 3-D Instruments precision pressure gauges. Temperatures were measured and controlled using platinum resistance temperature probes calibrated against the vapor pressure of water to within 0.01 K.

Vapor samples were analyzed by either gas chromatography or by means of a train of weighed collectors. In the train of collectors, HCl was used to remove NH_3, $CdSO_4$ was used to remove H_2S, Drierite was used to remove water, Ascarite was used to remove CO_2, and a wet test meter was used to measure inert gas volumes.

Liquid samples were allowed to flash to ambient temperature and pressure and the relative amounts of flash gas and flash liquid were measured. Flash gas volumes were measured using a gas buret or a wet test meter. Flash gas compositions were determined using a gas chromatograph.

Multiple analyses were made on vapor and liquid samples to ensure good precision.

Schematics of the measurement cells and equipment and more detailed experimental procedures are presented in references 1 and 2.

RESULTS

Measured results are given in attached tables and figures as follows:

Results	Tables	Figures
NH_3-CO_2-H_2S-H_2O VLE Data to 500°F	1 to 4	1
VLE Data with Inerts Present	5 to 8	2

Measured equilibrium data (1, 2) reported in Tables 1 to 8 and in Figures 1 and 2 give the temperature, pressure, liquid composition, vapor composition, and partial pressures or Henry's constants for each run. The Henry's constants were calculated as the ratio of the partial pressure of each inert component divided by the corresponding liquid mole fraction. In addition, calculated values based on the SWEQ correlation (3) are given for comparison purposes. Measurement problems were encountered during some of the runs which prevented direct

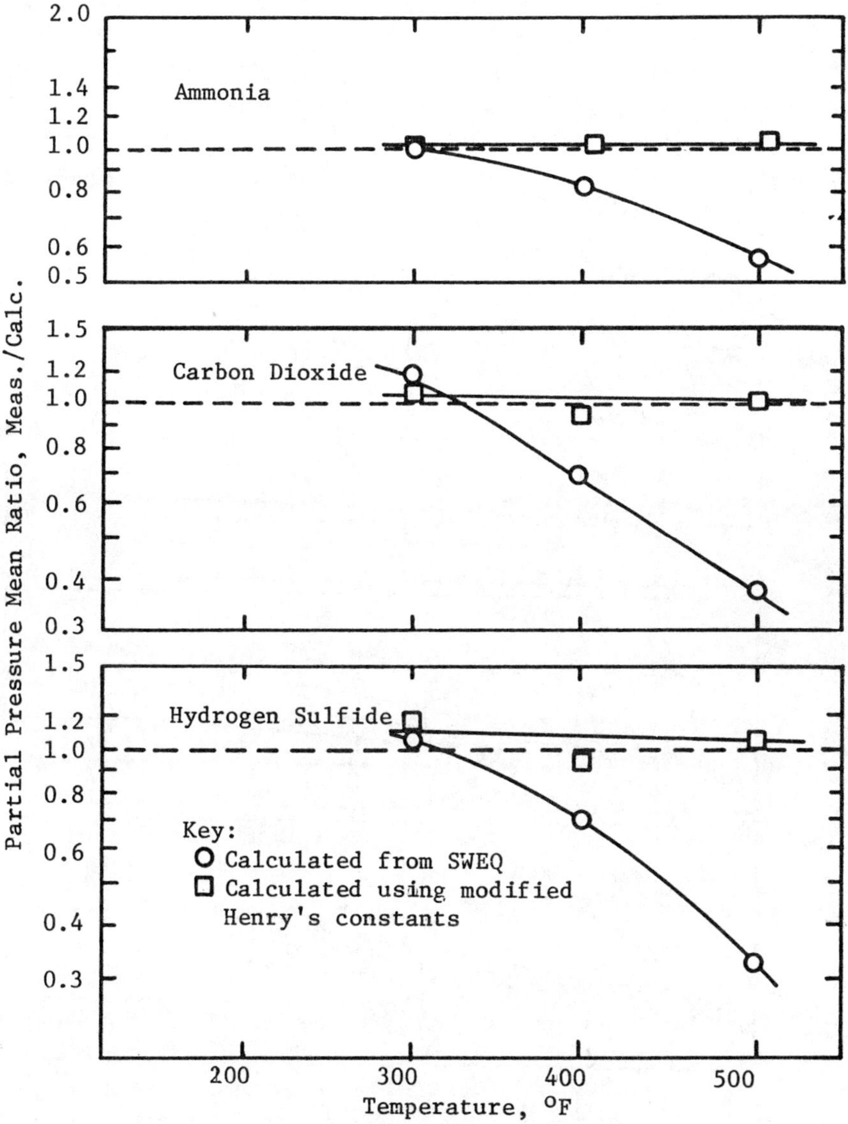

Figure 1. Comparison of measured over calculated partial pressures based on the SWEQ correlation.

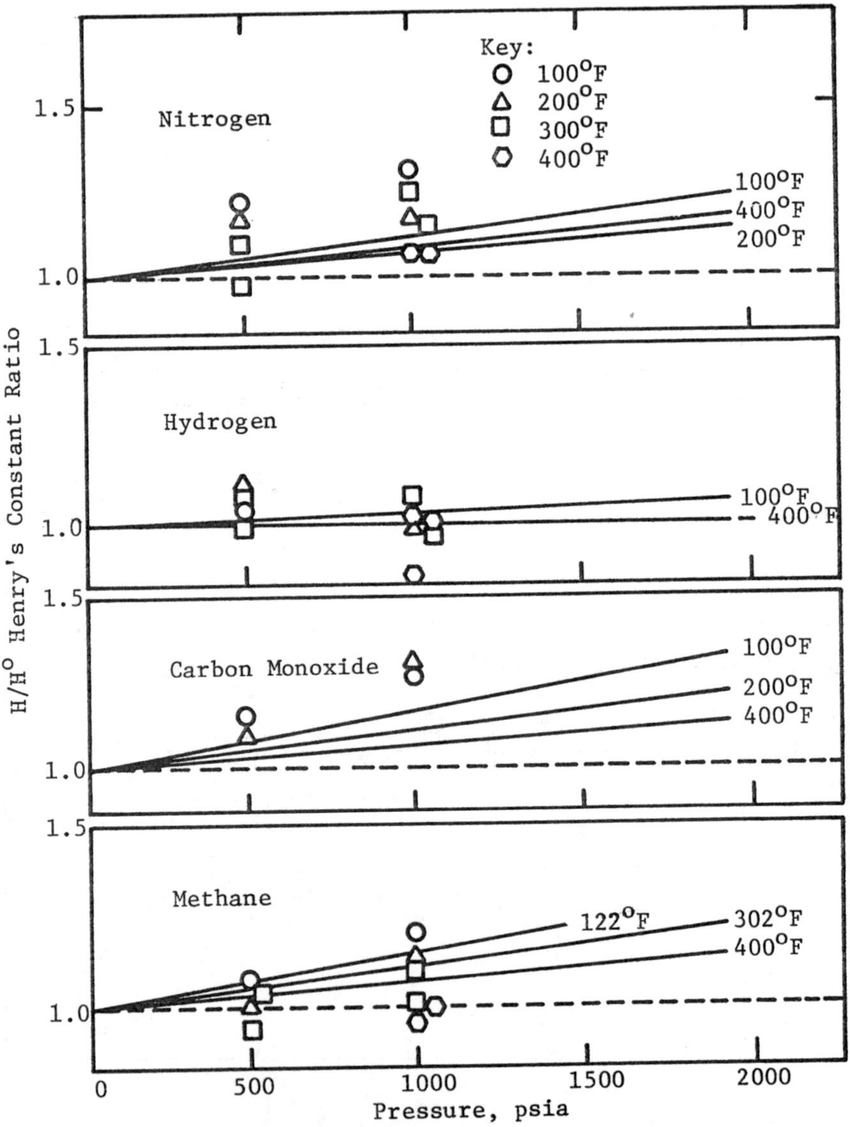

Figure 2. Effect of pressure on Henry's constant plotted as H/H° versus pressure.

Table 1
Ammonia-Carbon Dioxide-Hydrogen Sulfide-Water Quaternary at 200°F with a Comparison of Measured Versus Calculated Partial Pressures

Component	Composition Mole % Measured Liquid	Vapor	Partial Pressures, psia Meas.	Calc.(SWEQ)
Ammonia	2.00	6.09	1.50	1.41
Carbon Dioxide	0.57	33.99	8.38	7.45
Hydrogen Sulfide	0.51	17.45	4.30	4.70
Water	96.92	42.47	10.49	11.23
Total	100.00	100.00	24.67	24.79
Ammonia	2.06	13.25	2.20	2.10
Carbon Dioxide	0.26	8.80	1.45	1.64
Hydrogen Sulfide	0.51	11.85	1.96	2.45
Water	97.17	66.10	11.09	11.22
Total	100.00	100.00	16.70	17.41
Ammonia	17.76	35.52	23.44	17.32
Carbon Dioxide	3.84	17.36	11.46	8.56
Hydrogen Sulfide	4.08	32.65	21.55	19.04
Water	74.32	14.47[a)]	9.55	8.97
Total	100.00	100.00	66.00	53.88

a) Water too low to measure. Water concentration calculated by Raoult's law with correction for pressure enhancement.

Table 2
Ammonia-Carbon Dioxide-Hydrogen Sulfide-Water Quaternary at 300°F with a Comparison of Measured Versus Calculated Partial Pressures

Component	Composition Mole % Measured		Partial Pressures, psia	
	Liquid	Vapor	Meas.	Calc.(SWEQ)
Ammonia	2.00	6.81	8.86	9.63
Carbon Dioxide	0.24	27.14	35.31	41.42
Hydrogen Sulfide	0.49	13.78	17.93	17.92
Water	97.27	52.27	68.00	65.45
Total	100.00	100.00	130.10	134.42
Ammonia	2.12	8.33	11.74	9.92
Carbon Dioxide	0.29	30.87	43.52	53.45
Hydrogen Sulfide	0.53	14.44	20.37	20.95
Water	97.06	46.36	65.37	65.33
Total	100.00	100.00	141.00	149.65
Ammonia	17.32	15.69	96.02	76.50
Carbon Dioxide	4.19	43.83	268.24	489.47
Hydrogen Sulfide	3.82	29.12	178.21	173.62
Water	74.67	11.36[a)]	69.53	52.18
Total	100.00	100.00	612.00	791.77

Table 2 (Continued)
Ammonia-Carbon Dioxide-Hydrogen Sulfide-Water Quaternary at 300°F with a Comparison of Measured Versus Calculated Partial Pressures

Component	Composition Mole % Measured Liquid	Composition Mole % Measured Vapor	Partial Pressures, psia Meas.	Partial Pressures, psia Calc.(SWEQ)
Ammonia	2.16	.336	4.54	4.03
Carbon Dioxide	1.44	67.80	915.21	818.68
Hydrogen Sulfide	1.73	23.84	321.84	254.39
Water	94.72	8.03[a)]	108.41	63.93
Total	100.00	100.006	1350.00	1141.03

a) Water too low to measure. Water concentration calculated by Raoult's law with correction for pressure enhancement.

Table 3
Ammonia-Carbon Dioxide-Hydrogen Sulfide-Water Quaternary at 400°F with a Comparison of Measured Versus Calculated Partial Pressures

Component	Composition Mole % Measured Liquid	Composition Mole % Measured Vapor	Partial Pressures, psia Meas.	Partial Pressures, psia Calc.(SWEQ)
Ammonia	2.03	6.56	32.47	39.0
Carbon Dioxide	0.26	28.55	141.3	245.2
Hydrogen Sulfide	0.57	15.50	76.71	119.0
Water	97.14	49.39	244.5	242.4
Total	100.00	100.00	494.98	645.0
Ammonia	2.14	5.06	31.37	40.1
Carbon Dioxide	0.44	35.86	222.3	434.5
Hydrogen Sulfide	0.83	20.16	125.0	192.9
Water	96.59	38.92	241.3	241.0
Total	100.00	100.00	619.97	908.5
Ammonia	17.30	14.70	426.30	355.48
Carbon Dioxide	3.43	49.77	1443.33	2729.77
Hydrogen Sulfide	3.56	17.83	517.07	763.88
Water	75.70	17.70[a)]	513.30	190.85
Total	100.00	100.00	2900.00	4039.99

Table 3 (Continued)
Ammonia-Carbon Dioxide-Hydrogen Sulfide-Water Quaternary at 400°F with a Comparison of Measured Versus Calculated Partial Pressures

Component	Composition Mole % Measured Liquid	Vapor	Partial Pressures, psia Meas.	Calc.(SWEQ)
Ammonia	1.80	1.46	30.37	33.11
Carbon Dioxide	1.44	56.03	1165.42	1550.38
Hydrogen Sulfide	1.84	21.70	451.36	626.81
Water	94.74	20.81[a)]	432.85	236.49
Total	100.00	100.00	2080.00	2446.79

a) Water too low to measure. Water concentration calculated by Raoult's law with correction for pressure enhancement.

Table 4
Ammonia-Carbon Dioxide-Hydrogen Sulfide-Water Quaternary at 500°F with a Comparison of Measured Versus Calculated Partial Pressures

Component	Composition Mole % Measured Liquid	Composition Mole % Measured Vapor	Partial Pressures, psia Meas.	Partial Pressures, psia Calc.(SWEQ)
Ammonia	2.12	5.89	70.39	129.44
Carbon Dioxide	.313	15.04	179.73	455.97
Hydrogen Sulfide	.523	8.63	103.13	344.52
Water	97.05	70.44	841.75	667.43
Total	100.00	100.00	1195.00	1597.35
Ammonia	2.12	4.86	78.00	131.14
Carbon Dioxide	.575	22.88	376.22	839.27
Hydrogen Sulfide	.883	13.24	212.50	629.43
Water	96.42	59.02	947.28	663.14
Total	100.00	100.00	1605.00	2262.98
Ammonia	1.86	2.31	66.30	121.81
Carbon Dioxide	2.13	38.04	1091.75	3087.10
Hydrogen Sulfide	1.97	15.56	446.57	2071.42
Water	94.04	44.09[a)]	1265.38	646.78
Total	100.00	100.00	2870.00	5927.11

a) Water too low to measure. Water concentration calculated by Raoult's law with correction for pressure enhancement.

Table 5
Vapor-Liquid Equilibrium Data for Sour Water with Inerts at 100°F with a Comparison of Measured Versus Calculated Partial Pressures and with Henry's Constants of the Inerts

Press. psia	Comp.	Composition Mole % Vapor	Liquid	Meas. Partial Press. psia	Calc. SWEQ Partial Press. psia	Henry's Constant psia
33.4	NH3	.85	2.11	.28	.30	-
	CO2	.065	.27	.022	.020	-
	H2S	.47	.53	.16	.14	-
	H2O	2.33	97.09	.78	.93	-
	CO	25.99	a)	8.68	-	-
	N2	9.99	-	3.34	-	-
	CH4	12.96	-	4.33	-	-
	H2	47.35	-	15.81	-	-
		100.00	100.00			
500	NH3	.067	2.11	.34	.30	-
	CO2	.0029	.27	.015	.020	-
	H2S	.028	.53	.14	.14	-
	H2O	.19	97.05	.93	.93	-
	CO	26.89	.0101 b)	134.5	-	b)
	N2	10.32	.0025	51.60	-	b)
	CH4	13.44	.0063	67.21	-	b)
	H2	49.06	.0255	245.30	-	b)
		99.9979	100.0046			

Table 5 (Continued) Vapor-Liquid Equilibrium Data for Sour Water with Inerts at 100°F with a Comparison of Measured Versus Calculated Partial Pressures and with Henry's Constants of the Inerts

Press. psia	Comp.	Composition Mole % Vapor	Composition Mole % Liquid	Meas. Partial Press. psia	Calc. SWEQ Partial Press. psia	Henry's Constant psia
500	NH_3	c)	2.12		-	-
	CO_2	.23	.27	1.2	-	-
	H_2S		.53		-	-
	H_2O		97.04		-	-
	CO	25.73	.0112	128.7	-	1.15×10^6
	N_2	11.49	.0033	57.5	-	1.74×10^6
	CH_4	13.19	.0077	66.0	-	$.86 \times 10^6$
	H_2	49.36	.0220	246.8	-	1.12×10^6
		100.00	100.0042			
1000	NH_3	c)	2.13		-	-
	CO_2	.10	.27	1.0	-	-
	H_2S		.53		-	-
	H_2O		96.98		-	-
	CO	26.29	.0211	262.9	-	1.25×10^6
	N_2	10.96	.0059	109.6	-	1.86×10^6
	CH_4	13.70	.0141	137.0	-	$.97 \times 10^6$
	H_2	48.95	.0453	489.5	-	1.08×10^6
		100.00	99.9964			

Table 5 (Continued)
Vapor-Liquid Equilibrium Data for Sour Water with Inerts at 100°F with a Comparison of Measured Versus Calculated Partial Pressures and with Henry's Constants of the Inerts

Press. psia	Comp.	Composition Mole % Vapor	Composition Mole % Liquid	Meas. Partial Press. psia	Calc. SWEQ Partial Press. psia	Henry's Constant psia
1000	NH3	.036	2.11	.36	.30	-
	CO2	.0012	.27	.012	.020	-
	H2S	.015	.53	.15	.14	-
	H2O	.093	97.01	.93	.93	-
	CO	26.93	.021 b)	269.3	-	b)
	N2	10.33	.0052	103.4	-	b)
	CH4	13.46	.013	134.6	-	b)
	H2	49.13	.047	491.3	-	b)
		99.9952	100.0062			

a) Flash Gas/Flash Liquid ratio was too small to adequately measure the liquid compositions of the inert gases.

b) Henry's constants of the inert compounds were not calculated for these runs because of problems with the liquid analysis of the inert components.

c) No vapor phase analysis was made for the sour water components for these Henry's constants reruns, but by taking the ratio of the saturation pressure over the total pressure the mole percent of the sour water components could be assigned.

Table 6
Vapor-Liquid Equilibrium Data for Sour Water with Inerts at 200°F with a Comparison of Measured Versus Calculated Partial Pressures and with Henry's Constants of the Inerts

Press. psia	Comp.	Composition Mole % Vapor	Composition Mole % Liquid	Meas. Partial Press. psia	Calc. SWEQ Partial Press. psia	Henry's Constant psia
500	NH3	.48	2.05	2.40	2.13	-
	CO2	.32	.25	1.61	1.54	-
	H2S	.47	.51	2.37	2.36	-
	H2O	2.24	97.15	11.21	11.23	-
	CO	24.60	.0080	123.0[a)]	-	a)
	N2	9.53	.0025	47.6	-	a)
	CH4	12.58	.0058	62.9	-	a)
	H2	49.76	.0229	248.9	-	a)
		100.00	99.9991			
500	NH3	.083	2.07	.41	.31	-
	CO2	28.46	1.62	142.3	155.4	-
	H2S	4.74	.54	23.72	30.47	-
	H2O	2.16	95.74	10.79	11.18	-
	CO	13.17	c)	65.85	-	-
	N2	5.01	-	25.07	-	-
	CH4	9.18	-	45.90	-	-
	H2	37.20	-	185.9	-	-
		100.003	99.97			

Table 6 (Continued)
Vapor-Liquid Equilibrium Data for Sour Water with Inerts at 200°F with a Comparison of Measured Versus Calculated Partial Pressures and with Henry's Constants of the Inerts

Press. psia	Comp.	Composition Mole % Vapor	Composition Mole % Liquid	Meas. Partial Press. psia	Calc. SWEQ Partial Press. psia	Henry's Constant psia
500	NH_3	b)	2.12		-	-
	CO_2	3.53	.26	17.7	-	-
	H_2S		.52		-	-
	H_2O		97.06		-	-
	CO	25.08	.0092	125.4	-	1.36×10^6
	N_2	12.05	.0030	60.3	-	2.01×10^6
	CH_4	12.23	.0064	61.2	-	$.96 \times 10^6$
	H_2	47.11	.0205	235.6	-	1.15×10^6
		100.00	99.9991			
1000	NH_3	b)	2.13		-	-
	CO_2	1.82	.26	18.2	-	-
	H_2S		.52		-	-
	H_2O		97.01		-	-
	CO	26.25	.0167	262.5	-	1.57×10^6
	N_2	11.20	.0056	112.0	-	2.00×10^6
	CH_4	13.06	.0124	130.6	-	1.05×10^6
	H_2	47.67	.0462	476.7	-	1.03×10^6
		100.00	100.0009			

Table 6 (Continued)
Vapor-Liquid Equilibrium Data for Sour Water with Inerts at 200°F with a Comparison of Measured Versus Calculated Partial Pressures and with Henry's Constants of the Inerts

Press. psia	Comp.	Composition Mole % Vapor	Liquid	Meas. Partial Press. psia	Calc. SWEQ Partial Press. psia	Henry's Constant psia
1000	NH3	.28	2.02	2.75	2.03	-
	CO2	.18	.26	1.75	1.71	-
	H2S	.26	.52	2.64	2.53	-
	H2O	1.12	97.13	11.20	11.23	-
	CO	24.58	.0162	245.8[a)]	-	a)
	N2	9.84	.0048	98.4	-	a)
	CH4	12.68	.0103	126.8	-	a)
	H2	51.06	.0411	510.6	-	a)
		100.00	100.0024			

a) Henry's constants of the inert components were not calculated for these runs because of problems with the liquid analysis of the inert components.

b) No vapor phase analysis was made for the sour water components for these Henry's constants reruns, but by taking the ratio of the saturation pressure over the total pressure the mole percent of the sour water components could be assigned.

c) The liquid analysis of the inerts in this run seem to be in error causing the partial pressures and the Henry's constants to also be in error.

Table 7
Vapor-Liquid Equilibrium Data for Sour Water with Inerts at 300°F with a Comparison of Measured Versus Calculated Partial Pressures and with Henry's Constants of the Inerts

Press. psia	Comp.	Composition Mole % Vapor	Composition Mole % Liquid	Meas. Partial Press. psia	Calc. SWEQ Partial Press. psia	Henry's Constant psia
500	NH3	1.86	2.11	9.32	10.23	-
	CO2	9.90	.25	49.52	41.80	-
	H2S	4.19	.50	20.96	17.87	-
	H2O	13.07	97.10	65.35	65.37	-
	CO	8.60	.0020[a)]	43.01	-	a)
	N2	7.89	.0019	39.46	-	a)
	CH4	11.69	.0053	58.45	-	a)
	H2	42.80	.0284	213.9	-	a)
		100.00	99.9976			
500	NH3	b)	2.16		-	-
	CO2 } H2S } H2O } (NH3–H2O combined)	29.19	.27	146.0	-	-
	H2S		.53		-	-
	H2O		97.00		-	-
	N2	8.42	.0036	42.1	-	1.17×10^6
	CH4	12.78	.0092	63.9	-	$.69 \times 10^6$
	H2	49.61	.0307	248.1	-	$.81 \times 10^6$
		100.00	100.0035			

Table 7 (Continued)
Vapor-Liquid Equilibrium Data for Sour Water with Inerts at 300°F with a Comparison of Measured Versus Calculated Partial Pressures and with Henry's Constants of the Inerts

Press. psia	Comp.	Composition Mole % Vapor	Composition Mole % Liquid	Meas. Partial Press. psia	Calc. SWEQ Partial Press. psia	Henry's Constant psia
500	NH3	1.57	2.02	7.84	7.96	-
	CO2	29.20	.60	146.0	152.5	-
	H2S	4.68	.52	23.40	29.45	-
	H2O	13.93	96.83	69.64	65.33	-
	N2	6.40	.0024	32.02	-	1.33×10^6
	CH4	10.09	.0065	50.44	-	$.78 \times 10^6$
	H2	34.13	.0227	170.6	-	$.75 \times 10^6$
		100.00	100.0016			
1000	NH3	.69	2.01	6.94	6.82	-
	CO2	29.13	.94	291.3	297.4	-
	H2S	2.72	.48	27.24	35.33	-
	H2O	6.69	96.49	66.94	65.24	-
	N2	7.57	.0056	75.72	-	1.35×10^6
	CH4	12.24	.0150	122.4	-	$.82 \times 10^6$
	H2	40.96	.0560	409.5	-	$.73 \times 10^6$
		100.00	99.9966			
1000	NH3	b)	2.17		-	-
	CO2	14.60	.27	146	-	-
	H2S		.54		-	-
	H2O		96.92		-	-
	N2	10.85	.0073	108.5	-	1.49×10^6
	CH4	16.23	.0216	162.3	-	$.75 \times 10^6$
	H2	58.32	.0728	583.2	-	$.80 \times 10^6$
		100.00	100.0017			

(In the last block NH3, CO2, H2S and H2O are bracketed together: vapor 14.60 and measured partial pressure 146 apply to the group.)

Table 7 (Continued)
Vapor-Liquid Equilibrium Data for Sour Water with Inerts at 300°F with a Comparison of Measured Versus Calculated Partial Pressures and with Henry's Constants of the Inerts

Press. psia	Comp.	Composition Mole % Vapor	Liquid	Meas. Partial Press. psia	Calc. SWEQ Partial Press. psia	Henry's Constant psia
1000	NH3	.99	2.09	9.91	10.86	-
	CO2	4.28	.18	42.80	24.79	-
	H2S	2.55	.43	25.47	12.44	-
	H2O	6.54	97.20	65.37	65.44	-
	CO	17.52	.0119[a)]	175.2	-	a)
	N2	9.95	.0064	99.5	-	a)
	CH4	14.19	.0163	141.9	-	a)
	H2	43.98	.0637	439.9	-	a)
		100.00	99.9983			

a) Henry's constants of the inert components were not calculated for these runs because of problems with the liquid analysis of the inert components.

b) No vapor phase analysis was made for the sour water components for these Henry's constants reruns, but by taking the ratio of the saturation pressure over the total pressure the mole percent of the sour water components could be assigned.

Table 8
Vapor-Liquid Equilibrium Data for Sour Water with Inerts at 400°F with a Comparison of Measured Versus Calculated Partial Pressures and with Henry's Constants of the Inerts

Press. psia	Comp.	Composition Mole % Vapor	Composition Mole % Liquid	Meas. Partial Press. psia	Calc. SWEQ Partial Press. psia	Henry's Constant psia
1000	NH3	3.27	2.13	32.70	39.7	-
	CO2	22.28	.45	222.8	451.8	-
	H2S	12.44	.84	124.5	198.8	-
	H2O	23.99	96.51	239.8	241.0	-
	N2	8.13	.0057[a)]	81.3	-	a)
	CH4	10.02	.0111	100.2	-	a)
	H2	19.87	.0571	198.7	-	a)
		100.00	100.0039			
1000	NH3	3.15	2.02	31.52	38.2	-
	CO2	27.15	.77	271.5	760.3	-
	H2S	8.18	.61	81.77	137.2	-
	H2O	23.87	96.52	238.7	241.3	-
	N2	4.62	.0062	46.21	-	$.75 \times 10^6$
	CH4	7.27	.0155	72.72	-	$.47 \times 10^6$
	H2	25.76	.0556	257.6	-	$.46 \times 10^6$
		100.00	99.9973			

Table 8 (Continued)
Vapor-Liquid Equilibrium Data for Sour Water with Inerts at 400°F with a Comparison of Measured Versus Calculated Partial Pressures and with Henry's Constants of the Inerts

Press. psia	Comp.	Composition Mole % Vapor	Composition Mole % Liquid	Meas. Partial Press. psia	Calc. SWEQ Partial Press. psia	Henry's Constant psia
1000	NH3	b)	2.14		-	-
	CO2		.37		-	-
	H2S	62.00 (NH3, CO2, H2S, H2O)	.81	620.0 (NH3, CO2, H2S, H2O)	-	-
	H2O		96.59		-	-
	N2	4.62	.0062	46.2	-	$.75 \times 10^6$
	CH4	6.87	.015	68.7	-	$.46 \times 10^6$
	H2	26.51	.071	265.1	-	$.37 \times 10^6$
		100.00	100.0022			

a) Henry's constants of the inert components were not calculated for these runs because of problems with the liquid analysis of the inert components.

b) No vapor phase analysis was made for the sour water components for these Henry's constants reruns, but by taking the ratio of the saturation pressure over the total pressure the mole percent of the sour water components could be assigned.

analyses for water in the vapor phase. These problems resulted either from a low concentration of water in the vapor phase, which made analyses by absorption and direct weight rather inaccurate, or from the need to add water during vapor sampling in order to prevent plugging of the sample line. The plugging was presumably caused by urea formation in the sample line at high CO_2 and NH_3 partial pressures in the cell.

Particular runs where a direct analysis for water in the vapor phase was not determined are noted in Tables 1 to 4 by a superscript "a" adjacent to the vapor analysis for water and a footnote is given which explains the problem. In these cases, the water concentration in the vapor phase was estimated based on the assumption of "ideal" liquid behavior with a correction applied for the effect of pressure on the enhancement of the water volatility. This pressure enhancement effect was calculated from the virial equation of state using only the second virial coefficient. Table 9 gives a comparison of measured and calculated water enhancement ratios for the points in Tables 1 to 8 where the vapor concentration was measured. As is shown in Table 9, the average deviation between the measured and calculated ratios is $\pm 4.4\%$; but the measured values tend to

Table 9
Pressure Enhancement Ratio of Water Volatility, Comparison of Measured and Calculated Values

Run	Temp.	Press.	Mole Percent				Enhancemnt Ratio $Py/P^\circ x$		%
no.	°F	psia	NH3	CO2	H2S	H2O	Meas.	Calc.	Dev.[a)]
1	300	120	19.85	-	21.19	58.96	1.11	1.04	6.7
2	400	454	13.95	-	25.77	60.28	1.16	1.11	4.5
3	500	1268	11.87	-	23.47	64.66	1.27	1.25	1.6
5	400	695	4.02	-	53.94	42.05	1.22	1.20	1.7
6	500	1615	4.61	-	42.86	52.53	1.30	1.36	-4.4
7	500	1195	5.89	15.04	8.63	70.44	1.26	1.20	5.0
8	500	1605	4.86	22.88	13.24	59.02	1.43	1.36	5.1
12	300	120	20.95	18.74	-	60.31	1.13	1.04	8.7
13	400	505	13.45	34.01	-	52.54	1.11	1.13	-1.8
14	500	1115	12.16	21.11	-	66.73	1.13	1.18	-4.2
					Ave. Deviation[b)]				4.4%
					Ave. Bias[c)]				+2.3%

a) Percent Deviation = 100 (Meas. -Calc.)/Calc.
b) Ave. Deviation = $(\Sigma_i |d_i|)/n$
c) Ave. Bias = $(\Sigma_i d_i)/n$

be about 2.3% higher than the calculated values as shown by the average bias of $\pm 2.3\%$. These calculated values appeared sufficiently close to the measured values that the calculated enhancement ratios were used without modification to give the estimated water concentrations when no value was measured.

From thermodynamics, the enhancement of water volatility due to pressure effects is given by the following equation:

$$\text{Enhancement} = Py/P^{\circ}x\gamma = \phi^{\circ}\delta \,/\, \phi\delta^{\circ} \qquad (1)$$

In this case the activity coefficient is assumed to be unity because of the high concentrations of water in the liquid phase. The fugacity coefficients were calculated from the virial equation using only the second virial coefficient as follows:

$$Z = 1 + \frac{\Sigma_i \Sigma_j y_i y_j B_{ij}}{V} \qquad (2)$$

From this equation the fugacity coefficient is given as follows:

$$\ln \phi_i = 2\Sigma_j y_j B_{ij}/V - \ln Z \qquad (3)$$

In this equation the volume is calculated from the pressure by solving for V in equation 2; the result is as follows:

$$V = \frac{RT + \sqrt{(RT)^2 + 4RTPB_m}}{2P} \qquad (4)$$

where $$B_m = \Sigma_i \Sigma_j y_i y_j B_{ij} \qquad (5)$$

Equations 3 and 4 require values for the virial coefficients B_{ij}. These were calculated from the correlation published by C. Tsonopoulos (4) which involves the use of critical temperature, critical pressure, and acentric factor of the pure materials. These values were taken from Reid et al (5). For cross coefficients, k_{ij} values were assumed which appeared to be consistent with values recommended by Tsonopoulos. These values are summarized in Table 10.

DISCUSSION

The data given in Tables 1 to 8 are being used for development of an up-dated correlation model similar to the SWEQ correlation (3). This work is now in progress and is expected to be completed in 1984. The existing SWEQ correlation is inadequate at high temperatures as is shown in Figure 1 where the ratio of measured over calculated partial pressures of NH_3, CO_2, and H_2S are plotted versus temperature. Also shown in Figure 1 are ratios using Henry's constants in the SWEQ model for NH_3, CO_2, and H_2S based on work in this paper and on binary measurements

Table 10
Assumed k_{ij}'s and Partial Molar Volumes Used for Calculating Pressure Enhancement Ratio of Water

	k_{ij}			
Component	NH_3	CO_2	H_2S	H_2O
NH_3	0	0	0	-.2[b)]
CO_2	0	0	0	.14[a)]
H_2S	0	0	0	.14[b)]
H_2O	-.2[b)]	.14[a)]	.14[b)]	0
Molar Volume cc/mole	26.7	57.0	34.2	18.0

a) Based on recommended value published by C. Tsonopoulos

b) Estimated

which have been recently reported to the Gas Processors Association (6, 7). At 300°F, both models are in good agreement with the measured data; but at higher temperatures the SWEQ model diverges significantly while the modified SWEQ model does not. This seems to indicate that the major error in the SWEQ model at high temperatures is in the Henry's constants of NH_3, CO_2, and H_2S. The SWEQ model was only developed for temperatures up to 248°F, so these errors in the model at higher temperatures where an extrapolation is necessary are not surprising. This serves as an example of the hazards associated with the extrapolation of a correlation beyond the data base used to develop the correlation.

The volatility of the inert components from sour water mixtures are compared in Figure 2 with binary Henry's constant data for the pure compounds in water (6, 7) where the solid lines represent the binary data and points represent data in this paper. The data are plotted as the ratio of H/H° where H° is based on the binary data. These plots show some scatter in the measured Henry's constants in this paper which are mainly due to errors in the analyses of the inert components in the liquid phase which were at very low concentrations. Nevertheless, the scatter is not serious; and one can conclude that the Henry's constants of the inert components in sour water at the measured conditions are virtually the same as the Henry's constants of these same compounds in pure water at the same temperatures and pressures. Thus, the effects of the sour water components appear to be small.

SUMMARY AND CONCLUSIONS

Equilibrium measurements have been made for sour water systems with inert gases present at temperatures to 400°F and pressures to 1000 psia. The partial pressures of the sour water components have been compared with the SWEQ correlation. The effect of the inert gases on the volatilities of these components has been evaluated. The effect of the sour water components on the Henry's constants of the inert gases has also been examined.

The major conclusions from these measurements can be summarized as follows:

1. Measured partial pressures of NH_3, CO_2, and H_2S diverge significantly from predicted values based on the earlier SWEQ correlation at temperatures above 300°F. The earlier SWEQ correlation was developed for temperatures up to 248°F.

2. When new Henry's constant data which have recently been measured on NH_3, CO_2, and H_2S in pure water are used to modify the SWEQ correlation, then the agreement is significantly improved. This tends to show that there is consistency between the multicomponent data in this paper and recently measured binary Henry's constant data.
3. Henry's constant data of inerts in sour water systems appear to be virtually the same as in pure water at the conditions studied.
4. These data will be useful for the development of an up-dated correlation model similar to the SWEQ correlation.

NOTATION

x = liquid-phase mole fraction
y = vapor-phase mole fraction

B = virial coefficient
P = total pressure
P° = vapor pressure of pure component
R = gas constant
T = absolute temperature
V = molar volume
Z = compressibility ratio

Greek letters

γ = activity coefficient
δ = Poynting effect in mixture from zero pressure to mixture saturation pressure
δ° = Poynting effect of pure component from zero pressure to vapor pressure
ϕ = fugacity coefficient in mixture
ϕ° = fugacity coefficient of pure component

Subscripts

i = component i
j = component j
m = mixture

REFERENCES

1. Owens, J. L., J. R. Cunningham, and G. M. Wilson, "Vapor-Liquid Equilibria for Sour Water Systems at High Temperatures", Research Report RR-65, Gas Processors Association, Tulsa, OK, August 1983.

2. Wilson, G. M., J. L. Owens, J. R. Cunningham, "Vapor-Liquid Equilibria for Sour Water Systems With Inert Gases Present" Research Report RR-52, Gas Processors Association, Tulsa OK, April 1982.

3. Wilson, G. M., "A New Correlation of NH_3, CO_2, and H_2S Volatility Data from Aqueous Sour Water Systems," Publication 955, American Petroleum Institute, Washington DC, March 1978. SWEQ Sour Water Equilibria Correlation.

4. Tsonopoulos, C. Adv. in Chem. Series, No. 182 Equations of State in Engineering and Research, Am. Chem. Society, p 143-162 (1979).

5. Reid,R. C., J. M. Prausnitz, and T. K. Sherwood, The Properties of Gases and Liquids, Third Edition, Apppenix A, McGraw-Hill (1977).

6. Gillespie, P. C., G. M. Wilson, "Vapor-Liquid Equilibrium Data on Water-Substitute Gas Components: N_2-H_2O, H_2-H_2O, CO-H_2O, H_2-CO-H_2O, and H_2S-H_2O," Research Report RR-41, Gas Processors Association, Tulsa OK, April 1980.

7. Gillespie, P. C., G. M. Wilson, "Vapor-Liquid and Liquid-Liquid Equilibria: H_2O-CH_4, H_2O-CO_2, H_2O-H_2S, H_2O-nC_5, and H_2O-CH_4-nC_5," Research Report RR-48, Gas Processors Association, Tulsa OK, April 1982.

ACKNOWLEDGEMENT

This work was sponsored by the Gas Processors Association, 1812 First Place, Tulsa, OK 74103.

28

Ammonia Removal and Recovery by the PHOSAM-W Process

R. D. Rice
Director, Services-Chemicals (Retired)
USS Engineers and Consultants, Inc.

J. V. Busa
Director, Technical Services (Chemicals)
USS Engineers and Consultants, Inc.

Abstract:

U. S. Steel's ammonia recovery technology, developed and widely used in by-product coke plants, has gained acceptance in other industries, especially synthetic fuels. The paper gives an overview of the USS PHOSAM-W Process and its application, both actual and potential.

Recovery and upgrading of fossil fuels produces ammonia together with acid gases such as hydrogen sulfide, hydrogen cyanide, carbon and sulfur dioxides, etc. The nitrogen and sulfur in coal, peat, oil shale and tar sands, and also in crude oil, pose problems of product and environmental quality control that concern us all.

Ammonia is not an acid gas but it travels with the acid gases. Ammonia is strongly basic, and therefore it has a chemical affinity for H_2S and CO_2, with which it forms salts in water solution. Because of this affinity, separation of ammonia from acid gases by physical means, such as distillation, is a difficult proposition. The problem is complicated by the fact that CO_2 is absorbed in and released from water solutions more slowly than H_2S. Other troublesome aspects of CO_2 are its tendency to form ammonium salts from the vapor phase which can be deposited in heat exchangers and piping, and to form corrosive ammonium carbamate in ionic solution.

The PHOSAM-W process provides a unique solution to the problem of separating ammonia from other by-products of fossil fuel conversion processes. It is a chemical absorption process, not sensitive to the presence of acid gases, organic acids such as phenols and fatty acids and neutral or weakly basic organic compounds which may be present. The process enables production of pure liquid anhydrous ammonia, which is readily transported and marketed co-mingled with synthetic ammonia from other sources. Additionally, the process enables the production of ammonia and acid gas-free water for reuse or biological treatment, and of an ammonia-free acid gas stream.

The parent process was developed by United States Steel Corporation in the sixties, and is employed in about twenty coke plants worldwide. The PHOSAM process solves the problem of separating ammonia from gas and liquors produced in the conversion of coal to metallurgical coke. In the coking process, ammonia must be removed from coke-oven gas, to avoid problems of corrosion and deposition of solids in the processing, distribution, and burning of the coke-oven gas as fuel. The traditional process involves washing the coke-oven gas with a solution of ammonium sulfate containing free sulfuric acid, thereby producing ammonium sulfate crystals as the by-product. This traditional process has the disadvantage of producing a solid by-product that must be dried and stored for the seasonal fertilizer market; furthermore, a by-product that consumes three tons of expensive sulfuric acid for each ton of ammonia. Ammonium sulfate frequently fetches less income than the cost of the acid alone, not to mention process operating costs.

An alternative process which has been used in coke plants involves water-washing the coke-oven gas and burning the ammonia vapor from distillation of the wash water. However, this process inevitably removes and burns a large portion of the H_2S, present in the coke-oven gas, releasing SO_2 into the atmosphere. Therefore, the water-wash process is banned from new coke plant installations in Europe, and has rarely been applied in American coke plants.

A ten-year research effort by U. S. Steel resulted in the development of the parent process for coke plants, and of the PHOSAM-W process for applications outside the coking industry. The same fundamental data, proprietary to U. S. Steel, are used in the design of both processes. Both processes work with gaseous feeds, or vapors produced by simple stripping of sour waters.

The chemistry of the process is simple:

$$NH_4H_2PO_4 + NH_3 \rightleftarrows (NH_4)_2HPO_4$$

is the reaction of absorption; its reverse reaction, accomplished by heating, strips out the absorbed ammonia. Actually, the reaction between monoammonium phosphate and diammonium phosphate is not carried to completion in either direction. Solutions are actually varied between molar ratios (NH_3/H_3PO_4) of 1.2-1.4 in lean solution and 1.7-2.0 in rich solution, and the concentration of the mixed salts is kept well below the point at which crystals could form.

The solution travels in a closed circuit between an absorber, where ammonia is selectively absorbed from a gas or vapor stream, and a stripper, where the absorbed ammonia is stripped out and the lean solution is regenerated. The phosphate ion, which is the active ingredient and the key to selective absorption of ammonia, is not degraded in the process, therefore, there is no blowdown. The stripped-out ammonia is refined by fractionation in a separate column.

The advantage of a chemisorption system, over a physical absorption system lies in its selectivity for ammonia. Essentially, only ammonia is absorbed from the myriad of inorganic and organic components present in the feed gas or vapor; the small amounts of acid gases, acidic organics, and neutral compounds which are co-absorbed, are thoroughly eliminated from the product ammonia by other elements of the process.

When acid gas is the feed to a Claus process sulfur plant (or a sulfuric acid plant), removal of NH_3 will yield economies in capital and operating costs of the downstream plant. These economies stem from the elimination of that portion of the air required to burn the ammonia, thereby reducing equipment sizes and power requirements. Secondary benefits are elimination of nitrogen oxides from the product and the tail gas, and the avoidance of ammonium sulfate formation in the catalyst. In the case of an existing sulfur plant which requires expansion to handle high sulfur or high nitrogen crudes, addition of a PHOSAM-W plant may so increase the capacity of the existing sulfur plant that the need for expansion is eliminated.

The flowsheet for a hypothetical plant is shown in Figure 1. In this example, there are two feeds and two absorbers.

One feed is an overhead vapor from a sour water stripper containing about 10 percent NH_3, and an equal volume of H_2S and CO_2 in any proportion. The balance is water vapor with traces of HCN and volatile organics. Ammonia is removed by countercurrent contact with an ammonium phosphate solution, at a pressure of 35-70 kPa (5-10 psig) and a temperature of 90-100°C. Absorption

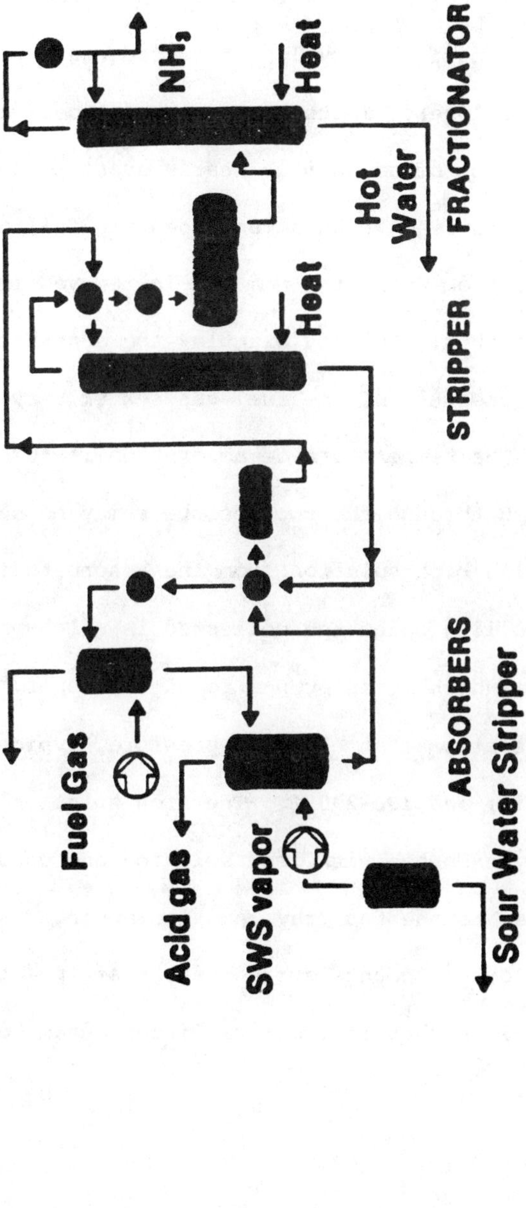

Figure 1. Example of USS PHOSAM-W Process.

efficiency may be in the range of 95-99 percent, but any ammonia remaining in the exit vapor can be recovered by partial condensation, recycling the condensate to the sour water stripper. In the presence of excess acid gas the condensation of ammonia is nearly quantitative.

The second feed is an ammonia-bearing fuel gas, rich in hydrogen, from which the ammonia is removed in a separate absorber. In this example, the absorber operates at the pressure of the fuel gas system, say 350 kPa (50 psig). The temperature of absorption is 50°C. The solution path through the two absorbers may be series or parallel. The rich solution from the absorbers is heated to the boiling point and collected in a low-pressure vessel, from which it is pumped to the stripper. The stripper operates at an elevated pressure, typically 1300 kPa (190 psig) and 180-200°C. The rich solution, preheated by exchange with lean solution and condensing vapor, is fed to the top tray, and the absorbed ammonia is stripped out by countercurrent steam stripping. The heat source is either indirect or direct steam, or a combination.

Regenerated lean solution returns to the absorbers through heat exchangers to complete the solution circuit. Physical losses of phosphate ion, due to pump and valve leaks, are made up from time to time by adding phosphoric acid purchased as 75 or 85 percent furnace grade acid.

Since the phosphate inventory is fixed, it is necessary to control the water balance so that water is neither accumulated nor depleted in the circuit. This is accomplished quite simply by observing the total solution volume in the absorber sumps, and adjusting the various heating and cooling steps to control the water balance.

In regenerating the lean solution in the stripper, an overhead vapor is produced consisting mainly of ammonia and water, typically about 20 percent ammonia. The vapor is condensed, without subcooling, in a two-stage condenser, and the condensed aqua ammonia is accumulated in the fractionator feed tank. From this tank, which provides surge capacity, the aqua ammonia is pumped to the fractionator.

The fractionator is a distillation column producing pure liquid ammonia overhead and ammonia-free water bottoms. It is normally designed to operate at a high enough pressure to condense the ammonia reflux and product at the maximum cooling water temperature. This avoids any need for an ammonia compressor in the process; liquid ammonia is directly produced to storage tanks for sale.

The fractionator bottoms stream is the only waste stream from the process. Its normal disposition is by recycle to the sour water stripper where it flashes to provide part of the stripping steam; its unflashed portion joins the sour water stripper bottoms.

BY-PRODUCT VALUES

Ammonia produced by the process will meet commercial or fertilizer grade specifications, and is readily saleable. As a rule, the by-product ammonia production from a coke plant, oil refinery, or a synfuels operation will be intermingled with synthetic ammonia produced by others and moved in the existing normal commercial channels. Ammonia is a commodity, transported by tank truck, tank car, barge, and interstate pipeline. The by-product ammonia producer will find his market among the large producers, distributors, or users of ammonia within a reasonable shipping radius of his plant.

Typical fertilizer or commercial grade ammonia specifications are shown in Table 1. For preliminary economic evaluation, the wholesale or co-manufacturers price for such ammonia may be estimated at $165 per metric ton ($150 per short ton), less freight. Thus, a plant producing 50 metric tons per day of ammonia would generate an income of about $8000 per day to defray the cost of operation.

In special cases, it may be possible to market a portion of the ammonia as a premium grade at a premium price. The process is capable of producing ammonia of exceptional quality, as shown in Table 2. Sufficient separation power is designed into the fractionator to produce ammonia containing as little as 100 ppm (0.01 percent) water, and correspondingly low concentrations of other impurities. If there exists a premium market for a portion of the production, it can be served by segregating the special ammonia in a separate tank.

Table 1
Commercial Grade Ammonia Specifications

Ammonia content	99.5% minimum
Water	0.2-0.5% weight
Oil	5 ppm maximum

Table 2
Premium Quality Ammonia Typical Specifications

Ammonia content	99.98% min.
Water	150 ppm max.
Oil	3 ppm max.
H_2S	1 ppm max.
CO_2	3 ppm max.
Noncondensible gas	0.2 cc/g. max.
Appearance	Clear, colorless

Table 3 lists some of the uses to which recovered anhydrous ammonia has been put successfully. A few of these uses require a premium grade of ammonia.

SPECIFIC APPLICATIONS

The parent process has found wide application in coking plants round the world, dating back to U. S. Steel's first commercial use in 1968 at the coke plant at Clairton, Pennsylvania (Figure 2). This plant is still in operation and has been expanded several times. It has also been operated at a severely reduced coking rate, and has been augmented by a second plant built specifically for waste water treatment.

The PHOSAM-W process has not fared as well as its parent process in terms of actual units in operation, owing to the abandonment of so many synthetic fuel projects in the past few years. Nearly all of the proposed synfuels ventures, which expected significant quantities of ammonia, have evaluated the process versus its alternatives and found it lower in capital and operating costs, as well as having the inherent reliability of a chemisorption system over physical separations based on

Table 3
Uses for Anhydrous Ammonia

Fertilizers
Ammonium Nitrate
Urea
Ammonium Phosphates
Direct Injection into the Soil
Chemicals, Plastics and Synthetic Fiber
Acrylonitrile for ABS and Fibers
Caprolactam for Nylon
Melamine for Laminating Resins
Miscellaneous
Ammonium Nitrate for Blasting
Nitric Acid
Refrigeration

Figure 2. PHOSAM Plant at U.S. Steel's Clairton Works.

relative volatility. To date, eight of these ventures progressed to the point of securing the license and the basic process engineering.

Let us comment on a few of these which are of particular and current interest. Sasol, of South Africa, is the largest exemplar of coal gasification and liquefaction via the Lurgi and Fischer-Tropsch processes, respectively. In the original plant at Sasolburg, Sasol One has, for twenty-five years, recovered ammonia from gas liquors as by-product ammonium sulfate by the conventional process common to coke-oven plants. The production rate is approximately 600 metric tons per day of sulfate, representing 150 metric tons per day of ammonia. In the two larger plants at Secunda, Sasol Two and Three recover anhydrous ammonia by a competitive (distillation) process which went into operation in the past few years. Sasol has secured a license, however, and has made an in-depth study of the process as an alternative to replace the aging ammonium sulfate plant at Sasol One. Under present market and economic conditions, Sasol has deferred decision to replace the sulfate plant, which can be kept in operation for awhile longer. However, they have advised that for application at their Sasol One plant their study shows lower investment costs and lower operating costs for the PHOSAM-W process compared to the competitive process.

In oil refineries, such processes as hydrodesulfurization, fluid catalytic cracking, and residual hydrocracking, produce sour water containing both H_2S and NH_3. Commonly, sour water is stripped to low levels of these pollutants, and the vapors are sent to a sulfur plant. But the ammonia content of the acid gas feed to the sulfur plant is crucial to the capacity, economy and operability of that plant. Therefore, one should consider ammonia removal and recovery not only for the "cash crop" of by-product ammonia it will produce, but also for its beneficial effects on the sulfur plant.

The first PHOSAM-W plant to go on stream in a refinery is at the Bakersfield, California refinery of Tosco Corporation. This unit, including two new sour water strippers, solved an environmental problem by enabling the recycle of process water to the hydrotreaters. This unit operated successfully and the ammonia was sold to a local distributor. The process can also be retrofitted to existing sour water strippers without any difficulty.

In the field of synthetic fuels, the process is an important part of two major projects in being. In the first phase of Union Oil Company of California's shale oil retorting and upgrading facilties at Parachute Creek, Colorado there is a plant with an ammonia capacity of 40 metric tons per day. The plant was commissioned in November 1983, and was placed in standby condition until the retort and upgrading plants were able to go into full operation.

The Great Plains Gasification Associates' project near Beulah, North Dakota, is the nation's largest synfuels project. It will go into operation in 1984, producing 125 million cubic feet of high-Btu gas daily from North Dakota lignite, via Lurgi gasification and methanation. The associated PHOSAM-W facility will strip dephenolized gas liquor and produce up to 115 metric tons per day of ammonia.

In conclusion, the PHOSAM-W process is generally accepted as the method of choice for the removal and recovery of ammonia from process streams. Those concerned with treatment of acid gases and sour waters should consider the process whenever such streams may be expected to contain ammonia. The technology is thoroughly grounded in practice, as well as theory. It has an unequalled track record of many successful applications, all producing saleable liquid ammonia, over the past fifteen years.

Index